Finite Time Thermodynamics of Power and Refrigeration Cycles

Shubhash C. Kaushik • Sudhir K. Tyagi
Pramod Kumar

Finite Time Thermodynamics of Power and Refrigeration Cycles

Springer

Shubhash C. Kaushik
Centre for Energy Studies
Indian Institute of Technology
New Delhi, India

Sudhir K. Tyagi
Centre for Energy Studies
Indian Institute of Technology
New Delhi, India

Pramod Kumar
Solid State Physics Laboratory (SSPL)
New Delhi, India

Co-published by Springer International Publishing, Cham, Switzerland, with Capital Publishing Company, New Delhi, India.

Sold and distributed in North, Central and South America by Springer, 233 Spring Street, New York 10013, USA.

In all other countries, except SAARC countries—Afghanistan, Bangladesh, Bhutan, India, Maldives, Nepal, Pakistan and Sri Lanka—sold and distributed by Springer, Tiergartenstr. 15, 69121 Heidelberg, Germany.

In SAARC countries—Afghanistan, Bangladesh, Bhutan, India, Maldives, Nepal, Pakistan and Sri Lanka—printed book sold and distributed by Capital Publishing Company, 7/28, Mahaveer Street, Ansari Road, Daryaganj, New Delhi 110 002, India.

Jointly published with Capital Publishing Company, New Delhi, India.

ISBN 978-3-319-62811-0 ISBN 978-3-319-62812-7 (eBook)
https://doi.org/10.1007/978-3-319-62812-7

Library of Congress Control Number: 2017954920

Printed on acid-free paper

This Springer imprint is published by Springer Nature
The registered company is Springer International Publishing AG
The registered company address is: Gewerbestrasse 11, 6330 Cham, Switzerland

Foreword

I am pleased to introduce the book on *Finite Time Thermodynamics of Power and Refrigeration Cycles* by my learned colleagues Prof. Shubhash C. Kaushik, Dr. Sudhir K. Tyagi, and Dr. Pramod Kumar. The book presents the basic concepts of finite time thermodynamics and its application to heat engine and heat pump cycles, used in power generation and space conditioning. The book incorporates the research by Prof. Kaushik and his associates in the last 15 years or so. The expertise of the authors in research and teaching in this emerging field of finite time thermodynamics is reflected in the book.

The book is suitable for study by B. Tech./M. Tech./research students and can be used as a reference by researchers and others. I wish the book all the success in enlightening numerous readers.

Mahendra Singh Sodha

Mahendra Singh Sodha
Padma Shri

Centre for Energy Studies (CES),
Indian Institute of Technology Delhi,
New Delhi
Former Vice-Chancellor DAVV Indore,
Lucknow University, and Bhopal University

Preface

Thermodynamics is one of the most interesting and important branches of thermal science and engineering, which plays a very crucial role in the design, development, and performance evaluation of thermal energy conversion systems including heat engine, refrigeration, and airconditioning systems.

Finite time thermodynamics (FTT) is a special branch of applied thermodynamics which deals mainly with external and internal irreversibilities associated with thermal energy conversion systems. With the increasing interest in power generation and space conditioning, FTT has become a very useful tool for the design and evaluation of these systems and gained a larger scope in the academic and scientific communities. In recent decades, it has been an interesting topic of discussion in many national and international conferences, but such material is scant and a comprehensive review on advanced systems is still lacking. Thus it was felt that there was a need for a comprehensive book on FTT of advanced and modified power and refrigeration cycles. The authors were inspired to write this book and are hopeful that it will be very useful for the research, design, and development of thermal energy conversion systems.

This book is mainly based on the research work carried out by the authors covering power and refrigeration cycles including basic, modified, and advanced versions of Carnot, Rankine, Brayton, Stirling, Ericsson, vapour compression, and vapour absorption cycles. However, relevant research papers and review articles were also consulted for the completeness of the book.

The content of the proposed book has been divided into 12 chapters. The first chapter presents the introduction of finite time thermodynamics, while the second chapter deals with its applications to the Carnot and Rankine power cycle. Chapters 3, 4, and 5 deal with the modified, advanced, and complex Brayton power cycles, respectively. The analysis of Stirling and Ericsson power cycles is presented in Chap. 6. Chapter 7 presents the finite time analysis of vapour compression refrigeration cycle, and Chap. 8 deals with the cascaded refrigeration cycle, while the Rankine cycle-operated vapour compression refrigeration cycles are presented in Chap. 9. Chapters 10 and 11 present the analysis and parametric study of Brayton and Stirling/Ericsson refrigeration cycles, whereas three heat reservoir-based (cooling and heating) cycles are given in Chap. 12 for a typical set of operating conditions. Finally, a list of references has been provided for further consultation to the readers.

New Delhi, India Shubhash C. Kaushik
New Delhi, India Sudhir K. Tyagi
New Delhi, India Pramod Kumar

Acknowledgements

The authors gratefully acknowledge the help from the following:

- Director/deputy directors of the Indian Institute of Technology Delhi for providing the necessary facilities and financial assistance in writing this book at the Centre for Energy Studies (CES) through the QIP/CEP Programme
- Prof. M.S. Sodha, former vice-chancellor of DAVV Indore, Lucknow University, and Bhopal University, former deputy director/acting director of IIT Delhi, and former head and presently honorary visiting professor of CES, IIT Delhi, for his overall guidance for writing the foreword of this book in time
- Prof. F. Meunier, director at LIMSI CNRS and professor at CNAM, Paris (France), for introducing the field of finite time thermodynamics and second law analysis to the authors
- Professors Jincan Chen and Lingen Chen of the People's Republic of China for their suggestions, reviews, and friendly associations with the authors and various stimulating discussions
- Prof. S. Prasad, former director of IIT Delhi and emeritus professor of the Electrical Engineering Department, for his encouragement from time to time
- Prof. V. Ramgopal Rao, present director of IIT Delhi, for his encouragement and moral support to the authors
- Professors P.L. Dhar and Sanjeev Jain of the Mechanical Engineering Department, IIT Delhi, for various academic interactions, suggestions, and stimulating discussions
- Professors N.D. Kaushik and N.K. Bansal for their cooperation and help in this endeavour
- Head, Centre for Energy Studies (Prof. V. Dutta), and other colleagues, including Professors R.P. Sharma, L.M. Das, and T.C. Kandpal and Drs. K.A. Subramanian, Vamsi Krishna, and Dibakar Rakshit, for their help and suggestions from time to time
- Research associates and students, viz. A. Khaliq, K. Manjunath, and V.V. Tyagi

- A.K. Pandey, N. Singh, A. Sharma, S. Manikandan, Rajesh, Ranjana, and others for their help and cooperation to this end
- Energy experts in the scientific community and various organizations all over the globe whose work has been consulted, referred, and used in writing this book
- Capital Publishing Company for timely and efficiently publishing this book
- Last but not the least, the family members (wife and children) of the authors for their high patience, cooperation, and everlasting moral support
- And above all, the Almighty God for giving intellect and strength to write this book with wisdom

New Delhi, India — Shubhash C. Kaushik
New Delhi, India — Sudhir K. Tyagi
New Delhi, India — Pramod Kumar

Contents

About the Authors

Shubhash C. Kaushik received his Ph.D. in plasma science from IIT Delhi after receiving his distinguished Master's degree in electronics from Meercut University, Meercut (UP). His research fields include the activities in plasma science and thermal science and engineering, energy conservation and heat recovery, solar refrigeration and airconditioning, solar architecture, and thermal storage and power generation. He has made significant contributions in these fields as evident by his more than 400 research publications in journals/conferences of repute at national and international levels. Dr. Kaushik has also completed several sponsored and consultancy projects from various government and private agencies. He has guided 50 Ph.D. theses and 75 M.Tech. projects and has authored several books. Dr. Kaushik has also been a postdoctoral fellow at Queensland University, Brisbane, Australia; visiting professor at LES-IIM-UNAM, Mexico; and Marie Skłodowska-Curie visiting fellow of the European Commission, Paris, France. Prof. Shubhash C. Kaushik has recently been awarded as top academic research performer (first rank holder) in all of India in the subject area of energy based on his research publication citations for the last 5–10 years and H-index as per Scopus data reported in NSTMIS-DST(GOI), New Delhi (2015).

Sudhir K. Tyagi is working as associate professor at the Centre for Energy Studies, IIT Delhi, and earlier he has worked as director/Scientist E at Sardar Swaran Singh National Institute of Renewable Energy, Kapurthala, an autonomous institution of the Ministry of New and Renewable Energy, Government of India. He has also worked as assistant professor at SMVDU, Katra, Jammu and Kashmir. Dr. Tyagi has worked as invited scientist at the Korea Institute of Energy Research, South Korea, and as postdoctoral fellow at the Hong Kong Polytechnic University, Hong Kong; Zhejiang University, Hangzhou; and Xiamen University, Xiamen, China, during 2002–2008. He has made a significant contribution in R&D activities as evident by more than 150 publications in journals/conferences of repute. He has guided 5 Ph.D. theses and 10 M.Tech. projects.

Pramod Kumar received his Ph.D. degree in the research field of finite time thermodynamics from IIT Delhi in 2003 and continued research work as postdoctoral fellow at IIT Delhi in the research field of exergy analysis. He had earlier completed his M.Sc. in physics, with specialization in electronics, from Meerut University in 1997 and also qualified in GATE-98 in physics and joint CSIR-UGC NET examination in December 2002 in physical sciences. His research areas include finite time thermodynamics, refrigeration and airconditioning systems, and energy and exergy analysis. Dr. Kumar joined the Defence Research and Development Organisation (DRDO), Ministry of Defence, directly as Scientist 'C' in 2005 at Naval College of Engineering, INS Shivaji, Lonavala, Pune, and taught at undergraduate level for about 4 years. In August 2008, he joined Solid State Physics Laboratory (SSPL) Delhi and is currently working as Scientist E. He has received the Commanding-in-Chief (C-in-C) Award – Commendation by Vice Admiral, Indian Navy, for his specific teaching ability. He has also received the Technology Group Award three times for his contribution at SSPL Delhi. He is a life member of the Solar Energy Society of India. He has made a significant contribution in R&D activities as evident by more than 25 research publications in journals/conferences of repute at national and international levels. He has also contributed towards the completion of four R&D projects and five technical reports and guided a couple of students at M.Tech./B.Tech. level. He is an expert reviewer in many national/international journals of repute.

Nomenclature

A	Heat exchanger area (m^2)
C	Heat capacitance rates (kW/K)
Cp	Specific heat (kJ/kg-k)
C-A	Curzon–Ahlborn
AE	Conductance on absorber side (kW/K)
CE	Conductance on condenser side (kW/K)
COP	Coefficient of performance
EE	Conductance on evaporator side (kW/K)
GE	Conductance on generator side (kW/K)
HP	Heat pump
h	Specific enthalpy (kJ/kg)
k	Conductance (kW\K)
L	Lagrangian operator
LMTD	Log Mean Temperature Difference
M	Mach number
M	Mass (kg)
$\dot{m}$	Mass flow rates (kg/s)
NTU	Number of transfer units
P	Power input/output (kW)
P_{c}	Condensation pressure (MPa)
P_{e}	Evaporation pressure (MPa)
P_{H}	Heating load (or output heat power) (kW)
P_{L}	Cooling load (or input heat power) (kW)
Q	Heat transfer (kJ)
$\dot{Q}$	Heat transfer rate (kW)
Ro	Gas constant (kJ/kmol·K)
$R_{\Delta S}, R'_{\Delta S}, R_{\mathrm{K-T}}$	Internal irreversibility parameters
R/AC/HP	Refrigeration/airconditioning/heat pump
r_c	Pressure/volume ratio

S	Entropy (kJ/K)
s	Specific entropy (kJ/K-kg)
t, t_{cycle}	Total/cycle time (s)
T	Temperature (K)
U	Overall heat transfer coefficient kW/(m^2–k)
V, V_s	Speed (m/s)
VC	Vapour compression
VCR	Vapour compression refrigeration VA Vapour absorption
VAR	Vapour absorption refrigeration
W	Work input/output (kJ)
x	Dryness fraction or the quality of the mixture

Subscripts

a	Absorber side
A	Absorber
c	Condenser side/compression
cy	Cycle
C	Condenser/heat sink side
C-A	Curzon–Ahlborn
e	Evaporator side
ea	Entropic average
E	Evaporator/heat source side
f	Fluid
g	Generator side
G	Generator
h	Heat source reservoir side
H	High temperature/hot side reservoir
in	Input
l	Sink/cold side
L	Low temperature/cold side reservoir
min	Minimum
max	Maximum
p	Pressure
s	Sound/source
wf	Working fluid
1, 2	Inlet, outlet
opt	Optimal
o	Output/optimal
r	Volume
w	Warm/hot side

Superscripts

HP	Heat pump
o, a	Environment/ambient
Ref	Refrigeration cycle
Rev.	Reversible
i	Initial
f	Final

Greeks

ε	Effectiveness/overall performance
λ μ	Lagrangian multipliers
γ	Specific heat ratio
η	Thermal efficiency
α, β	Thermal conductance (kW/K)
λ	Pressure/volume ratio

Chapter 1
General Introduction and the Concept of Finite Time Thermodynamics

1.1 Background

Thermodynamics is an important branch of Applied Thermal Science and Engineering in which the effects of heat and mass transfer, transformation of energy, etc. are studied. It is based on two basic laws, i.e. the first law and second law of thermodynamics. The first law of thermodynamics is one of the most fundamental laws and is of special importance for the energy conversion processes and also called as the 'Law of Conservation of Energy'. According to this law, energy is always conserved, neither can it be produced nor it can be destroyed; however, it can be converted from one form to another. All the energy that goes into the system comes out somewhere, in some form or the other which is true for all energy conversion systems. Thus, the first law of thermodynamics can be used in the form of energy conservation equation to solve the problem of various devices like heat engines, refrigerators, airconditioners, heat pumps, and other energy conversion systems. However, it tells nothing about how much heat can be converted into useful work and also it does not place restrictions on the conversion of heat into work and vice versa. Thus, as far as the first law of thermodynamics is concerned, all the heat transferred from a heat source to a heat engine could conceivably be converted into useful work, which is not true as shown by the experiments. It is also impossible to predict from the first law of thermodynamics whether a process is physically possible or not. Until now it has been assumed that energy change in thermodynamics can proceed in either direction, which is not true.

However, through rigorous experimental and hypothetical observations, it was predicted that a directional law should exist which would tell whether the particular process may occur or not and later on it came into existence as the second law of thermodynamics. Thus, the second law of thermodynamics provides the means of measuring the energy degradation which is also known as the 'Law of Degradation of Energy'. In other words, it explains that the quantity of energy is conserved but the quality of energy is degraded. Thus, the understanding and application of the

S.C. Kaushik et al., *Finite Time Thermodynamics of Power and Refrigeration Cycles*,
https://doi.org/10.1007/978-3-319-62812-7_1

second law of thermodynamics enable to investigate the problems of more efficient energy conversion and utilization processes. The second law of thermodynamics, in its fundamental form, cannot be used in energy utilization; therefore, some of its derived principles are to be used in energy conversion and conservation. Some of the considerations are given as below:

1. Whenever heat energy is used to do work, some of its part must be lost to the surroundings, and it is never available for a useful work/purpose. For instance, a part of energy is being used and the rest is wasted/lost.
2. The maximum possible amount of energy available in power-producing device such as an engine or turbine can be calculated, and hence, the best possible efficiency of the plant can be determined and compared with an actual installation.
3. The maximum possible amount of energy required to produce a given amount of refrigeration effect can be calculated and may be compared to that of an actual operating system.

Some of the relevant processes involved in understanding the second law analysis are reversible, endoreversible, and irreversible processes. Any process that occurs without leaving any change of initial state when reversed is called a reversible process and the process which cannot be reversed in such a way is called an irreversible process, while an endoreversible process lies somewhere between the two mentioned above. A reversible process is an ideal case which is seldom achieved in real practice, and hence, the minimization of irreversible effects is sought in the interest of efficient energy conversion and conservation techniques.

In order to understand the thermodynamic processes associated with heat engine and refrigeration systems, it is worthwhile to understand the difference between equilibrium and non-equilibrium processes, reversible and irreversible thermodynamics, and quasi-static and finite time thermodynamics. A process in which no external force is acting on the system is said to be in mechanical equilibrium. The state of a system in equilibrium is defined in terms of properties like pressure, temperature, density, and so on. Also when a system moves from one equilibrium process to another, it may do so either through a nonequilibrium process or a quasi-equilibrium process. A quasi-equilibrium process is one in which the deviation from equilibrium is minimal and all the states through which the system passes during quasi-equilibrium process may be considered as equilibrium state. On one hand, if the weights on the systems as shown in Fig. 1.1a are removed one by one, then the piston shown in Fig. 1.1b moves very slowly without deviating much from the equilibrium state, and hence, the process can be considered as quasi-equilibrium process. On the other hand, if all the weights are removed simultaneously, the piston rises rapidly and the system would not be in equilibrium at any time during this change of state. A reversible process is the idealized version of a quasi-equilibrium process. When the number of weights on the piston are made infinitely large and are removed one by one, then the process can be reversed at each step without bringing any change in the system or the surroundings. In other words, it

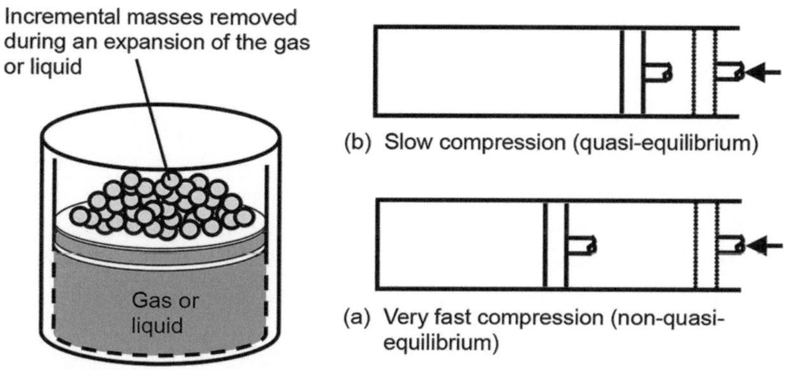

Fig. 1.1 Schematic of quasi-equilibrium and non-quasi-equilibrium processes (Source: Internet)

would take infinite time to remove or add infinite number of weights one by one to carry out a reversible process. Thus, a reversible process can be defined as the one, which can be reversed and in doing so no change either in the system or the surroundings takes place; however, it takes infinite time to execute such a process, and hence, seldom achieved in real practice.

One of the most important insights of thermodynamics has been the capacity of this discipline to furnish natural limits on the effectiveness with which one can carry out real processes. Much of the strong motivation for the creation of thermal science was the desire to make steam and gas engines as efficient as possible. The main emphasis from the time of Carnot and the tradition carried on by Clausius and Kelvin and more rigorously by Caratheodory was using thermodynamics as a tool to find limits on work, heat transfer, and efficiency. But after Gibbs, the focal point of the subject moved away from the process variables of heat and work towards the state variables. The Gibbsian view only became possible with the recognition of the existence of the state variables 'energy and entropy'. The earlier approach is far more in the spirit of Carnot, Clausius, and Kelvin. The Carnot–Clausius–Kelvin view emphasized the interaction of the thermodynamic system with its surroundings, while the Gibbsian view makes the properties of the system, the state variables themselves, dominant.

Thus, the Carnot–Clausius–Kelvin formulation is the one in which the impossibility of perpetual motion machines is an important physical postulate. Most important for us is the formulation in which the role of thermodynamic postulate becomes one of the setting natural limits on the process of variable as of heat and work. Traditionally, thermodynamic potentials are defined without any regard to time and thus achieve their rates as limits when the corresponding limiting processes proceed reversibly and, therefore, infinitely slowly in Carnot–Clausius–Kelvin approach, and it is these limits on process variables that we really want to know.

A number of reversible physical effects cause a loss of energy available for doing work or cause an increase in energy required to produce a given amount of refrigeration effect both cooling and heating. Such effects cannot be avoided, but should be reduced to a minimum. Some effects are as follows:

(a) *Temperature difference*: Large temperature difference causes greater loss during heat transfer; therefore, temperature difference should be kept as small as possible, e.g. between the source, sink, and the devices, viz. the heat engine or the refrigeration cycle.
(b) *Friction*: Friction causes loss of useful energy and, therefore, should be minimized. For example, regular cleaning of tube walls prevents scale build-up. Thus, if fluid friction is less, then less energy will be lost in pumping power.
(c) *Rapid expansion*: The adiabatic expansion of a high-pressure fluid to a low pressure, sometimes called throttling, is a process that wastes some energy available in the high-pressure fluid which could have been used to do some work.
(d) *Mixing*: Mixing of fluids of different temperatures will result in a loss of the useful available energy. In all these processes, irreversibilities are due to the entropy generation mechanism.

The term 'entropy' is a physical property of substances related to energy utilization and conservation and is a measure of the energy that is not available to do work. For a fluid with isentropic nature, the amount of work required by the compressor will be the minimum or work done by the turbine will be the maximum. The constant entropy (isentropic) process is an ideal reversible process which can never take place in practice, and hence, in a real process where work is required, the entropy increases and efforts should be made to minimize this increase. Among the important topics in thermodynamics has been the formulation of criteria for comparing the performance of real and ideal processes. The early tradition was carried out by Clausius, Kelvin, and others using thermodynamics as a tool to find the limit on work, heat transfer, efficiency, and the coefficient of performance of various energy conversion systems.

In classical thermodynamics, the heat transfer processes are reversible and are conducted infinitely slowly so that the working fluid comes into thermal equilibrium with the surroundings. The contemporary classical thermodynamics gives a fairly complete description of equilibrium states and reversible processes. However, in real engineering world, laws of thermodynamics are all conceived about irreversible processes and emphasize on the interaction of a thermodynamic system with its surroundings. In general, irreversible thermodynamic problems are too complex to solve exactly. Thus the classical equilibrium problems can be extended to quasi-static processes which happens in finite time, produce entropy, and provide a better approximation to real processes than that could be provided by equilibrium thermodynamics.

In this regard, a systematic study of different thermal energy conversion systems along with the application of finite time thermodynamics has been made considering external as well as internal irreversibilities. The 'missing link' in the treatments of different authors, which create confusion in the minds of the readers, has also been discussed, and appropriate gaps have been filled with relevant studies, in order to make the present study more comprehensive and coherent. In the present study we have done an appraisal of the performance analysis of different cycles, such as Carnot, Rankine, Brayton, Stirling, Ericsson, etc. for both power generation and space airconditioning applications, besides the vapour compression and vapour absorption cycles under different sets of operating conditions.

Thus the present book is an attempt to understand the basic concept of finite time thermodynamics for visualizing various losses in terms of irreversibilities/entropy generation associated with different energy conversion systems. Both external and internal (heat and mass transfer) irreversibilities play important roles in the deviation of actual system efficiency from the ideal system efficiency. The concept of 'finite time thermodynamics' deals with external (thermal coupling) irreversibilities, while the concept of exergy or available energy or availability is confined to internal component irreversibilities. However, in the ideal/reversible Carnot cycle, the heat transfer at infinitesimal temperature difference implied an infinitely slow rate of heat transfer or infinite heat transfer area, both of which are not realistic. From practical point of view, the heat transfer to and from the cycles should be completed in finite time, and hence, it is desirable to achieve maximum power output in case of heat engine and the minimum power input and/or the maximum cooling/heating load in RAC/HP systems, rather than the maximum/minimum work output/input to and from these cycles.

In power generation/consumption systems, it is desirable to study the net exergy (available energy) content of the input source(s) and how much of it one can extract/consume efficiently. For that, one has to go into the depth of the basic concept of classical thermodynamics. The traditional engineering design and analysis of energy systems, in practice, rarely proceeded beyond the first law of thermodynamics. However, it does not give the satisfactory explanations concerning qualitative transformations, so this methodology is incomplete, and hence, the second law analysis has to be embodied in. At present, a considerable conceptual progress has been made in this field, but a lot need to be done on the qualification of various thermodynamic losses. At most, all engineering applications, the implications of the thermodynamic laws, reveal a close relationship between the irreversibility (entropy generation) and one-way destruction of available energy. This relationship is very important in the field of engineering thermodynamics, as the interest is on accomplishing maximum power out of the available energy and minimum power in or maximum heating/cooling effect, while the losses can be measured in terms of entropy generation for heat engine and RAC/HP systems, respectively. Thus, any thermodynamic analysis of any system remains incomplete unless the second law analysis becomes a part of that analysis and thus the deviation is explained in terms of various entropy generation/irreversibilities occurring during different thermodynamic processes.

In real engineering world, we find that the actual changes in enthalpy and free energy in a process rarely approach the ideal thermodynamic enthalpy and free energy changes for that process. Typically, the actual expenditure of enthalpy and free energy as fuel and other inputs tend to be several times more than the ideal thermodynamic limit. Sometime we even expand energy on processes that could return less energy output. Thus energetics of thermodynamic systems has to be considered from the point of view of energy conservation. It is tempting to use the differences of actual and ideal requirements of energy, enthalpy, free energy, or availability as an index of how a process could be improved. Any realistic thermodynamics challenges this position on the grounds that nobody wants to operate a system reversibly. This discussion is intended to meet the challenge by providing an extension of conventional thermodynamics that will give limits on process variables carried out in limited or finite time intervals. This discipline is known as finite time thermodynamics or finite temperature difference thermodynamics and has grown very fast in the field of 'Thermal Science and Engineering'.

The goal of finite time thermodynamics is a means to evaluate the ideal limits on heat and work of processes operating at finite rates. One approach is to require processes to take place in an arbitrary but fix time interval. We then carry the analysis further to determine the optimal interval in which a process should be carried out in order to optimize power or any index of optimality. The subsequent events of the subject, however, turned away from the reversible processes and focused on equilibrium systems. This happened even though many questions, which gave the subject its original impetus, remained unanswered, such as:

- What is the minimum energy (ΔE) required by a given machine to produce a given work (W) in a given time (Δt)?
- What is the maximum work (W) that can be produced by a particular machine in a given time (Δt) utilizing a given amount of energy (ΔE)?
- What is the most efficient way to run a machine at a given thermodynamic process in finite time to produce maximum power?

The classical thermodynamics does not try to answer these questions, although it does give fairly complete description of the equilibrium states and the reversible processes. The only fact it tells us about real processes is that they always produce less work and more entropy than the corresponding reversible processes. The reversible processes are, however, possible in the limit of infinite time. But no one wants to produce finite work in infinite time and/or to run the machine infinitely slowly, and in practice, real engines seldom attain more than 30–40% of reversible efficiency.

1.2 Development of Irreversible Thermodynamics

A system is said to have a reversible process if at the conclusion of the process it leaves no change anywhere within the system or with the surrounding. A reversible process is hypothetical which can never be attained in practice but can be approximated as close as reversible process. Such processes are defined in the limit of infinitely slow execution. So the real thermodynamic process should proceed in finite time or limited time which causes a change between the system and its surrounding and is known as irreversible process. All real/actual processes are irreversible which means there is some entropy generation, which is a measure of irreversibility. Thus irreversibility can be defined as lost opportunities to do work. It represents the entropy generation that could have been converted into work but was not. The smaller the irreversibility associated with the cycle, the greater the work that will be produced by the heat engine cycle (or smaller work will be consumed by the refrigeration cycle). To improve the performance of thermodynamic systems, one should reduce the primary source of irreversibility associated with each component of the systems. For a closed system, the irreversibility is the difference of the reversible and actual work, i.e.

$$I = W_{\text{rev}} - W_{\text{actual}} \tag{1.1}$$

For energy-producing devices, like heat engines, the actual output work (W_{actual}) will be less, while, for energy-consuming devices like RAC and HP systems, the actual input work (W_{actual}) will be more than the reversible input work, i.e.

$$\text{For Heat Engine Cycles} \quad I = W_{\text{rev}} - W_{\text{actual}} \tag{1.2}$$

$$\text{For RAC and HP Cycles} \quad I = W_{\text{actual}} - W_{\text{rev}} \tag{1.3}$$

The irreversible thermodynamics in general deals with irreversibility of any kind in the system. The finite time thermodynamics deals with external irreversibility due to source/sink reservoirs, while the internal irreversibility deals with thermodynamic processes (including friction, etc.) within the cycle. Irreversibility in any system cannot be removed completely; however, it can be minimized up to some extent. The irreversible thermodynamics has become an increasing powerful tool for energy conversion systems. Thus the concept of finite time thermodynamics came into existence after the novel work of Curzon and Ahlborn (1975) which not only answer the above questions but also solved the problems of heat transfer and energy conversion systems.

1.3 Concept of Finite Time Thermodynamics

Heat is a kind of energy, which can be transferred from one body to another because of temperature difference between them, but finite temperature difference makes the process irreversible. Therefore, the heat transfer process approaches a reversible one as the temperature difference between two bodies reaches zero. However, transfer of finite amount of heat through infinitesimal temperature difference would take infinite time or infinite heat transfer area. As we know from the heat transfer theory, the amount of heat (Q) transfer between two bodies is proportional to the temperature difference, the contact area, and the time taken, i.e.

$$Q \propto A\ \Delta T \Delta t \quad \text{or} \quad Q = U\ A \Delta T \Delta t \tag{1.4}$$

where U is the overall heat transfer coefficient, A is the heat transfer area, ΔT is the temperature difference between the bodies, and Δt is the time taken in the process.

For infinitesimal temperature difference [i.e. $\Delta T \rightarrow 0$].

Then either $A \rightarrow \infty$ or $\Delta t \rightarrow \infty$ or $U \rightarrow \infty$.

But due to the finite conductivity of materials U will be finite, thus the only possibility is that either $A \rightarrow \infty$ or $\Delta t \rightarrow \infty$, again:

1. If $A \rightarrow \infty$ means the heat exchanger area is infinite, the heat engine/heat transfer unit becomes economically unviable, since the heat exchanger units are quite expensive.
2. If $\Delta t \rightarrow \infty$ means it takes infinite time to produce the finite amount of work, the power (which is work per unit time, i.e. $P = W/\Delta t \rightarrow 0$ as $\Delta t \rightarrow \infty$) will tend to zero.

Thus, the very purpose of power generation is defeated, since no engineer wants to design such type of machine whose output power is zero. Also the prime object of the machine is to produce power, which means finite amount of work in finite time rather than finite work in infinite time. This means we need finite heat transfer in finite time, which produces irreversibility. In the real engineering world, every machine has some power, which means the production of finite amount of work in finite time. This requires that there should be a finite temperature difference between the working fluid and the external reservoirs. Thus, irreversible heat transfer due to finite temperature difference is known as 'finite time thermodynamics' and/or 'finite temperature difference thermodynamics'. Curzon and Ahlborn (1975) proposed that due to finite conductivity of materials, the heat transfer to and from the heat engine is flowing through a finite temperature difference. They applied the idea in Carnot heat engine whose efficiency was found lower than the well-known Carnot efficiency.

$$\eta_{\text{Carnot}} = 1 - \frac{T_{\text{L}}}{T_{\text{H}}} \text{ while } \eta_{\text{C-A}} = 1 - \sqrt{\frac{T_{\text{L}}}{T_{\text{H}}}} \Rightarrow \eta_{\text{C-A}} < \eta_{\text{Carnot}} \tag{1.5}$$

where T_{L} and T_{H} are the temperatures of sink and source reservoirs, respectively. The main object of finite time thermodynamics is to understand irreversible finite time processes and to establish bonds on efficiency and maximum power for such processes. Further, it seeks to establish general operating principle for system, which serves as a model for real processes. The real processes can be associated with two kinds of irreversibilities, as below:

- External irreversibilities
- Internal irreversibilities

The external irreversibilities are due to finite temperature difference and the direct heat leak loss from source to sink (heat engines) and/or from sink to source (RAC and HP systems) and can be found using finite time thermodynamics methodology. The internal irreversibilities are due to non-isentropic compression/ expansion, friction, and entropy generation in the machine and can be determined using exergy analysis and irreversible thermodynamics. Thus the concept of exergy should be used to review the performance of energy conversion systems, because exergy is the maximum possible work that can be extracted from a given system in a given state by a process, which brings it into equilibrium with environment. Thus exergy analysis is the true mission to calculate the actual performance of real energy conversion devices.

1.4 Application of Finite Time Thermodynamics

There are three main power cycles being used in power generation sector, and different options are characterized by the type of the working fluid and/or the thermodynamic cycles followed by a particular heat engine. The power cycles mainly used in power generation sector are, namely, Rankine cycle, Brayton cycle, and Stirling cycle. The Rankine cycle is most commonly used for power generation and basically a steam/vapour power cycle. This cycle comprises the constant pressure heat addition and rejection processes along with the isentropic expansion and pumping processes and utilizes the working fluids, which change the phase during the heat addition processes to provide essentially isothermal heat addition and rejection. The other power cycles, such as Brayton, Stirling, and Ericsson cycles, are gas power cycles and utilize the gas or air as the working fluid and are preferred for higher temperature applications. On one hand, the Brayton cycle combines the adiabatic compression and expansion along with the isobaric heat addition and rejection processes, while on the other hand, the Stirling cycle combines the isothermal heat addition and rejection along with the isochoric compression and expansion processes. Whereas the Ericsson cycle combines the isobaric compression and expansion along with the isothermal heat addition and rejection processes, respectively.

In the present book, finite time thermodynamic analysis of various cycles, viz. Rankine, Brayton, Stirling, and Ericsson, vapour compression, vapour absorption, etc. have been carried in detail with different possible options from a simple to a more complex version for both power generation and refrigeration applications. All these three cycles offer higher performance, and hence, attention on these cycles is increasing nowadays. From the analysis presented in this book, one can understand and explain the reasons for the difference between the actual and ideal performance of different cycles for power generation and space airconditioning applications.

1.5 Conclusion

In this chapter the basic background of thermodynamics and its relation along with the limitation of classical thermodynamics with the Carnot–Clausius–Kelvin and Gibbs approach was summarized. The difference along with the concept of various processes such as reversible, endoreversible, irreversible, equilibrium, and quasi-static equilibrium and non-quasi-static equilibrium along with the development of finite time thermodynamics is given in detail. Also, the concept and development of irreversible thermodynamics along with the brief mathematical formulation for the possible applications of finite time thermodynamics are given.

Chapter 2
Finite Time Thermodynamic Analysis of Carnot and Rankine Heat Engines

2.1 Introduction

We know that heat engine is a power-producing device which absorbs heat from a high-temperature reservoir (heat source) and converts some of its part into mechanical (shaft) work and rejects the rest to a low-temperature reservoir (heat sink). All power-producing devices (heat engines) operate on cycles utilizing either gas or vapour, as the working fluid, and are known by their name. For example, if a heat engine utilizes gas as the working fluid, it is known as the gas power cycle such as Carnot, Brayton, Stirling, Ericsson, Braysson, etc. Similarly, if a heat engine utilizes vapour as the working fluid, it is known as the vapour power cycle such as Rankine cycle. The study and analysis of power cycles are very exciting and are an important part of the engineering thermodynamics, because they involve many complicated aspects and, hence, are very difficult to analyse completely. Some of these complicated aspects are pressure drop, friction, heat loss, entropy generations due to various causes, and the lack of sufficient time for the establishment of equilibrium condition, during each cycle, leading to more difficult, complex, and incomplete analysis of power cycles. To overcome these difficulties associated with power cycles, some idealistic assumptions are made. When a cycle is free from all irreversibilities and complexities, it is called an ideal cycle, and the processes followed by such cycle are reversible processes and vice versa. Most of the power cycles are designed for the purpose of converting thermal energy into shaft (mechanical) work. The performance of a heat engine is expressed in terms of thermal efficiency, which is basically the ratio of network produced by the heat engine to that of total heat supplied and denoted by η_{th}, as below:

$$\eta_{th} = \frac{W_{net}}{Q_{total}} = \frac{w_{net}}{q_{total}} \tag{2.1a}$$

S.C. Kaushik et al., *Finite Time Thermodynamics of Power and Refrigeration Cycles*,
https://doi.org/10.1007/978-3-319-62812-7_2

Based on the reversible processes, Nicolas Sadi Carnot, a French engineer, designed a heat engine in 1824, which goes after his name. The Carnot cycle has the highest (theoretical) thermal efficiency among all the heat engines operating between the same temperature levels. In other words, no engine can be as efficient as the Carnot engine operating between two temperature levels. Then a general question arises in the mind of the readers that if Carnot cycle is the best possible cycle, why do we need to have other so-called ideal cycle? The answer to this question is related to hardware. In other words, it is very difficult to design a reversible Carnot cycle in real practice.

2.2 Ideal Carnot Cycle

The (ideal) Carnot cycle consists of four reversible processes such as isothermal heat absorption, isentropic expansion, isothermal heat rejection, and isentropic compression as shown on the *T–s* diagram of Fig. 2.1a and on the *P–V* diagram of Fig. 2.1b, respectively.

The Carnot cycle can also be executed in a closed system like a piston–cylinder device or a steady-flow system utilizing two turbines and two compressors as shown in Fig. 2.1c and having a gas or vapour as the working fluid. As mentioned above, Carnot cycle is the most efficient cycle among all possible cycles that can be executed between two reservoirs having temperatures T_H and T_L of heat source and sink, respectively. The thermal efficiency of Carnot heat engine is the maximum efficiency that can never be achieved in practice for any real cycle and is defined as:

$$\eta_{\text{Carnot}} = 1 - \frac{T_L}{T_H} \tag{2.1b}$$

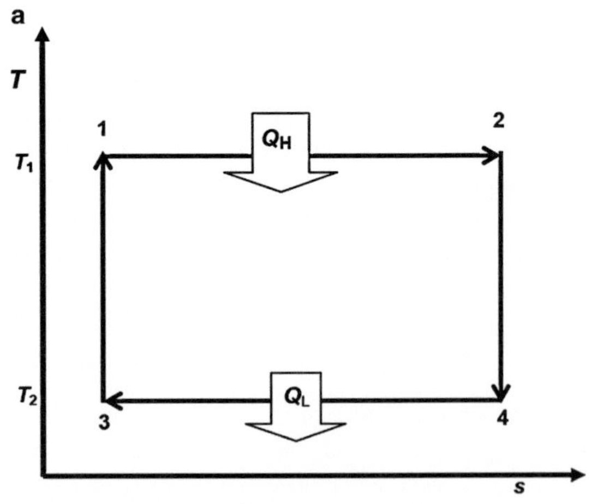

Fig. 2.1a *T–s* diagram of ideal Carnot cycle

Fig. 2.1b *P–V* diagram of ideal Carnot cycle

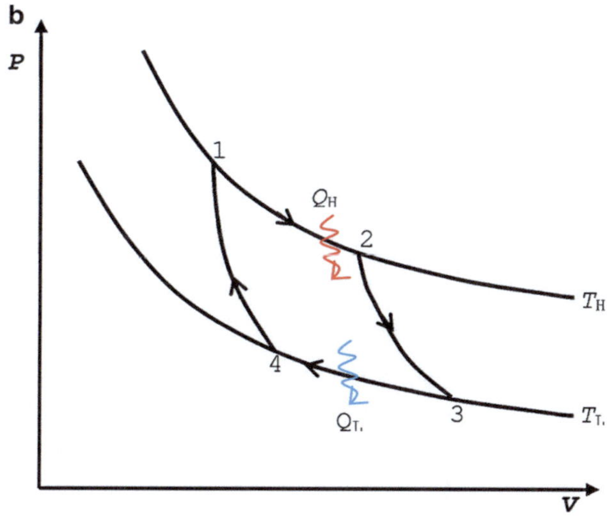

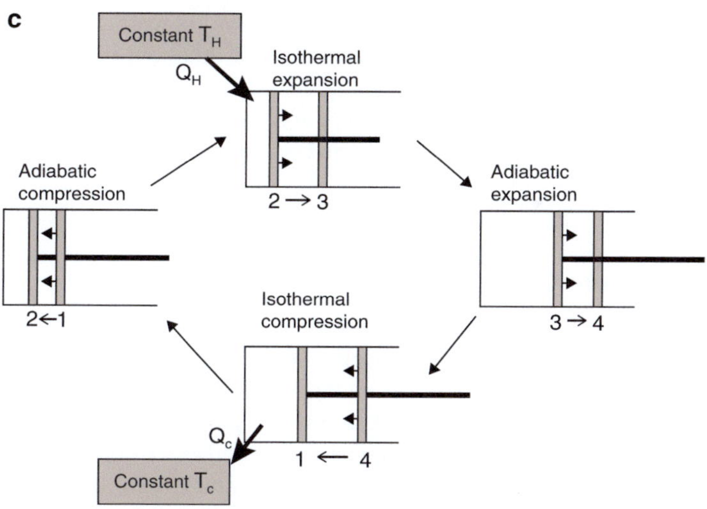

Fig. 2.1c Carnot power cycle with piston–cylinder arrangement

where T_H and T_L are, respectively, the absolute temperatures of the heat source and heat sink. However, the reversible isothermal heat transfer is very difficult to achieve in real practice because it requires infinite heat transfer area and/or infinite heat addition time and both of these are not realistic, because:

- If it requires infinite heat exchanger area, the heat engine becomes economically unviable, as the heat exchanger units are quite expensive and hence, the very purpose of power generation is defeated.

- If it takes infinite time to produce the finite amount of work, then the power which is work per unit time tends to zero which is again of no use in real engineering problems.

However, in practical situations, we need finite heat transfer in finite time or finite amount of work in finite time which produces irreversibility. Hence the new discipline of finite time thermodynamics or finite temperature difference thermodynamics came into existence after the novel work of Curzon and Ahlborn (1975). They applied the concept of finite time thermodynamics to Carnot heat engine that was further extended by Andreson et al. (1984), Bejan (1993, 1995, 1996a, b), Berry et al. (1999), Chambadal (1957), Chen (1994), Chen and Andresen (1995), Chen and Schouten (1998), Chen et al. (2003), Chen and Yan (1989), Chen et al. (1995, 1997, 1999a, b), Hoffmann et al. (1997), Kodal et al. (2000, 2002), Leff (1987), Medina et al. (1996), Moran and Shapiro (2007), Novikov (1957), Sahin et al. (1996, 1998), Sahin and Kodal (1999), Salamon and Nitzan (1981), Tyagi et al. (2010), Wu (1995), Wu et al. (1996, 1999), Ibrahim et al. (1991), Wu and Kiang (1992), Kumar (2000), Kaushik and Tyagi (2002), and others. Further, Tyagi et al. (2002–2008) applied the concept of finite time thermodynamics to different heat engines such as Brayton, Stirling, Ericssion, Braysson, and refrigeration cycles for power generation, space airconditioning, etc. Besides, they also have made a thorough systematic study of the application of finite time thermodynamics to different heat engines and space airconditioning systems including the concept of external and internal irreversibilities along with the finite heat capacities of external reservoirs.

2.3 Finite Time Carnot Cycle

Curzon–Ahlborn in 1975 applied the concept of finite time thermodynamics to Carnot heat engine with infinite heat capacitance rates of source and sink reservoirs and the time of two adiabatic processes was taken directly proportional to the time of two isothermal processes. They assume that due to finite conductivity of materials, the heat engine is working between the working fluid temperatures (T_w and T_c) rather than between the source and sink temperatures (T_H and T_L) where $T_H > T_w$ and $T_L < T_c$ and depends on the output of the machine (Fig. 2.2a). They optimized the power output with respect to the working fluid temperatures (T_w and T_c) and find the expression for maximum power output and computed the corresponding thermal efficiency, which is different than the well-known 'Carnot' efficiency. They gave the theoretical model of Carnot heat engine by assuming that heat fluxes through the vessel containing the working substance are proportional to the temperature difference across the walls of the vessel in the isothermal stage. There are two different cases of such a system, one in which it is assumed that the temperature of the heat source and sink remain constant viz. the heat source and heat sink are of infinite heat capacity (Fig. 2.2b), two in which it changes with time viz. the heat capacity of source and sink reservoirs are finite, respectively. The two different cases are mentioned as below.

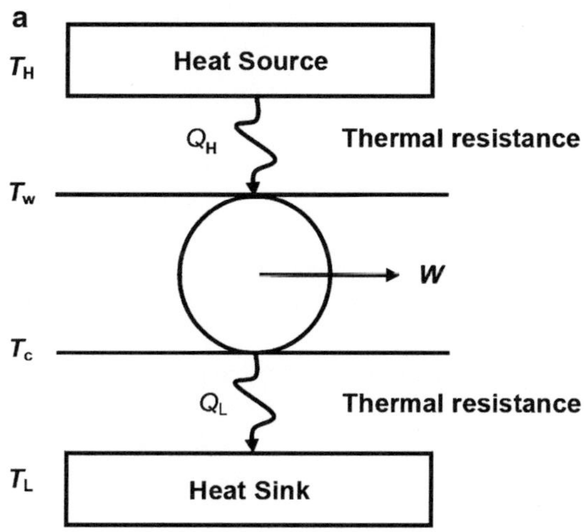

Fig. 2.2a Schematic of finite time Carnot cycle

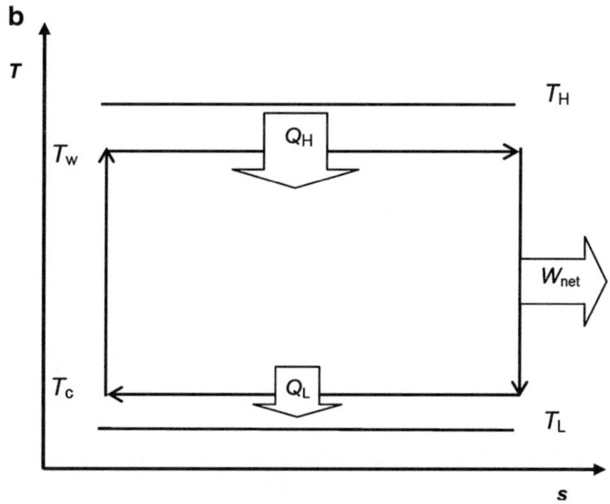

Fig. 2.2b Finite time Carnot cycle on *T–s* diagram

2.3.1 *Infinite Heat Capacity*

Let $\dot{Q}_H$ and $\dot{Q}_L$ are the heat transfer rates between the heat source and the heat engine and between the heat engine and the heat sink and are given by:

$$\dot{Q}_H = \alpha(T_H - T_w) \tag{2.2}$$

$$\dot{Q}_L = \beta(T_c - T_L) \tag{2.3}$$

where α and β are the constants, depending on the thickness and thermal conductivity of the walls. If the isothermal expansion and compression last for t_H and t_L seconds, respectively, then Q_H and Q_L are given by:

$$Q_H = \alpha t_H (T_H - T_w)$$
$$Q_L = \beta t_L (T_c - T_L)$$

Curzon and Ahlborn (1975) also assume that the adiabatic expansion and compression are completely reversible and no heat exchange occurs with the surroundings.

Again, using the second law of thermodynamics for a reversible cycle, yields:

$$\frac{Q_H}{T_w} = \frac{Q_L}{T_c} \Rightarrow \frac{Q_H - Q_L}{T_w - T_c}$$

and the ratio of time taken for the two isothermal processes can be given as below:

$$\frac{t_H}{t_L} = \frac{\beta T_w (T_c - T_L)}{\alpha T_c (T_H - T_w)}$$

The power output of the cycle is given by:

$$P = \frac{(Q_H - Q_L)}{(t_H + t_L + t_{adiabatic})} \tag{2.4}$$

where $t_{adiabatic}$ is the time taken to complete the two adiabatic processes and proportional to the time of two isothermal processes. Thus the power output will be:

$$P = \frac{(Q_H - Q_L)}{\gamma (t_H + t_L)} = \frac{\alpha \beta x y (T_H - T_L - x - y)}{[\alpha x T_L + \beta y T_H + xy(\alpha - \beta)]} \tag{2.5}$$

where $x = (T_H - T_w)$ and $y = (T_c - T_L)$, from Eq. (2.5), we see that power (P) is function of two variables x and y. Thus using Eq. (2.5) and its optimal condition, yields:

$$\beta T_H y (T_H - T_L - x - y) = x[\beta T_H y + \alpha T_L x + xy(\alpha - \beta)] \tag{2.6}$$
$$\alpha T_L x (T_H - T_L - x - y) = y[\beta T_H y + \alpha T_L x + xy(\alpha - \beta)] \tag{2.7}$$

Solving Eqs. (2.6) and (2.7) for x and y, gives:

$$y = x \sqrt{\frac{\alpha T_L}{\beta T_H}} \tag{2.8}$$

Solving Eq. (2.6) with the help of Eq. (2.8) by substituting $\mu = x/T_H$, yields:

$$\left[1-\frac{\alpha}{\beta}\right]\mu^2 - 2\left[\left(\frac{\alpha T_L}{\beta T_H}\right)^{1/2}+1\right]\mu + \left[1-\frac{T_L}{T_H}\right] = 0 \tag{2.9}$$

Thus, Eq. (2.9) is second-degree equation in μ whose solution is:

$$\mu = \frac{\left[\left(\frac{\alpha T_L}{\beta T_H}\right)^{1/2}+1\right] \pm \sqrt{\left[\left(\frac{\alpha T_L}{\beta T_H}\right)^{1/2}+1\right]^2 - \left[1-\left(\frac{\alpha}{\beta}\right)\right]\left[1-\left(\frac{T_L}{T_H}\right)\right]}}{\left[1-\left(\frac{\alpha}{\beta}\right)\right]} \tag{2.10}$$

which on solving, gives:

$$\mu = \frac{x}{T_H} = \frac{\left[1-\sqrt{\frac{T_L}{T_H}}\right]}{\left[1+\frac{\alpha}{\beta}\right]} \tag{2.11}$$

and

$$\frac{y}{T_L} = \frac{\left[\sqrt{\frac{T_H}{T_L}}-1\right]}{\left[1+\frac{\beta}{\alpha}\right]} \tag{2.12}$$

Substituting Eqs. (2.11) and (2.12) into Eq. (2.5), the optimum power output is:

$$P_{max} = \frac{\alpha\beta}{\gamma}\left[\frac{(\sqrt{T_H}-\sqrt{T_L})}{(\sqrt{\alpha}+\sqrt{\beta})}\right]^2 \tag{2.13}$$

and the corresponding thermal efficiency:

$$\eta_m = 1-\sqrt{\frac{T_L}{T_H}} = \eta_{C-A} \tag{2.14}$$

1. The most striking feature of the above results is that the efficiency does not depend on the heat transfer coefficients α and β, as in the case of ideal Carnot engine, where it depends only on the temperatures of the heat reservoirs.
2. These results serve as a much better guide to the performance of real heat engines.
3. The efficiency values at the maximum power have been compared with the observed efficiency of the actual power plants as given in Table 2.1.

Table 2.1 Observed efficiencies of real heat engines (Curzon and Ahlborn 1975)

Source	T_H (°C)	T_L	Carnot efficiency (%)	C–A efficiency (%)	Observed efficiency (%)
Coal-fired steam plant	565	25	64.44	40.37	36
CANDU PHW nuclear reactor	300	25	47.99	27.88	30
Geothermal steam plant	250	80	32.50	17.84	16

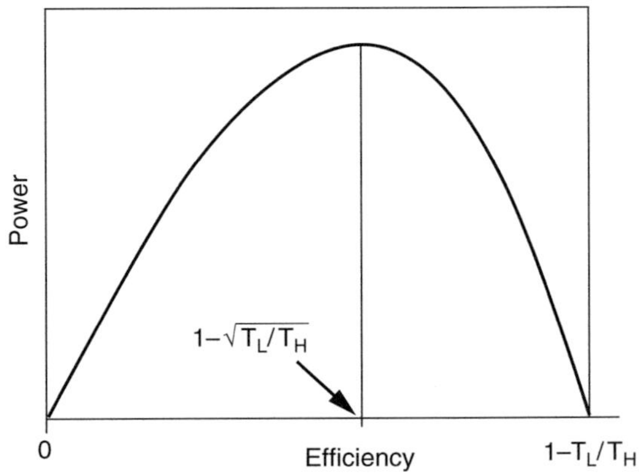

Fig. 2.3 Power output vs. thermal efficiency of a Carnot cycle (Ibrahim et al. 1991)

From Table 2.1, it can be seen that the observed efficiencies are much closer to Curzon–Ahlborn efficiency than the Carnot efficiency. Ibrahim et al. (1991) mathematically proved the observations made by Curzon and Ahlborn (1975), and they plotted the graph between power output and thermal efficiency of a Carnot cycle. They observed that the power output increases with the thermal efficiency, attains its peak at Curzon–Ahlborn point and then decreases sharply, tend to zero as the thermal efficiency approaches to Carnot efficiency. The graph plotted by Ibrahim et al. (1991) for infinite heat capacity of external reservoirs is also given below (Fig. 2.3) to make the readers clear about the trend of power output vs. thermal efficiency.

2.3.2 Alternative Derivation of Curzon–Ahlborn Efficiency

Let us consider a heat engine having heat transfers between the system and thermal reservoirs and work output shown schematically in Fig. 2.2a and on *T–s* diagram Fig. 2.2b.

Using first law of thermodynamics, net power output is given by:

$$P = \dot{Q}_H - \dot{Q}_L = \eta_{th}\dot{Q}_H \tag{2.15}$$

Heat transfer rates are defined in earlier Sect. 2.3.1. If working fluid temperatures are very much close to the value of corresponding thermal reservoirs temperature then heat transfer rates will be close to zero and heat engine will be externally reversible, thermal efficiency will be highest equal to Carnot efficiency but our desired output power (work output per unit time) will be zero. Again, if the temperature difference between the working fluid and the corresponding reservoir is too large, then the heat transfer rates will be high but the thermal efficiency will be low, in the limiting case will be zero and, hence, the power output will also be zero. So there must be an optimal set of working fluid temperatures in between these two limiting cases that produces the maximum output power. Now, using the second law of thermodynamics for a reversible cycle yields:

$$\frac{\dot{Q}_{\mathrm{H}}}{T_{\mathrm{w}}} = \frac{\dot{Q}_{\mathrm{L}}}{T_{\mathrm{c}}} \tag{2.16}$$

Using Eq. (2.16), power output can be written:

$$P = \eta_{\mathrm{th}}\dot{Q}_{\mathrm{H}} = \left(1 - \frac{T_{\mathrm{c}}}{T_{\mathrm{w}}}\right) C_{\mathrm{H}}(T_{\mathrm{H}} - T_{\mathrm{w}}) \tag{2.17}$$

Using Eqs. (2.2) and (2.3) and Eq.(2.16), we have:

$$T_{\mathrm{c}} = T_{\mathrm{L}}\left[1 - \frac{\alpha}{\beta}\left(\frac{T_{\mathrm{H}}}{T_{\mathrm{w}}} - 1\right)\right]^{-1} \tag{2.18}$$

Substituting Eq. (2.18) in Eq. (2.17), power output becomes the function of only one working fluid temperature T_w. Now, to find out the optimal working fluid temperatures, we take derivative of power output w.r.t. 'T_w' and equate to zero.

$$\frac{dp}{dT_{\mathrm{w}}} = 0 \text{ gives}$$

$$\left(\frac{T_{\mathrm{c}}}{T_{\mathrm{w}}} - \frac{dT_{\mathrm{c}}}{dT_{\mathrm{w}}}\right)\alpha\left(\frac{T_{\mathrm{H}}}{T_{\mathrm{w}}} - 1\right) - \alpha\left(1 - \frac{T_{\mathrm{c}}}{T_{\mathrm{w}}}\right) = 0 \tag{2.19}$$

$$\text{where } \frac{dT_{\mathrm{c}}}{dT_{\mathrm{w}}} = -T_{\mathrm{L}}\left[1 - \frac{\alpha}{\beta}\left(\frac{T_{\mathrm{H}}}{T_{\mathrm{w}}} - 1\right)\right]^{-2} \frac{\alpha T_{\mathrm{H}}}{\beta T_{\mathrm{w}}^2} \tag{2.20}$$

After substituting Eqs. (2.18) and (2.20) in Eq. (2.19), we get the optimal working fluid temperatures given as below:

$$(T_{\mathrm{w}})_{\mathrm{opt}} = \frac{\alpha}{\alpha + \beta} T_{\mathrm{H}} + \frac{\beta}{\alpha + \beta}\sqrt{T_{\mathrm{H}} T_{\mathrm{L}}} \tag{2.21}$$

$$(T_c)_{opt} = \frac{\beta}{\alpha+\beta} T_L + \frac{\alpha}{\alpha+\beta} \sqrt{T_H T_L} \tag{2.22}$$

Using Eqs. (2.21) and (2.22), optimal thermal efficiency becomes:

$$(\eta_{th})_{opt} = 1 - \frac{(T_c)_o}{(T_w)_o} = 1 - \frac{\beta T_L + \alpha \sqrt{T_H T_L}}{\alpha T_H + \beta \sqrt{T_H T_L}}$$

which reduces to:

$$(\eta_{th})_{opt} = 1 - \sqrt{\frac{T_L}{T_H}} = \eta_{C-A} \tag{2.23}$$

Equation (2.23) is optimal thermal efficiency corresponding to the maximum power output and also known as Curzon–Ahlborn efficiency, which is independent of heat transfer coefficients but depends only on the reservoirs temperatures and always less than the corresponding Carnot efficiency.

2.3.3 *Finite Heat Capacity*

If the Carnot heat engine is having finite heat capacity of heat source and sink as shown in Fig. 2.4, then the heat transfer rates to and from the cycle will be proportional to the Log Mean Temperature Difference (LMTD), as below:

$$\dot{Q}_H = \frac{Q_H}{t_H} = U_H A_H (\text{LMTD})_H \tag{2.24}$$

$$\dot{Q}_L = \frac{Q_L}{t_L} = U_L A_L (\text{LMTD})_L \tag{2.25}$$

where $(\text{LMTD})_H$ and $(\text{LMTD})_L$ are defined as:

$$(\text{LMTD})_H = \frac{[(T_{H1} - T_w) - (T_{H2} - T_w)]}{\ln\left[\frac{(T_{H1} - T_w)}{(T_{H2} - T_w)}\right]}$$

$$(\text{LMTD})_L = \frac{[(T_c - T_{L1}) - (T_c - T_{L2})]}{\ln\left[\frac{(T_c - T_{L1})}{(T_c - T_{L2})}\right]}$$

Also from the property of external fluids, the heat transfer rates to and from the cycle will be:

$$\dot{Q}_H = \frac{Q_H}{t_H} = \dot{m}_H C_{pH} (T_{H1} - T_{H2}) \tag{2.26}$$

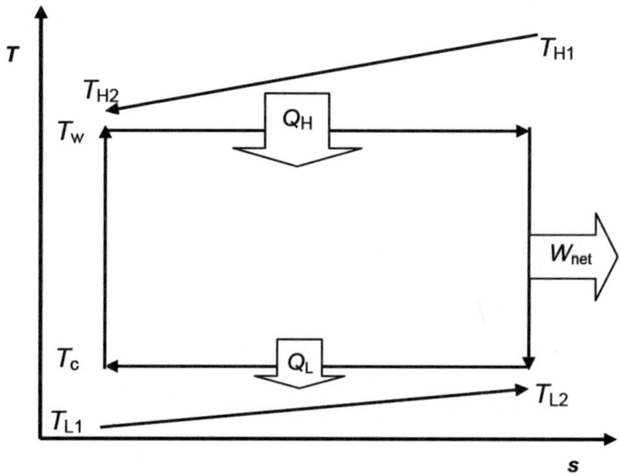

Fig. 2.4 Finite time Carnot cycle with finite heat capacity

$$\dot{Q}_{\mathrm{L}} = \frac{Q_{\mathrm{L}}}{t_{\mathrm{L}}} = \dot{m}_{\mathrm{L}} C_{\mathrm{pL}} (T_{\mathrm{L2}} - T_{\mathrm{L1}}) \tag{2.27}$$

where t_{H} and t_{L} are the heat addition and rejection times, $\dot{m}_{\mathrm{H}}$ and $\dot{m}_{\mathrm{L}}$ are the mass flow rates and C_{pH} and C_{pL} are the specific heats of the source and sink reservoir fluids, respectively. Using Eqs. (2.24), (2.25), (2.26), and (2.27), we have:

$$T_{\mathrm{H2}} = T_{\mathrm{w}} + (T_{\mathrm{H1}} - T_{\mathrm{w}}) \exp\left[-\frac{U_{\mathrm{H}} A_{\mathrm{H}}}{\dot{m}_{\mathrm{H}} C_{\mathrm{pH}}}\right] \tag{2.28}$$

$$T_{\mathrm{L2}} = T_{\mathrm{c}} + (T_{\mathrm{c}} - T_{\mathrm{L1}}) \exp\left[-\frac{U_{\mathrm{L}} A_{\mathrm{L}}}{\dot{m}_{\mathrm{L}} C_{\mathrm{pL}}}\right] \tag{2.29}$$

Neglecting the time of two adiabatic processes, which are quite fast, the total cycle time t_{cy}, will be:

$$t_{\mathrm{cy}} = (t_{\mathrm{H}} + t_{\mathrm{L}}) = \frac{Q_{\mathrm{H}}}{\dot{m}_{\mathrm{H}} C_{\mathrm{pH}} (T_{\mathrm{H1}} - T_{\mathrm{H2}})} + \frac{Q_{\mathrm{L}}}{\dot{m}_{\mathrm{L}} C_{\mathrm{pL}} (T_{\mathrm{L2}} - T_{\mathrm{L1}})} \tag{2.30}$$

Substituting Eqs. (2.28) and (2.29) into Eq. (2.30) yields:

$$t_{\mathrm{cy}} = \frac{Q_{\mathrm{H}}}{\mathrm{HE}(T_{\mathrm{H1}} - T_{\mathrm{w}})} + \frac{Q_{\mathrm{L}}}{\mathrm{LE}(T_{\mathrm{c}} - T_{\mathrm{L1}})} \tag{2.31}$$

where HE and LE are given as below:

$$\mathrm{HE} = \dot{m}_{\mathrm{H}} C_{\mathrm{pH}} \left[1 - \exp\left(-\frac{U_{\mathrm{H}} A_{\mathrm{H}}}{\dot{m}_{\mathrm{H}} C_{\mathrm{pH}}}\right)\right]$$

$$\mathrm{LE} = \dot{m}_{\mathrm{L}} C_{\mathrm{pL}} \left[1 - \exp\left(-\frac{U_{\mathrm{L}} A_{\mathrm{L}}}{\dot{m}_{\mathrm{L}} C_{\mathrm{pL}}} \right) \right]$$

Using the first law of thermodynamics gives:

$$W = Q_{\mathrm{H}} - Q_{\mathrm{L}} \tag{2.32}$$

Also, the second law of thermodynamics gives:

$$\frac{Q_{\mathrm{H}}}{T_{\mathrm{w}}} = \frac{Q_{\mathrm{L}}}{T_{\mathrm{c}}} \tag{2.33}$$

From Eqs. (2.32) and (2.33), we have:

$$Q_{\mathrm{H}} = \frac{W T_{\mathrm{w}}}{(T_{\mathrm{w}} - T_{\mathrm{c}})} \tag{2.34}$$

$$Q_{\mathrm{L}} = \frac{W T_{\mathrm{c}}}{(T_{\mathrm{w}} - T_{\mathrm{c}})} \tag{2.35}$$

Using Eqs. (2.34), (2.35), and (2.31), we have:

$$t_{\mathrm{cy}} = \left[\frac{W T_{\mathrm{w}}}{\mathrm{HE}(T_{\mathrm{H1}} - T_{\mathrm{w}})(T_{\mathrm{w}} - T_{\mathrm{c}})} + \frac{W T_{\mathrm{c}}}{\mathrm{LE}(T_{\mathrm{c}} - T_{\mathrm{L1}})(T_{\mathrm{w}} - T_{\mathrm{c}})} \right] \tag{2.36}$$

The power output will be:

$$P = \frac{\mathrm{HE\,LE} xy (T_{\mathrm{H1}} - T_{\mathrm{L1}} - x - y)}{[\mathrm{HE} x T_{\mathrm{L1}} + \mathrm{LE} y T_{\mathrm{H1}} + xy(\mathrm{HE} - \mathrm{LE})]} \tag{2.37}$$

where $x = (T_{\mathrm{H1}} - T_{\mathrm{h}})$ and $y = (T_{\mathrm{l}} - T_{\mathrm{L1}})$. It can be seen from Eq. (2.37) that P is function of x and y because T_{H1}, T_{L1}, HE, and LE are constants for a particular set of operating parameters; thus optimizing P w.r.t. x and y gives:

$$T_{\mathrm{w}} = C_1 \sqrt{T_{\mathrm{H1}}} \,\mathrm{and}\, T_{\mathrm{c}} = C_1 \sqrt{T_{\mathrm{L1}}} \tag{2.38}$$

where

$$C_1 = \left[\frac{\sqrt{\mathrm{HE} T_{\mathrm{H1}}} + \sqrt{\mathrm{LE} T_{\mathrm{L1}}}}{\sqrt{\mathrm{HE}} + \sqrt{\mathrm{LE}}} \right] \tag{2.39}$$

Substituting Eqs. (2.38) and (2.39) into Eq. (2.37) the maximum power output of the engine will be:

$$P_{\max} = \frac{\mathrm{HE}\,\mathrm{LE}\left[\sqrt{T_{\mathrm{H1}}} - \sqrt{T_{\mathrm{L1}}}\right]^2}{\left[\sqrt{\mathrm{HE}} + \sqrt{\mathrm{LE}}\right]^2} \tag{2.40}$$

while the thermal efficiency at maximum power output will be:

$$\eta_{\mathrm{m}} = 1 - \sqrt{\frac{T_{\mathrm{L1}}}{T_{\mathrm{H1}}}}(\approx \eta_{\mathrm{C-A}}) \tag{2.41}$$

1. It is observed that the thermal efficiency at maximum power output depends only on the inlet temperature of the reservoirs.
2. For the case of heat engine operating between heating and cooling fluids with infinite heat capacity ($m_{\mathrm{H}}\ C_{\mathrm{pH}}$ and $m_{\mathrm{L}}\ C_{\mathrm{pL}}$ tends to infinity) and the time taken for the two adiabatic processes is directly proportional to that of the two isothermal processes, then the temperatures of the external fluids will remain constant and the expressions for maximum power output and the corresponding thermal efficiency will be equal to that obtained by Curzon and Ahlborn (1975), given in Eq. (2.14).

2.4 Special Cases

As it is clearly seen from the above two cases, the performance of the Carnot cycle changes slightly with the assumptions made to evaluate its performance. There can be several more cases where the assumptions can be chosen according to the simplicity of the expressions. It is also important to note that efficiency of a Carnot cycle depends on the source and sink temperatures only, as can be seen from Eqs. (2.14) and (2.41), respectively.

Time of Two Adiabatic Processes

Following the same analysis and neglecting the time of two adiabatic processes, Wu (1988) analysed the Carnot heat engine taking into account the overall heat transfer coefficient and contact area between the heat engine and the reservoirs and gave the expressions for the maximum power output and the corresponding thermal efficiency as below:

$$P_{\max} = U_{\mathrm{H}}A_{\mathrm{H}}U_{\mathrm{L}}A_{\mathrm{L}}\left[\frac{\left(\sqrt{T_{\mathrm{H}}} - \sqrt{T_{\mathrm{L}}}\right)}{\left(\sqrt{U_{\mathrm{H}}A_{\mathrm{H}}} + \sqrt{U_{\mathrm{L}}A_{\mathrm{L}}}\right)}\right]^2 \tag{2.42}$$

and

$$\eta = 1 - \sqrt{\frac{T_{\mathrm{L}}}{T_{\mathrm{H}}}} = \eta_{\mathrm{C-A}} \tag{2.43}$$

where $U_H A_H$ and $U_L A_L$ are the overall heat transfer coefficient-area product for source side and sink side, respectively. If the expression given in Eq. (2.42) is compared with that of given in Eq. (2.13) for P_{max}, then $\alpha = U_H A_H$ and $\beta = U_L A_L$.

Again, if the internal irreversibility associated with the cycle due to number of reasons is also considered, the cycle is known as irreversible Carnot cycle. The irreversible Carnot cycle is a cycle where both external and internal irreversibilities are associated with the cycle as given below.

2.5 Irreversible Carnot Cycle

The Curzon and Ahlborn (1975) did not consider any internal irreversibility in their cycle model. However, every cycle has both internal and external irreversibilities associated with it, which is a complex analysis as mentioned earlier. Keeping this in mind, few authors have applied the concept of internal as well as external irreversibility (Wu and Kiang 1992; Kaushik 1999; Kumar 2000) and proposed a Carnot cycle model assuming the heat transfer rates to and from the cycle which also includes both the external and internal irreversibility in the Curzon–Ahlborn model as can be seen on the *T–s* diagram of Fig. 2.5. The external irreversibilities are due to finite temperature difference between the cycle and its external reservoirs, while the internal irreversibilities are due to various reasons such as friction, non-isentropic compression and expansion process, pressure drop, etc. and neglecting the time taken in the two adiabatic processes. Kumar (2000) included

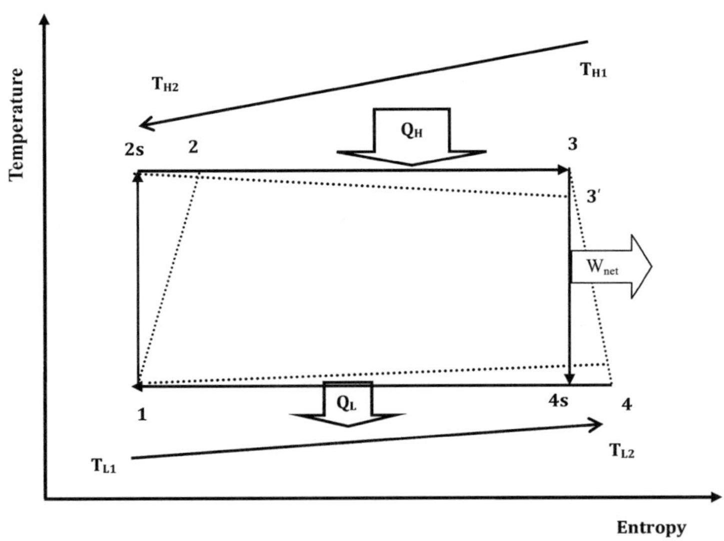

Fig. 2.5 Finite time irreversible Carnot cycle

a total irreversibility parameter (R_{K-T}) which is defined as the ratio of entropy generation during heat rejection to that of during heat addition which is more than unity. Let $\dot{Q}_H$ and $\dot{Q}_L$ are, respectively, the heat transfer rates to and from the heat engine during the two isothermal processes, following earlier trends, as below:

$$\dot{Q}_H = k_H(T_{H1} - T_w) = T_w(s_3 - s_2) \tag{2.44}$$

$$\dot{Q}_L = k_L(T_c - T_{L1}) = T_c(s_4 - s_1) \tag{2.45}$$

where k_H and k_L are the conductance on the source and sink side, respectively, and the s's are entropy values at the respective state points. Also from the first and second laws of thermodynamics, we have:

$$P = \dot{Q}_H = \dot{Q}_L = k_H(T_{H1} - T_w) - k_L(T_c - T_{L1}) \tag{2.46}$$

and

$$\int \frac{dQ}{T} = \frac{\dot{Q}_H}{T_w} - \frac{\dot{Q}_L}{T_c} \leq 0 \tag{2.47}$$

If we define the irreversibility parameter R_{K-T} which is the ratio of entropy generation during heat rejection to that of during heat addition and greater than unity for a real cycle, as below:

$$\frac{\dot{Q}_L}{T_c} \bigg/ \frac{\dot{Q}_H}{T_w} = \frac{(s_4 - s_1)}{(s_3 - s_2)} = R_{K-T} > 1 \tag{2.48}$$

$$R_{K-T}\frac{\dot{Q}_H}{T_w} = \frac{\dot{Q}_L}{T_c} \tag{2.49}$$

Substituting Eq. (2.49) into Eq. (2.46) and maximizing with respect to T_w and T_c as done in the previous section, we obtained the expressions for maximum power output and the corresponding thermal efficiency, as given below:

$$P_{max} = \frac{k_H k_L \left[\sqrt{T_{H1}} - \sqrt{R_{K-T} T_{L1}}\right]^2}{(R_{K-T} k_H + k_L)} \tag{2.50}$$

and

$$\eta_m = 1 - \sqrt{\frac{R_{K-T} T_{L1}}{T_{H1}}} \tag{2.51}$$

The cycle irreversibility parameter (R_{K-T}) appears in both the equations for maximum power output and the corresponding thermal efficiency.

Since the value of R_{K-T} is always more than unity for a real cycle, the power delivered by the cycle is always lesser than that of the Curzon–Ahlborn model. The expressions for the maximum power and the corresponding thermal efficiency will reduce to the endoreversible cycle with finite heat capacity mentioned above if R_{K-T} equals to unity. Further, if the external reservoirs are infinitely large the present case tends to the Curzon–Ahlborn case, as below:

$$P_{\max} = \frac{k_H k_L \left[\sqrt{T_{H1}} - \sqrt{T_{L1}}\right]^2}{(k_H + k_L)} \tag{2.52}$$

and

$$\eta_m = 1 - \sqrt{\frac{T_{L1}}{T_{H1}}} \tag{2.53}$$

Further, if the irreversibility parameter tends to the temperature ratio of the external reservoirs viz. the ratio of the source temperature to that of the sink temperature, the power output of the cycle tends to zero. This corresponds to the Carnot case where heat source and heat sink are of infinite heat capacity, i.e. as $R_{K-T} \to T_H/T_L$ and $P \to 0$. Again, if R_{K-T} equals to unity and the reservoirs are of infinite heat capacity, the above expressions reduce to:

$$P_{\max} = \frac{k_H k_L \left[\sqrt{T_H} - \sqrt{T_L}\right]^2}{(k_H + k_L)} \tag{2.54}$$

and

$$\eta_m = 1 - \sqrt{\frac{T_L}{T_H}} = \eta_{C-A} \tag{2.55}$$

Thus, the cycle model presented by Kumar (2000) is more general, and most of the cycle models analysed so far can be derived from this particular model as a special case, and hence, this makes this particular model of special importance.

As it is well understood from experimental and analytical observations, the Carnot cycle cannot be a practical cycle by any means, yet it will be the unique model that represents the upper limit of thermal efficiency that a real power cycle can never achieve. However, difficulties with Carnot cycle may be eliminated by replacing the isothermal processes to that of an isobaric processes using the working fluid which change its phase while heating or cooling at constant temperature, and hence, the concept of Rankine cycle/vapour power cycle comes into existence during the 1820s.

2.6 Ideal Rankine Cycle

The Rankine cycle was proposed and evaluated mathematically by William John Macquorn Rankine, a Scottish Civil Engineer and Mathematician and Professor at Glasgow University which was named after him (1820–1872). The Rankine cycle is the fundamental thermodynamic underpinning of the steam engine. The simple Rankine cycle comprises four main components, namely, boiler, turbine, condenser, and feed pump. In an ideal Rankine cycle, there are four sequential operations, namely, isobaric heating, isentropic expansion, isobaric heat rejection, and isentropic pumping, as shown on the schematic diagram of Fig. 2.6a and *T–s* diagram of Fig. 2.6b, respectively. It is well understood that the actual Rankine cycle is not as efficient as the ideal cycle, and hence, the quest for higher thermal efficiencies has resulted in some innovative modifications to the basic vapour power cycle, such as lowering the condenser pressure, superheating the turbine inlet steam in the boiler, increasing the boiler pressure, use of regeneration, inclusion of feed water heater and reheater, and so on. However, the details about these modified Rankine cycles can be found in different basic text books and engineering thermodynamics. In this part of the chapter, the finite time thermodynamic analysis of Rankine cycle will be presented as the main objective of this book.

2.7 Finite Time Rankine Cycle

After the novel work of finite time thermodynamics by Curzon and Ahlborn (1975), it was extended further by others for Carnot and Rankine cycles using different modifications. As discussed earlier, a major objective of finite time thermodynamics is to understand irreversible finite time process and to establish a bound on

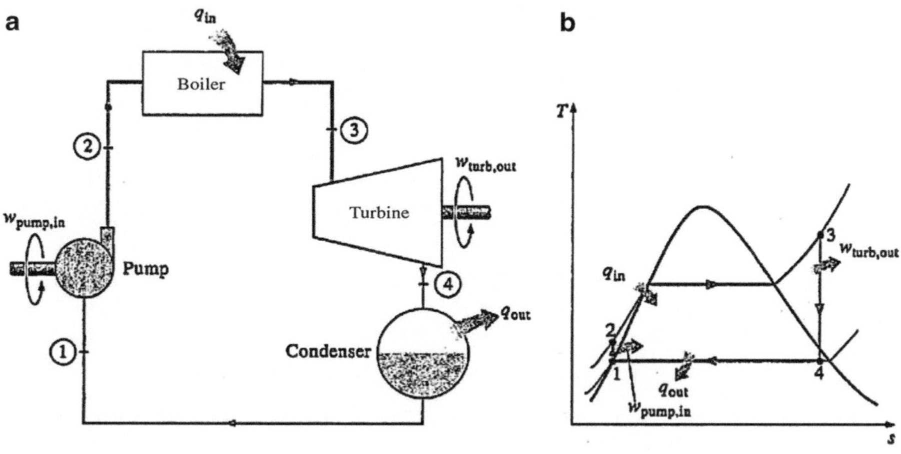

Fig. 2.6 (**a**) The schematic and (**b**) *T–s* diagrams of an ideal Rankine cycle

efficiency and the maximum power for such processes. The system adopted is a reversible Rankine heat engine, which is connected to the heating and cooling fluids with infinite/finite heat capacity rates. The heat transfer processes which occur across finite temperature differences are irreversible as shown on the temperature–entropy diagram of Fig. 2.7a. Heat is transferred from the heating fluid to the Rankine heat engine, which converts a fraction of heat into work and rejects the rest to the heat sink of finite/infinite heat capacity of cooling fluid. If the heat source and sink are of finite extent, their temperatures will change as a result of heat transfer process, while it will remain constant otherwise. In the present section, both the finite and infinite heat capacity of external reservoirs will be studied similar to that of Carnot cycle presented earlier.

Unlike Carnot cycle, the Rankine cycle has variation in the working fluid temperature during heat addition, which is basically due to the change of phase of the working fluid during heat addition and heat rejection. However, during heat rejection, the working fluid changes only phase and, hence, remains at almost constant temperature, while during heat addition process from the source, the working fluid undergoes different phases. For example, subcooled liquid, saturated liquid, liquid–vapour mixture, saturated vapour, and finally superheated vapour, as a result, the temperature of the working fluid during heat addition changes and has three different zones as can be seen from Fig. 2.7a.

Due to the reasons mentioned above, the Rankine cycle has a different temperature–entropy diagram than that of a Carnot cycle, and hence, the analysis of Rankine cycle is slightly different than that of the Carnot cycle by making some assumption and defining the entropic average temperature (T_a) which is also shown on the T–s diagram of Fig. 2.7b. The entropic average temperature can be defined as:

$$T_a = \Delta Q/\Delta s \tag{2.56}$$

Fig. 2.7a T–s diagram of finite time Rankine cycle

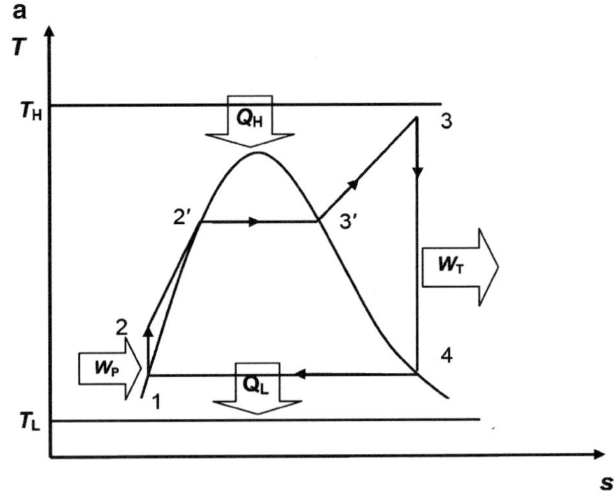

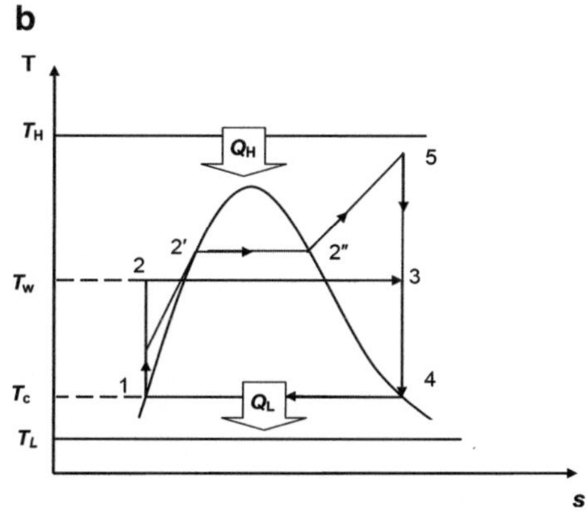

Fig. 2.7b *T–s* diagram of Carnot equivalent Rankine cycle

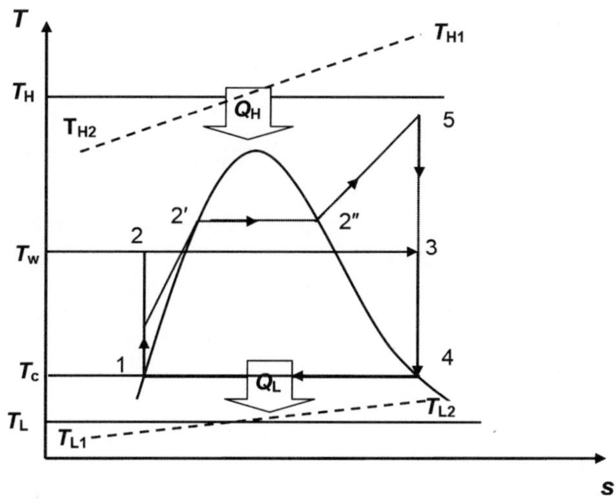

Fig. 2.8 *T–s* diagram of Carnot equivalent Rankine cycle

where Q and s represent the heat and entropy, respectively. Thus the entropic average temperature of heat addition and rejection, T_w and T_c become:

$$T_w = (h_3 - h_2)/(s_3 - s_2) \tag{2.57}$$

$$T_c = (h_4 - h_1)/(s_4 - s_1) \tag{2.58}$$

where h's denote the specific enthalpy values at the respective state points. Thus the Rankine heat engine becomes an equivalent Carnot heat engine operating between the source and sink of infinite heat capacity with working fluid temperatures T_w and T_c as shown in Fig. 2.8. Thus, with the above modifications, there are now two cases for a Rankine cycle, as below.

2.7.1 Alternatively Connected Rankine Cycle

The heat engine operates in a mode with fixed time, t_{cy} allotted for each cycle. Thus, after time t_{cy} has leashed, the working fluid returns to the initial state. We assume that the inlet temperatures of a heat source and heat sink (T_{H1}, T_{L1}), the heat conductance ($U_H A_H$, $U_L A_H$) of the heat exchangers, and the heat capacity rates ($\dot{m}_H C_{pH}$, $\dot{m}_L C_{pL}$) of the heat source and the heat sink are fixed but are otherwise arbitrary. Therefore, the temperature distributions of the heating and cooling fluids are not constant through the heat exchangers as shown in Fig. 2.8. The analysis of Rankine cycle is similar to that of a Carnot cycle, so the infinite and finite heat capacities of external reservoirs have been indicated on the same diagram. The formulations for the two cases may be left to readers to understand the difference and similarity between the two different cases; besides, the comparative analysis of the two cases mentioned above may be done in the similar fashion, separately.

The rate of heat flow from the heat source to the heat engine is proportional to the Log Mean Temperature Difference $(\text{LMTD})_H$ and is equivalent to the decreasing rate of heat input from the heating fluid. If the heat input tests for time t_H, per cycle, the rate of heat input is:

$$\dot{Q}_H = Q_H/t_H = U_H A_H (\text{LMTD})_H = \dot{m}_H C_{pH}(T_{H1} - T_{H2}) \tag{2.59}$$

where $(\text{LMTD})_H$ is the log mean temperature difference (LMTD) at the source side and is given as below:

$$(\text{LMTD})_H = \frac{[(T_{H1} - T_w) - (T_{H2} - T_w)]}{\ln \frac{(T_{H1} - T_w)}{(T_{H2} - T_w)}} \tag{2.60}$$

Similarly, for condensation, the rate of heat flow from the heat engine to the heat sink is given by:

$$\dot{Q}_L = Q_L/t_L = U_L A_L (\text{LMTD})_L = \dot{m}_L C_{pL}(T_{L2} - T_{L1}) \tag{2.61}$$

where $(\text{LMTD})_L$ is the log mean temperature difference at the sink side and is given as below:

$$(\text{LMTD})_L = \frac{[(T_c - T_{L1}) - (T_c - T_{L2})]}{\ln \frac{(T_c - T_{L1})}{(T_c - T_{L2})}} \tag{2.62}$$

where $(UA)_H$ and $(UA)_L$ are the overall heat transfer coefficient-area products on the source- and sink-side heat exchangers, while $\dot{m}_H C_{pH}$ and $\dot{m}_L C_{pL}$ are the products of mass flow rate times the specific heat of the external fluids in the source and sink reservoirs, respectively. If t_H and t_L are, respectively, the time taken during the heat addition and rejection processes, the total time taken per cycle will be:

$$t_{\mathrm{cy}} = t_{\mathrm{H}} + t_{\mathrm{L}} + t_{12} + t_{34} \tag{2.63}$$

where t_{12} and t_{34} are the times required in two isentropic processes and are negligible as compared to the time of two isobaric processes. Thus the cycle time becomes:

$$t_{\mathrm{cy}} = t_{\mathrm{H}} + t_{\mathrm{L}} \tag{2.64}$$

Using Eqs. (2.59) and (2.64) yields:

$$t_{\mathrm{cy}} = \frac{Q_{\mathrm{H}}}{\dot{m}_{\mathrm{H}} C_{\mathrm{pH}} (T_{\mathrm{H1}} - T_{\mathrm{H2}})} + \frac{Q_{\mathrm{L}}}{\dot{m}_{\mathrm{L}} C_{\mathrm{pL}} (T_{\mathrm{L2}} - T_{\mathrm{L2}})} \tag{2.65}$$

Using Eqs. (2.59) and (2.61) yields:

$$T_{\mathrm{H2}} = T_{\mathrm{w}} + (T_{\mathrm{H1}} - T_{\mathrm{w}}) \exp\left[-\frac{U_{\mathrm{H}} A_{\mathrm{H}}}{\dot{m}_{\mathrm{H}} C_{\mathrm{pH}}}\right] \tag{2.66}$$

$$T_{\mathrm{L2}} = T_{\mathrm{c}} - (T_{\mathrm{c}} - T_{\mathrm{L1}}) \exp\left[-\frac{U_{\mathrm{L}} A_{\mathrm{L}}}{\dot{m}_{\mathrm{L}} C_{\mathrm{pL}}}\right] \tag{2.67}$$

Substituting Eqs. (2.66) and (2.67) into Eq. (2.65) yields:

$$t_{\mathrm{cy}} = \frac{Q_{\mathrm{H}}}{\mathrm{HE}(T_{\mathrm{H1}} - T_{\mathrm{w}})} + \frac{Q_{\mathrm{L}}}{\mathrm{LE}(T_{\mathrm{c}} - T_{\mathrm{L1}})} \tag{2.68}$$

where HE and LE are given as below:

$$\mathrm{HE} = \dot{m}_{\mathrm{H}} C_{\mathrm{pH}} \left[1 - \exp\left(-\frac{U_{\mathrm{H}} A_{\mathrm{H}}}{\dot{m}_{\mathrm{H}} C_{\mathrm{pH}}}\right)\right] \tag{2.69}$$

$$\mathrm{LE} = \dot{m}_{\mathrm{L}} C_{\mathrm{pL}} \left[1 - \exp\left(-\frac{U_{\mathrm{L}} A_{\mathrm{L}}}{\dot{m}_{\mathrm{L}} C_{\mathrm{pL}}}\right)\right] \tag{2.70}$$

The first law and second laws of thermodynamics for this particular model yield:

$$W = Q_{\mathrm{H}} - Q_{\mathrm{L}} \tag{2.71}$$

$$\frac{Q_{\mathrm{H}}}{T_{\mathrm{w}}} = \frac{Q_{\mathrm{L}}}{T_{\mathrm{c}}} \tag{2.72}$$

From Eqs. (2.68) and (2.72), we have:

$$Q_{\mathrm{H}} = \frac{W T_{\mathrm{w}}}{(T_{\mathrm{w}} - T_{\mathrm{c}})} \tag{2.73}$$

$$Q_{\mathrm{L}} = \frac{WT_{\mathrm{c}}}{(T_{\mathrm{w}} - T_{\mathrm{c}})} \tag{2.74}$$

From Eqs. (2.68, 2.73 and 2.74), we have:

$$t_{\mathrm{cy}} = \left[\frac{WT_{\mathrm{w}}}{\mathrm{HE}(T_{\mathrm{H1}} - T_{\mathrm{w}})(T_{\mathrm{w}} - T_{\mathrm{c}})} + \frac{WT_{\mathrm{c}}}{\mathrm{LE}(T_{\mathrm{c}} - T_{\mathrm{L1}})(T_{\mathrm{w}} - T_{\mathrm{c}})}\right] \tag{2.75}$$

The power output will be:

$$P = \frac{W}{t_{\mathrm{cy}}} = \left[\frac{T_{\mathrm{w}}}{\mathrm{HE}(T_{\mathrm{H1}} - T_{\mathrm{w}})} + \frac{T_{\mathrm{c}}}{\mathrm{LE}(\mathrm{T_c} - \mathrm{T_{L1}})}\right]^{-1} (T_{\mathrm{w}} - T_{\mathrm{c}}) \tag{2.76}$$

or

$$P = \frac{\mathrm{HE\,LE}\,xy\,(T_{\mathrm{H1}} - T_{\mathrm{L1}} - x - y)}{[x\mathrm{HE}T_{\mathrm{L1}} + y\mathrm{LE}T_{\mathrm{H1}} + xy(\mathrm{HE} - \mathrm{LE})]} \tag{2.77}$$

where $x = (T_{\mathrm{H1}} - T_{\mathrm{w}})$ and $y = (T_{\mathrm{c}} - T_{\mathrm{L1}})$. It can be seen from Eq. (2.77) that P is function of x and y only because T_{H1}, T_{L1}, HE, and LE are constants for a typical set of operating parameters. Thus, using Eq. (2.77) and its extremal condition yields:

$$T_{\mathrm{w}} = C_1\sqrt{T_{\mathrm{H1}}} \text{ and } T_{\mathrm{c}} = C_1\sqrt{T_{\mathrm{L1}}} \tag{2.78}$$

where C_1 is given as below:

$$C_1 = \left[\frac{\sqrt{\mathrm{HE}T_{\mathrm{H1}}} + \sqrt{\mathrm{LE}T_{\mathrm{L1}}}}{\sqrt{\mathrm{HE}} + \sqrt{\mathrm{LE}}}\right] \tag{2.79}$$

Substituting Eqs. (2.78) and (2.79) into Eq. (2.77), the maximum power output of the engine will be:

$$P_{\mathrm{max}} = \frac{\mathrm{HE\,LE}\left[\sqrt{T_{\mathrm{H1}}} - \sqrt{T_{\mathrm{L1}}}\right]^2}{\left[\sqrt{\mathrm{HE}} + \sqrt{\mathrm{LE}}\right]^2} \tag{2.80}$$

and the thermal efficiency at maximum power output will be:

$$\eta_{\mathrm{m}} = 1 - \sqrt{\frac{T_{\mathrm{L1}}}{T_{\mathrm{H1}}}} \tag{2.81}$$

From Eqs. (2.80) and (2.81), it is clear that the optimum power and the corresponding thermal efficiency of the Rankine cycle are also same as for the Carnot heat engine. It is also found from these equations that the optimum power and the corresponding thermal efficiency depend on the initial temperatures and the heat capacity of the heating and cooling fluids as well as on the heat transfer coefficients. Although they also depend on the heat engine model and the assumption made for a typical cycle, yet the final expressions remains the same.

Again, if the external reservoirs are of infinite heat capacity, the expressions for the maximum power output and the corresponding thermal efficiency will reduce to:

$$P_{\max} = \frac{k_{\mathrm{H}} k_{\mathrm{L}} \left[\sqrt{T_{\mathrm{H1}}} - \sqrt{T_{\mathrm{L}}}\right]^2}{\left[\sqrt{k_{\mathrm{H}}} + \sqrt{k_{\mathrm{L}}}\right]} \tag{2.82}$$

$$\eta_{\mathrm{m}} = 1 - \sqrt{\frac{T_{\mathrm{L}}}{T_{\mathrm{H}}}} \tag{2.83}$$

where k_{H} and k_{L} are the conductance on the source and sink side, respectively. This is the similar case to that of Carnot cycle where the time of two adiabatic processes has been neglected as given in Eqs. (2.42) and (2.43).

2.7.2 Continuously Connected Rankine Cycle

For such an engine, the time taken for heat transfer to the heat engine is same, as the time required for the heat transfer from the heat engine, i.e.

$$t_{\mathrm{H}} = t_{\mathrm{L}} \tag{2.84}$$

In such a case, the power output may be written as:

$$P = \mathrm{HE}(T_{\mathrm{H1}} - T_{\mathrm{w}}) \left[\frac{T_{\mathrm{w}}(\mathrm{HE} + \mathrm{LE}) - \mathrm{HE}T_{\mathrm{H1}} - \mathrm{LET}_{\mathrm{L1}}}{(\mathrm{HE} + \mathrm{LE})T_{\mathrm{w}} - \mathrm{HET}_{\mathrm{H1}}}\right] \tag{2.85}$$

From Eq. (2.85) it is seen that P is a function of single variable T_{w}. Thus maximizing 'P' w. r. t. T_{w} yields:

$$T_{\mathrm{w}} = C\sqrt{T_{\mathrm{H1}}}\,\mathrm{and}\,T_{\mathrm{c}} = C\sqrt{T_{\mathrm{L1}}} \tag{2.86}$$

where C is given as below:

$$C = \left[\frac{\mathrm{HE}\sqrt{T_{\mathrm{H1}}} - \mathrm{LE}\sqrt{T_{\mathrm{L1}}}}{(\mathrm{HE} + \mathrm{LE})}\right] \tag{2.87}$$

The maximum power output and the corresponding thermal efficiency become:

$$P_{\max} = \frac{\mathrm{HE}\,\mathrm{LE}\left(\sqrt{T_{\mathrm{H1}}} - \sqrt{T_{\mathrm{L1}}}\right)^2}{\mathrm{HE} + \mathrm{LE}} \tag{2.88}$$

and

$$\eta = 1 - \sqrt{\frac{T_{L1}}{T_{H1}}} \tag{2.89}$$

The expressions given in Eqs. (2.88) and (2.89) are similar to those given in Eqs. (2.50) and (2.51) where $HE = k_H, LE = k_L$ and $R_{K-T} = 1.0$, which is also a case of Carnot cycle continuously connected with the finite heat capacity of external reservoirs. However, the thermal efficiency remains the same as in the previous (alternatively connected) case, and this is also a similar expression as derived for the Carnot heat engine. Hence, this equation can also be applied for the performance evaluation of other operational heat engines, in general, and the Rankine cycle, in particular.

2.8 Irreversible Rankine Cycle

So far a Rankine heat engine has been evaluated with different cases of continuously and alternatively connected with external reservoirs assuming the external irreversibility due to finite temperature difference during the heat transfer processes. However, since the expansion in the turbine and pumping in the feed pump are not perfectly isentropic processes, irreversibilities are present within the heat engine. Thus, in real practice, one has to deal with both internal as well as external irreversibilities. The external irreversibility is due to finite temperature difference between the heat engine and the external reservoirs, while the internal irreversibilities are due to friction, entropy generation during the heat addition, rejection processes, etc.

The cycle model is shown in Fig. 2.9 with internal irreversibility due to various reasons mentioned above, and the entropic average temperature was applied in order to modify the Rankine heat engine to an equivalent Carnot cycle similar to the previous cases. Hence, the irreversible Rankine cycle becomes an equivalent Carnot cycle with internal irreversibilities operating between the temperatures T_w and T_c as shown on the *T–s* diagram of Fig. 2.9 with finite heat capacity of heat source and sink reservoirs. Let $\dot{Q}_H$ and $\dot{Q}_L$ be, respectively, the heat transfer rates to and from the heat engine during the two modified isothermal (isobaric) processes as below (Kaushik 1999; Wu and Kiang 1992):

$$\dot{Q}_H = k_H(T_{H1} - T_w) = T_w(s_3 - s_2) \tag{2.90}$$

$$\dot{Q}_L = k_L(T_c - T_{L1}) = T_c(s_4 - s_1) \tag{2.91}$$

where k_H and k_L are the conductance on the source and sink side, respectively, and *s* is the entropy values at the respective state points. Also from the first and second laws of thermodynamics, we have:

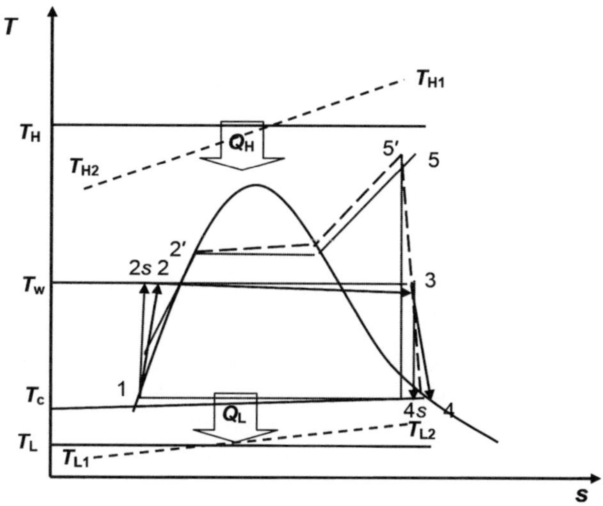

Fig. 2.9 *T–s* diagram of Carnot equivalent irreversible Rankine cycle

$$P = \dot{Q}_{\mathrm{H}} - \dot{Q}_{\mathrm{L}} = k_{\mathrm{H}}(T_{\mathrm{H1}} - T_{\mathrm{w}}) - k_{\mathrm{L}}(T_{\mathrm{c}} - T_{\mathrm{L1}}) \tag{2.92}$$

$$\int \frac{dQ}{T} = \frac{\dot{Q}_{\mathrm{H}}}{T_{\mathrm{w}}} - \frac{\dot{Q}_{\mathrm{L}}}{T_{\mathrm{c}}} \leq 0 \tag{2.93}$$

If we define the irreversibility parameter $R_{\mathrm{K-T}}$ which is the ratio of entropy generation during heat rejection to that of during heat addition and greater than unity for a real cycle as below:

$$\frac{\dot{Q}_{\mathrm{L}}}{T_{\mathrm{c}}} \bigg/ \frac{\dot{Q}_{\mathrm{H}}}{T_{\mathrm{w}}} = \frac{(s_4 - s_1)}{(s_3 - s_2)} = R_{\mathrm{K-T}} \geq 1 \tag{2.94}$$

$$R_{\mathrm{K-T}} \frac{\dot{Q}_{\mathrm{H}}}{T_{\mathrm{w}}} = \frac{\dot{Q}_{\mathrm{L}}}{T_{\mathrm{c}}} \tag{2.95}$$

Substituting Eq. (2.94) into Eq. (2.92) and maximizing '*P*' with respect to T_{h} and T_{l} as done in the previous section we obtained the expressions for maximum power output and the corresponding thermal efficiency as given below:

$$P_{\max} = \frac{k_{\mathrm{H}} k_{\mathrm{L}} \left[\sqrt{T_{\mathrm{H1}}} - \sqrt{R_{\mathrm{K-T}} T_{\mathrm{L1}}}\right]^2}{(k_{\mathrm{H}} R_{\mathrm{K-T}} + k_{\mathrm{L}})} \tag{2.96}$$

$$\eta_{\mathrm{m}} = 1 - \sqrt{\frac{R_{\mathrm{K-T}} T_{\mathrm{L1}}}{T_{\mathrm{H1}}}} \tag{2.97}$$

The cycle irreversibility parameter ($R_{\mathrm{K-T}}$) appears in both the equations for maximum power output and the corresponding thermal efficiency. The value of $R_{\mathrm{K-T}}$ is always more than unity, and hence, the power delivered by the cycle is always lesser than that of an endoreversible Rankine cycle mentioned in the earlier sections. For example, if R_{K-T} equals to unity, the expressions for the maximum power and the corresponding thermal efficiency will reduce to the endoreversible cycle case, as below:

$$P_{\max} = \frac{k_H k_L \left[\sqrt{T_{H1}} - \sqrt{T_{L1}}\right]^2}{(k_H + k_L)} \tag{2.98}$$

$$\eta_m = 1 - \sqrt{\frac{T_{L1}}{T_{H1}}} \tag{2.99}$$

Again, if the irreversibility parameter tends to the temperature ratio (sink to source) the power output tends to zero. This corresponds to the Carnot case where heat source and heat sink are of infinite heat capacity. Further, if the internal irreversibility parameter tends to unity and the heat capacity of the external reservoirs, then the expressions for the maximum and the corresponding thermal efficiency are given as:

$$P_{\max} = \frac{k_H k_L \left[\sqrt{T_H} - \sqrt{T_L}\right]^2}{(k_H + k_L)} \tag{2.100}$$

$$\eta_m = 1 - \sqrt{\frac{T_L}{T_H}} \tag{2.101}$$

Thus, this particular cycle model presented here in Eqs. (2.96) and (2.97) is more general, and most of the other models of Rankine cycle analysed so far can be derived directly from this particular model as a special case, and hence, this makes this particular model of special importance for a typical set of operating conditions mentioned above.

2.9 Conclusion

This chapter presents the finite time thermodynamic analysis of Carnot and Rankine cycles with external and internal irreversibilities for finite/infinite heat capacity of the external reservoirs. The external irreversibility is due to the finite temperature difference between the cycle and the external reservoirs, while the internal irreversibility is due to other losses such as friction, pressure drop, and non-isentropic compression/pumping and expansion processes in the compressor/pump and the turbine, respectively. Some special cases were also presented, and it was found that the cycle model presented by Kaushik (1999) is more general and very close to real heat engine for both Carnot and Rankine cycles as the other cases discussed by number of authors and can be derived directly as a special case of this particular cycle model.

Chapter 3
Finite Time Thermodynamic Analysis of Brayton Cycle

3.1 Introduction

The Brayton cycle was first proposed by George Brayton in 1870 for the reciprocating oil-burning engine and has been extensively used in power plants and aeroplanes since then. Today, it is used for gas turbines only where both the compression/expansion processes may operate on either in an open or a closed cycle. The Brayton cycle may be either an open or a closed cycle depending on the working fluid; e.g., for gas other than air, a closed cycle is desirable. In the open cycle, the atmospheric air is continuously drawn into the compressor, where it is compressed to a high pressure. The air then enters the combustion chamber where it is mixed with the fuel and combustion occurs, resulting in combustion products at an elevated temperature and pressure. The combustion product at high temperature and pressure then expands through the turbine and subsequently discharged to the atmosphere. Also, the turbine work developed is partly used to drive the compressor, while the remainder is available to generate electricity, to propel vehicle, and/or for other useful purposes (Fig. 3.1a).The open gas turbine cycle described here can be modelled as a closed cycle by utilizing the air standard assumption, as shown in Fig. 3.1b.

However, in the case of a closed Brayton cycle, the compression and expansion processes remain the same, but the combustion process is replaced by a constant pressure heat addition from a heat source, and the exhaust process is replaced by a constant pressure heat rejection to an external reservoir/heat sink as can be seen on the schematic diagram of Fig. 3.1b. In general, the temperature of the exhaust of the gas turbine is of the order of few hundred degrees Celsius and, hence, can be utilized for other useful applications such as combined power cycle, cogeneration, trigeneration, etc.; however, the details about these applications are not given in this book.

S.C. Kaushik et al., *Finite Time Thermodynamics of Power and Refrigeration Cycles*,
https://doi.org/10.1007/978-3-319-62812-7_3

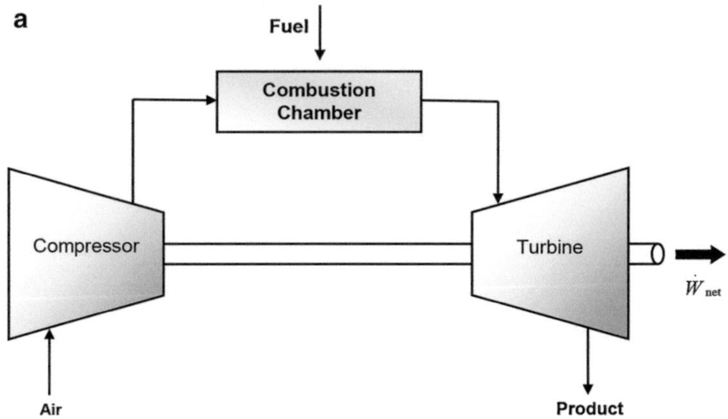

Fig. 3.1a Schematic diagram of an open cycle Brayton heat engine

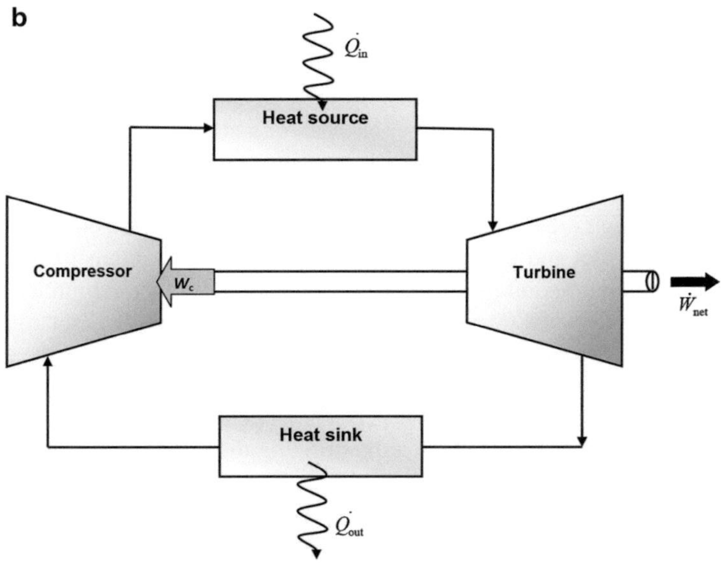

Fig. 3.1b Schematic diagram of closed cycle Brayton heat engine

3.2 Ideal Brayton Cycle

The ideal closed Brayton cycle consists of four reversible processes such as isobaric heat addition, isentropic expansion, isobaric heat rejection, and isentropic compression as shown on the *T–s* and *P–v* diagrams of Figs. 3.2a and 3.2b, respectively. Since all the four processes of a Brayton cycle are executed on steady-state flow devices, they must be analysed on steady-state flow process. When the changes in

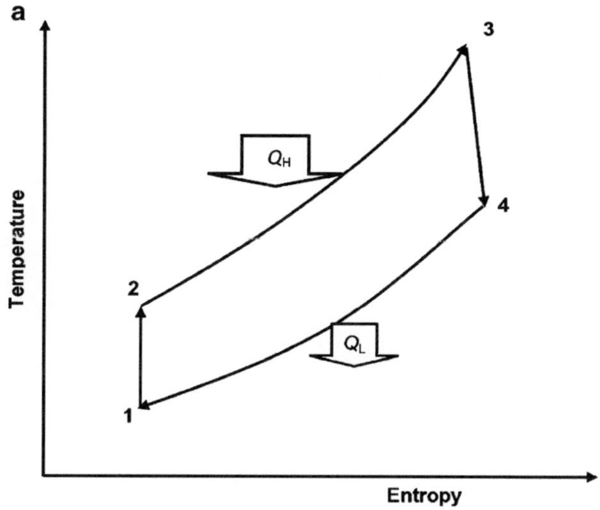

Fig. 3.2a *T–s* diagram of a closed Brayton cycle

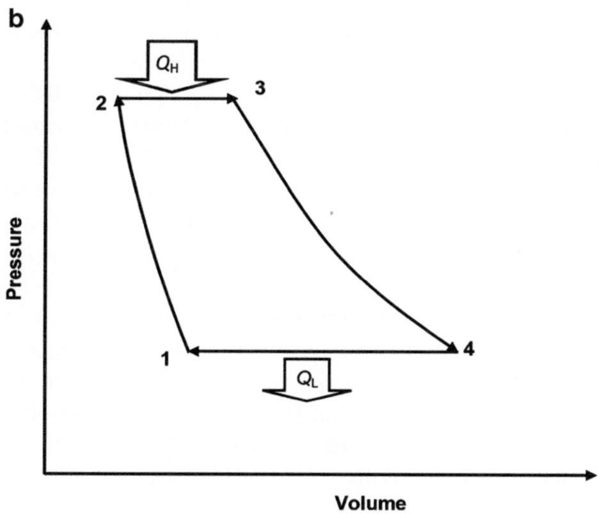

Fig. 3.2b *P–v* diagram of a closed Brayton cycle

kinetic and potential energies are neglected, the energy balance for steady-state flow process can be expressed as below:

$$\dot{Q}_{\mathrm{H}} = \dot{m}\, c_{\mathrm{p}}(T_3 - T_2) \tag{3.1}$$

$$\dot{Q}_{\mathrm{L}} = \dot{m}\, c_{\mathrm{p}}(T_4 - T_1) \tag{3.2}$$

The thermal efficiency of an ideal Brayton cycle can be expressed as:

$$\eta_{\mathrm{th}} = \frac{\dot{Q}_{\mathrm{H}} - \dot{Q}_{\mathrm{L}}}{\dot{Q}_{\mathrm{H}}} = \frac{\dot{m}\,c_{\mathrm{p}}(T_3 - T_2) - \dot{m}\,c_{\mathrm{p}}(T_4 - T_1)}{\dot{m}\,c_{\mathrm{p}}(T_3 - T_2)} = \frac{T_1(T_4/T_1 - 1)}{T_2(T_3/T_2 - 1)} \tag{3.3}$$

Also for an ideal Brayton cycle, $P_1 = P_4$ and $P_2 = P_3$ yields:

$$T_2/T_1 = \left(\frac{P_2}{P_1}\right)^{(\gamma-1)/\gamma} = T_3/T_4 = \left(\frac{P_3}{P_4}\right)^{(\gamma-1)/\gamma} \tag{3.4}$$

Using Eqs. (3.3) and (3.4) yields:

$$\eta_{\mathrm{th}} = 1 - \frac{1}{\left(R_{\mathrm{p}}\right)^{(\gamma-1)/\gamma}} \tag{3.5}$$

where R_p is the pressure ratio of the cycle and γ is the specific heat ratio of the gas/working fluid being used in the Brayton cycle. It is clear from Eq. (3.5) that the thermal efficiency of a Brayton cycle increases as the cycle pressure ratio and the specific heat ratio of the working fluid increase, which is also the case in practice, especially for simple Brayton cycle. Also the highest temperature in the cycle occurs at the end of the combustion process which is limited by the maximum temperature that the turbine blades can withstand. Nowadays, with the advancement in material technology, the upper limit has gone above 1500 °C. Thus, the upper limit of the maximum temperature of a gas turbine also limits the pressure ratio that can be used in the cycle. However, in modified and complex Brayton cycles, the upper limit of cycle pressure ratio is also influenced by other parameters, such as the effectiveness of regenerator, intercooler pressure ratio, the reheat pressure ratio, etc. There are two major areas for unique applications of gas turbine, one is aircraft propulsion and the other is stationary power generation in thermal power plant either stand-alone and/or in conjunction of steam power (Rankine) cycle for cogeneration applications. It is well known that a practical Brayton cycle is not as efficient as the classical Brayton cycle. To achieve the theoretical Brayton cycle efficiency, the isobaric heating and cooling processes of the cycle must be carried out infinitely slowly to ensure that the working fluid is in thermal equilibrium.

Thus the power output of the cycle approaches zero, since it requires an infinite time to produce a finite amount of work. On the other hand, it requires infinite heat exchanger area to transfer a finite amount of heat in finite time. Again, if it is the case, the system will be uneconomical as the heat exchanger components are quite expensive, and hence, the very purpose of power production is defeated. Thus, to obtain finite power/finite amount of work in finite time, the cycle should be speeded up and sizes of heat exchangers should be finite. However, if the heat engine speeds were infinitely fast, the heat would flow directly from source to sink, and hence, no mechanical work would be performed by the heat engine; thus, the power output would be again zero. Somewhere between these two limits, the heat engine has the maximum power output. Keeping the above aspects in mind, a number of authors have studied the Brayton cycle following Curzon and Ahlborn (1975) approach using the concept of finite time thermodynamics. The efficiency of the heat engine under the condition of maximum power output has been evaluated by many authors taking different assumptions and modifications in the Brayton cycle, like:

- Finite temperature difference between the heat engine and the external reservoirs, viz. finite time Brayton cycle
- Introducing a regenerator to utilize the high-temperature exhaust of the turbine to preheat the working fluid before it enters the combustion chamber, viz. a regenerative Brayton cycle

3.3 Finite Time Brayton Cycle

As mentioned above, a real Brayton cycle is not as efficient as an ideal one. However, a real cycle also consists of four processes mentioned above, yet there is a finite temperature difference between the cycle and the external reservoirs as shown on the *T–s* diagram of Fig. 3.3. Since all the four processes of a Brayton cycle are executed on steady-state flow devices, they must be analysed on steady-state flow process. Also, when the changes in kinetic and potential energies are neglected, the heat transfer rates to and from the cycle for a steady-state flow process can be expressed separately, for two different cases as below.

3.3.1 *Infinite Heat Capacity*

When the external reservoirs are of infinite heat capacity, the temperature (T_H) of the source and the sink (T_L) does not change with time. Using the heat transfer

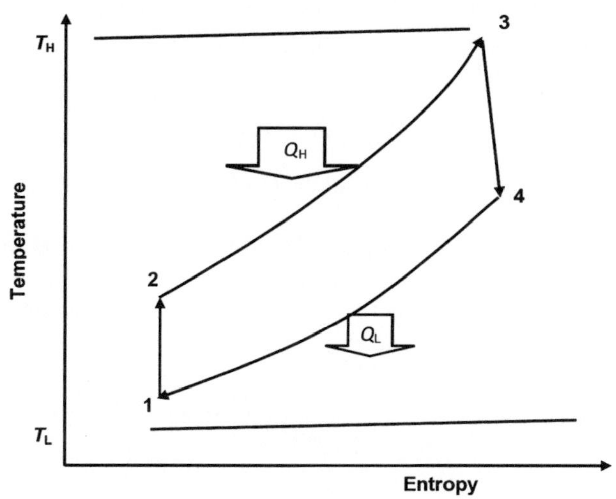

Fig. 3.3 *T–s* diagram of finite time Brayton cycle with infinite heat capacity

theory and neglecting the changes in the kinetic and potential energies for steady-state flow, the heat transfer rates to and from the cycle can be expressed, following Ibrahim et al. (1991) and Kaushik and Tyagi (2002), as below:

$$\dot{Q}_{\mathrm{H}} = C_{\mathrm{wf}}(T_3 - T_2) = (\mathrm{UA})_{\mathrm{H}}(\mathrm{LMTD})_{\mathrm{H}} \tag{3.6}$$

$$\dot{Q}_{\mathrm{L}} = C_{\mathrm{wf}}(T_4 - T_1) = (\mathrm{UA})_{\mathrm{L}}(\mathrm{LMTD})_{\mathrm{L}} \tag{3.7}$$

where C_{wf} is the heat capacitance rate of the working fluid, (UA) is the overall heat transfer coefficient-area product, while LMTD is the Log Mean Temperature Difference, on the source- and sink-side reservoirs, respectively, and given as below:

$$(\mathrm{LMTD})_{\mathrm{H}} = \frac{[(T_{\mathrm{H}} - T_2) - (T_{\mathrm{H}} - T_3)]}{\ln\left(\frac{(T_{\mathrm{H}} - T_2)}{(T_{\mathrm{H}} - T_3)}\right)} \tag{3.8}$$

$$(\mathrm{LMTD})_{\mathrm{L}} = \frac{[(T_4 - T_{\mathrm{L}}) - (T_1 - T_{\mathrm{L}})]}{\ln\left(\frac{(T_4 - T_{\mathrm{L}})}{(T_1 - T_{\mathrm{L}})}\right)} \tag{3.9}$$

Using Eqs. (3.6), (3.7), (3.8), and (3.9) yields:

$$\dot{Q}_{\mathrm{H}} = C_{\mathrm{wf}}(T_3 - T_2) - C_{\mathrm{wf}}\varepsilon_{\mathrm{H}}(T_{\mathrm{H}} - T_2) \tag{3.10}$$

$$\dot{Q}_{\mathrm{L}} = C_{\mathrm{wf}}(T_4 - T_1) - C_{\mathrm{wf}}\varepsilon_{\mathrm{L}}(T_4 - T_{\mathrm{L}}) \tag{3.11}$$

where ε_{H} and ε_{L} are the effectiveness of the source- and sink-side heat exchangers, respectively, and are given as below:

$$\varepsilon_{\mathrm{H}} = 1 - \exp^{\frac{(\mathrm{UA})_{\mathrm{H}}}{C_{\mathrm{wf}}}} \Rightarrow 1 - e^{-\mathrm{NTU_H}} \quad \text{and} \quad \varepsilon_{\mathrm{L}} = 1 - \exp^{\frac{(\mathrm{UA})_{\mathrm{L}}}{C_{\mathrm{wf}}}} \Rightarrow 1 - e^{-\mathrm{NTU_L}}$$

where NTU is defined as the number of transfer units on the hot- and cold-side heat exchangers, respectively (Ibrahim et al. 1991; Kumar 2000; Kaushik and Tyagi 2002). Again, using the first and second laws of thermodynamics for a closed Brayton cycle yields:

$$P = \dot{Q}_{\mathrm{H}} - \dot{Q}_{\mathrm{L}} = C_{\mathrm{wf}}\varepsilon_{\mathrm{H}}(T_{\mathrm{H}} - T_2) - C_{\mathrm{wf}}\varepsilon_{\mathrm{L}}(T_4 - T_{\mathrm{L}}) \tag{3.12}$$

$$C_{\mathrm{wf}} \ln(T_3/T_2) - C_{\mathrm{wf}} \ln(T_4/T_1) = 0 \tag{3.13}$$

Using the Lagrangian multiplier, the maximum power output and the corresponding thermal efficiency of a closed Brayton cycle coupled with infinite heat capacity (isothermal heat) reservoirs following Ibrahim et al. (1991) and Kaushik and Tyagi (2002) are given by:

$$P_{\max} = \frac{C_{\mathrm{wf}}\varepsilon_{\mathrm{H}}\varepsilon_{\mathrm{L}}\left(\sqrt{T_{\mathrm{H}}} - \sqrt{T_{\mathrm{L}}}\right)^2}{\varepsilon_{\mathrm{H}} + \varepsilon_{\mathrm{L}} - \varepsilon_{\mathrm{H}}\varepsilon_{\mathrm{L}}} \tag{3.14}$$

$$\eta_{\mathrm{m}} = 1 - \sqrt{\frac{T_{\mathrm{L}}}{T_{\mathrm{H}}}} \tag{3.15}$$

Thus, the maximum power output and the corresponding thermal efficiency of a finite time internal reversible Brayton cycle is similar to that of a Carnot cycle model given by Ibrahim et al. (1991), Wu and Kiang (1992), Sahin et al. (1998), and Kumar (2000). It is also important to note that the thermal efficiency of a finite time internal reversible Brayton cycle is also a function of the sink and source temperatures only, which shows that the model proposed by Curzon and Ahlborn (1975) for Carnot cycle is equally important and applicable to other cycles. Also, the efficiency of a Brayton cycle can be expressed in terms of pressure ratio, as below:

$$\eta_{\mathrm{m}} = 1 - \frac{1}{(R_{\mathrm{p}})^{(\gamma-1)/\gamma}} \tag{3.16}$$

where R_{p} is the pressure ratio of the cycle and γ is the specific heat ratio of the gas/working fluid being used in the Brayton cycle as given in Eq. (3.5). Thus, using Eqs. (3.15) and (3.16), it is observed that for the maximum power point, there exists a relation between the temperature and pressure ratios as below:

$$\sqrt{\frac{T_{\mathrm{L}}}{T_{\mathrm{H}}}} = \left(\frac{1}{R_{\mathrm{p}}}\right)^{(\gamma-1)/\gamma} = \left(\frac{P_{\mathrm{L}}}{P_{\mathrm{H}}}\right)^{(\gamma-1)/\gamma} = \left(\frac{P_1}{P_2}\right)^{(\gamma-1)/\gamma} \tag{3.17}$$

where $P_{\mathrm{L}}\,(=P_1)$ and $P_{\mathrm{H}}\,(=P_2)$ are, respectively, the pressure on the lower and higher side of the heat engine. In other words, P_{H} is the higher and P_{L} is the lower pressures of the Brayton cycle, working between the two reservoirs on source and sink side, respectively, as can be seen from Fig. 3.3. It is also important to note that the thermal efficiency of a Brayton cycle increases as the cycle pressure ratio and the specific heat ratio of the working fluid increase, which is also true in the case of temperature ratio, as can be seen from Eq. (3.16), especially for a simple Brayton cycle.

3.3.2 *Finite Heat Capacity*

When the external reservoirs are of finite heat capacity, the temperature of the source and the sink changes with time as can be seen from Fig. 3.4. Using the heat transfer theory and neglecting the changes in the kinetic and potential energies for steady-state flow, the heat transfer rates to and from the cycle can be expressed, following Ibrahim et al. (1991) and Kaushik and Tyagi (2002), as below:

$$\begin{aligned}\dot{Q}_{\mathrm{H}} &= C_{\mathrm{wf}}(T_3 - T_2) = \mathrm{UA_H}(\mathrm{LMTD})_{\mathrm{H}} = C_{\mathrm{H}}(T_{\mathrm{H1}} - T_{\mathrm{H2}}) \\ &= C_{\mathrm{H,min}}\varepsilon_{\mathrm{H}}(T_{\mathrm{H1}} - T_2)\end{aligned} \tag{3.18}$$

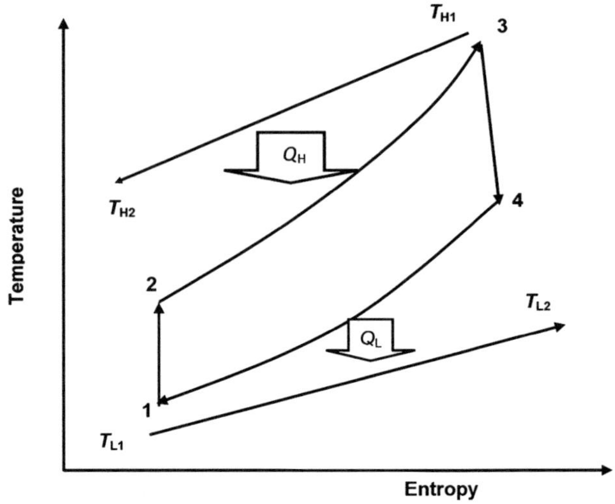

Fig. 3.4 *T–s* diagram of finite time Brayton cycle with finite capacity

$$\begin{aligned}\dot{Q}_L &= C_{wf}(T_4 - T_1) = UA_L(LMTD)_L = C_L(T_{L2} - T_{L1}) \\ &= C_{L,min}\varepsilon_L(T_4 - T_{L1})\end{aligned} \tag{3.19}$$

where $(LMTD)_H$ and $(LMTD)_L$ are the Log Mean Temperature Differences and ε_H and ε_L are the effectiveness of the source- and sink-side heat exchangers, respectively. Also $C_{H,min}$, $C_{L,min}$, $C_{H,max}$, and $C_{L,max}$ are the minimum and the maximum heat capacitance rates of the external fluids on the source and sink side, respectively. For counterflow heat exchangers, these parameters are given (Ibrahim et al. 1991; Kaushik and Tyagi 2002; Kays and London 1964) as below:

$$\begin{aligned}(LMTD)_H &= \frac{(T_{H1} - T_2) - (T_{H2} - T_3)}{\ln\left(\frac{(T_{H1}-T_2)}{(T_{H2}-T_3)}\right)} \text{ and } (LMTD)_L \\ &= \frac{(T_4 - T_{L2}) - (T_1 - T_{L1})}{\ln\left(\frac{(T_4-T_{L2})}{(T_1-T_{L1})}\right)}\end{aligned} \tag{3.20}$$

$$\varepsilon_H = \frac{1 - e^{-NTU_H\left(1-\frac{C_{H,min}}{C_{H,max}}\right)}}{1 - \frac{C_{H,min}}{C_{H,max}} e^{-NTU_H\left(1-\frac{C_{H,min}}{C_{H,max}}\right)}} \text{ and } \varepsilon_L = \frac{1 - \exp^{-NTU_L\left(1-\frac{C_{L,min}}{C_{L,max}}\right)}}{1 - \frac{C_{L,min}}{C_{L,max}} \exp^{-NTU_L\left(1-\frac{C_{L,min}}{C_{L,max}}\right)}} \tag{3.21}$$

$$\begin{pmatrix} C_{H,min} = \min(C_{wf}, C_H) \\ C_{H,max} = \max(C_{wf}, C_H) \end{pmatrix} \text{ and } \begin{pmatrix} C_{L,min} = \min(C_{wf}, C_L) \\ C_{L,max} = \max(C_{wf}, C_L) \end{pmatrix} \tag{3.22}$$

Using the first and second laws of thermodynamics for a closed internal reversible (endoreversible) Brayton cycle (Ibrahim et al. 1991; Kumar 2000) yields:

$$P = \dot{Q}_H - \dot{Q}_L = C_{H,min}\varepsilon_H(T_{H1} - T_2) - C_{L,min}\varepsilon_L(T_4 - T_{L1}) \tag{3.23}$$

$$C_{\text{wf}} \ln\left(T_3/T_2\right) - C_{\text{wf}} \ln\left(T_4/T_1\right) = 0 \tag{3.24}$$

Using the Lagrangian multiplier similar to the previous case of infinite heat capacity of external reservoirs, the maximum power output and the corresponding thermal efficiency of a closed Brayton cycle coupled with finite heat capacity reservoirs following Ibrahim et al. (1991) and Kumar (2000) are given by:

$$P_{\max} = \frac{C_{\text{H,min}}\varepsilon_{\text{H}} C_{\text{L,min}}\varepsilon_{\text{L}}\left(\sqrt{T_{\text{H1}}} - \sqrt{T_{\text{L1}}}\right)^2}{C_{\text{H,min}}\varepsilon_{\text{H}} + C_{\text{L,min}}\varepsilon_{\text{L}} - C_{\text{H,min}}\varepsilon_{\text{H}} C_{\text{L,min}}\varepsilon_{\text{L}}/C_{\text{wf}}} \tag{3.25}$$

$$\eta_{\text{m}} = 1 - \sqrt{\frac{T_{\text{L1}}}{T_{\text{H1}}}} \tag{3.26}$$

Thus, the maximum power output and the corresponding thermal efficiency of a finite time internal reversible Brayton cycle is similar to that of a Carnot cycle model given by Ibrahim et al. (1991), Wu and Kiang (1992), Sahin et al. (1998), and Kumar (2000). Again, the thermal efficiency of a finite time internal reversible Brayton cycle is also a function of the sink and source temperatures only. However, this cycle model is more general, and the results obtained earlier can be derived directly from this particular model by setting the sink and source temperatures as constant. Also it is important to note that the optimum value of the heat capacitance rate of different fluids is always bounded by the relation, as below:

$$C_{\text{L}} \geq C_{\text{wf}} \geq C_{\text{H}} \tag{3.27}$$

When the heat capacitance rate of the external fluids is equal to each other and the heat capacitance rate of the working is equal to the heat capacitance rate of the source side, viz. $C_{\text{L}} = C_{\text{H}}$ and $C_{\text{wf}}/C_{\text{H}} = 1.0$, the maximum power of the Brayton cycle asymptotically approaches to the maximum power output of the Carnot cycle as C_{wf} increases (Ibrahim et al. 1991).

3.4 Further Modifications in Brayton Cycle

The Brayton (gas turbine) cycle has experienced phenomenal progress and growth since its first successful development in the 1930s. The early gas turbines built in the mid-1990s were having poor efficiency of about 17–18% due to various reasons such as low compressor, turbine efficiencies, and lower turbine inlet temperature due to metallurgical limitations/constraints at that time. However, concrete efforts have been made by scientists and technologists to improve the efficiency of gas turbine with new modifications such as:

1. *Turbine inlet temperatures*: The primary approach to increase the turbine inlet temperature was taken into consideration to improve the overall performance of

gas turbine. A relatively high progress has been made in this regard, and the turbine inlet temperature has been increased from 540 °C in the 1940s to around 1500 °C at present. These changes were made possible by the development of new materials and innovative coating techniques for the important and critical components, such as coating the blades of the turbine with ceramic layers, cooling the turbine blades with the discharge air from the compressor, and so on.

2. *Enhanced efficiency of turbomachinery components*: The performance of Brayton cycle also suffered due to low efficiency of important components such as turbine and compressor. But due to advancement of computer-aided design and simulation techniques, it is possible to design high-performance turbine and compressor using these techniques. As a result, the performance of gas turbine cycle has improved greatly during the recent years.
3. *Modifications in the basic design*: With the advancement of scientific and technological knowledge in the area of power generation, material, simulation, and design, the efficiency of Brayton cycle has improved significantly in the last few decades. For example, the simple-cycle efficiency of early gas turbine has been doubled by incorporating few additional components such as regenerator, intercooler, reheater, isothermal heat addition, and so on. All the modifications made so far in the Brayton cycle have improved the cycle efficiency significantly, and some of them will be discussed one by one in the preceding chapters.

3.5 Irreversible Regenerative Brayton Cycle

As mentioned above, the Brayton cycle can be modified using a regenerator, which basically utilizes the high-temperature heat available in the exhaust of the turbine to preheat the working fluid before it enters the combustion chamber, thereby reducing the requirement of heat supplied by the fuel, and hence, the requirement of fuel also reduces. The regenerative closed cycle Brayton heat engine coupled with a finite heat capacity of source- and sink-side external fluids and having non-isentropic expansion and compression processes is shown on the schematic diagram of Fig. 3.5a along with the *T*–*s* diagram on Fig. 3.5b. The working fluid enters the compressor at state 4 and compressed up to state 1 non-isentropically and enters the regenerator where it is partially heated up to state 1R by the turbine exhaust. In an ideal/real regenerator, the working fluid leaves the regenerator at the temperature equal to/less than the turbine exhaust ($T_3 \geq T_{1R}$). The primary heat addition takes place between states 1 and 1R. The working fluid leaving the regenerator enters the hot-side heat source (combustion chamber) and heated up to state 2, while the external hot-side working fluid cooled from T_{H1} to T_{H2}. The working fluid then enters the turbine and expands up to state 3.

The turbine exhaust enters the regenerator where it transfers heat partly to the compressor outlet and then enters the cold-side heat exchanger of finite heat capacity whose temperature increases from T_{L1} to T_{L2}, and the working fluid is cooled up to state 4. The processes 4–1 s and 4–1 are, respectively, the ideal and real

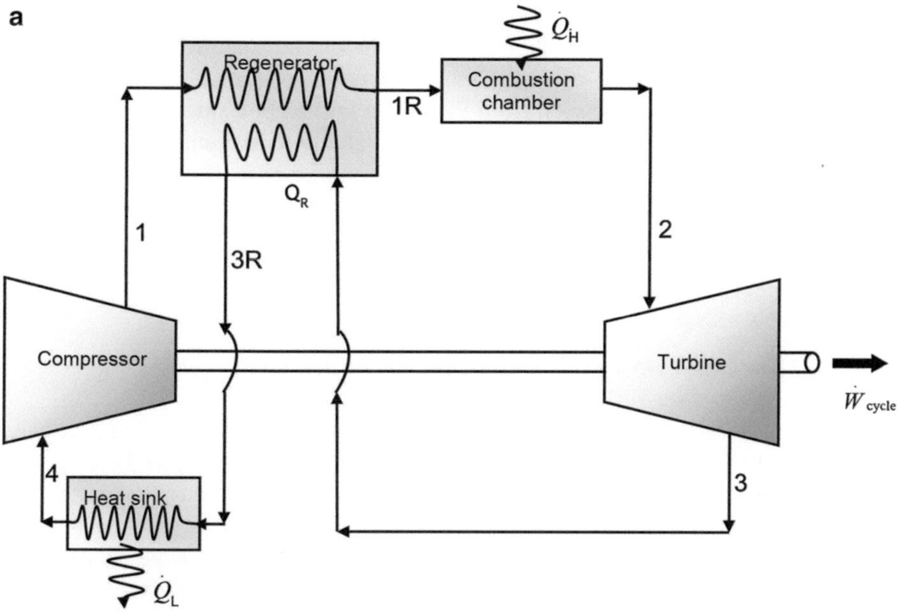

Fig. 3.5a Schematic diagram of regenerative Brayton cycle

Fig. 3.5b *T–s* diagram of an irreversible regenerative Brayton cycle

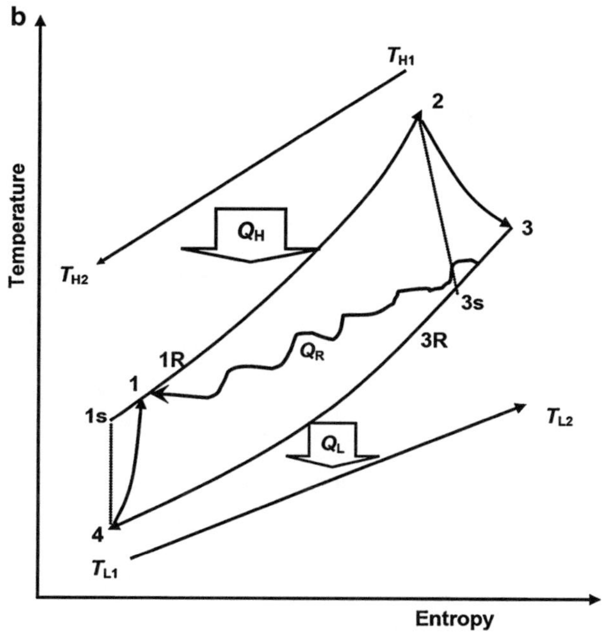

compression processes in the compressor. Similarly, 2–3 s and 2–3 are, respectively, the ideal and real expansion processes in the turbine. When the external reservoirs are of finite heat capacity, the temperature of the source and the sink changes with time as can be seen from Figs. 3.5a and 3.5b. Using the heat transfer theory and neglecting the changes in the kinetic and potential energies for steady-state flow, the different heat transfer rates can be expressed as (Ibrahim et al. 1991; Kumar 2000; Kaushik and Tyagi 2002) below:

$$\dot{Q}_{\mathrm{H}} = C_{\mathrm{wf}}(T_2 - T_{1\mathrm{R}}) = (\mathrm{UA})_{\mathrm{H}}(\mathrm{LMTD})_{\mathrm{H}} = C_{\mathrm{H}}(T_{\mathrm{H1}} - T_{\mathrm{H2}}) \tag{3.28}$$

$$\dot{Q}_{\mathrm{L}} = C_{\mathrm{wf}}(T_{3\mathrm{R}} - T_4) = (\mathrm{UA})_{\mathrm{L}}(\mathrm{LMTD})_{\mathrm{L}} = C_{\mathrm{L}}(T_{\mathrm{L2}} - T_{\mathrm{L1}}) \tag{3.29}$$

$$\dot{Q}_{\mathrm{R}} = C_{\mathrm{wf}}(T_3 - T_{3\mathrm{R}}) = C_{\mathrm{wf}}(T_{1\mathrm{R}} - T_1) = (\mathrm{UA})_{\mathrm{R}}(\mathrm{LMTD})_{\mathrm{R}} \tag{3.30}$$

where $(\mathrm{UA})_{\mathrm{H}}$, $(\mathrm{UA})_{\mathrm{L}}$, and $(\mathrm{UA})_{\mathrm{R}}$ are the heat transfer coefficient-area products; $(\mathrm{LMTD})_{\mathrm{H}}$, $(\mathrm{LMTD})_{\mathrm{L}}$, and $(\mathrm{LMTD})_{\mathrm{R}}$ are the Log Mean Temperature Differences; and ε_{H}, ε_{L}, and ε_{R} are the effectiveness of the hot-, sink-, and the regenerative-side heat exchangers, respectively.

For counterflow heat exchangers are given (Kays and London 1964; Bejan 1995; Ibrahim et al. 1991; Kumar 2000; Kaushik and Tyagi 2002) as below:

$$(\mathrm{LMTD})_{\mathrm{H}} = \frac{(T_{\mathrm{H1}} - T_2) - (T_{\mathrm{H2}} - T_{1\mathrm{R}})}{\ln\left(\frac{(T_{\mathrm{H1}} - T_2)}{(T_{\mathrm{H2}} - T_{1\mathrm{R}})}\right)} \tag{3.31}$$

$$(\mathrm{LMTD})_{\mathrm{L}} = \frac{(T_{3\mathrm{R}} - T_{\mathrm{L2}}) - (T_4 - T_{\mathrm{L1}})}{\ln\left(\frac{(T_{3\mathrm{R}} - T_{\mathrm{L2}})}{(T_4 - T_{\mathrm{L1}})}\right)} \tag{3.32}$$

$$\varepsilon_{\mathrm{H}} = \frac{1 - \mathrm{e}^{-NTU_{\mathrm{H}}\left(1 - \frac{C_{\mathrm{H,min}}}{C_{\mathrm{H,max}}}\right)}}{1 - \frac{C_{\mathrm{H,min}}}{C_{\mathrm{H,max}}}\mathrm{e}^{-NTU_{\mathrm{H}}\left(1 - \frac{C_{\mathrm{H,min}}}{C_{\mathrm{H,max}}}\right)}} \tag{3.33}$$

$$\varepsilon_{\mathrm{L}} = \frac{1 - \exp^{-NTU_{\mathrm{L}}\left(1 - \frac{C_{\mathrm{L,min}}}{C_{\mathrm{L,max}}}\right)}}{1 - \frac{C_{\mathrm{L,min}}}{C_{\mathrm{L,max}}}\exp^{-NTU_{\mathrm{L}}\left(1 - \frac{C_{\mathrm{L,min}}}{C_{\mathrm{L,max}}}\right)}} \tag{3.34}$$

$$\varepsilon_{\mathrm{R}} = \frac{NTU_{\mathrm{R}}}{1 + NTU_{\mathrm{R}}} \tag{3.35}$$

where *NTU*s are the number of transfer units on the respective-side heat exchangers. From Eqs. (3.28), (3.29), (3.30), (3.31), (3.32), (3.33), (3.34), and (3.35), the heat transfer rates from the source to the heat engine ($\dot{Q}_{\mathrm{H}}$), heat engine to the sink ($\dot{Q}_{\mathrm{L}}$), and within the regenerator ($\dot{Q}_{\mathrm{L}}$), respectively, are given as below:

$$\dot{Q}_{\mathrm{H}} = C_{\mathrm{wf}}(T_2 - T_{1\mathrm{R}}) = C_{\mathrm{H,min}}\varepsilon_{\mathrm{H}}(T_{\mathrm{H1}} - T_{1\mathrm{R}}) \tag{3.36}$$

$$\dot{Q}_{\mathrm{L}} = C_{\mathrm{wf}}(T_{3\mathrm{R}} - T_4) = C_{\mathrm{L,min}}\varepsilon_{\mathrm{L}}(T_{3\mathrm{R}} - T_{\mathrm{L}1}) \tag{3.37}$$

$$Q_{\mathrm{R}} = C_{\mathrm{wf}}(T_{1\mathrm{R}} - T_1) = C_{\mathrm{wf}}(T_3 - T_{3\mathrm{R}}) = C_{\mathrm{wf}}\varepsilon_{\mathrm{R}}(T_3 - T_1) \tag{3.38}$$

where $C_{\mathrm{H,min}}$ and $C_{\mathrm{H,max}}$ and $C_{\mathrm{L,min}}$ and $C_{\mathrm{L,max}}$ are, respectively, the minimum and the maximum capacitance rates of the external fluids on the source and sink side, while ε_{H}, ε_{L}, and ε_{R} are, respectively, the effectiveness of hot-, cold-, and regenerative-side heat exchangers and, for counterflow heat exchanger, are defined as below:

$$\begin{pmatrix} C_{\mathrm{H,min}} = \min(C_{\mathrm{wf}}, C_{\mathrm{H}}) \\ C_{\mathrm{H,max}} = \max(C_{\mathrm{wf}}, C_{\mathrm{H}}) \end{pmatrix} \text{and} \begin{pmatrix} C_{\mathrm{L,min}} = \min(C_{\mathrm{wf}}, C_{\mathrm{L}}) \\ C_{\mathrm{L,max}} = \max(C_{\mathrm{wf}}, C_{\mathrm{L}}) \end{pmatrix} \tag{3.39}$$

where $\mathrm{NTU_H}$, $\mathrm{NTU_L}$, and $\mathrm{NTU_R}$ are, respectively, the number of heat transfer units for different heat exchangers based on the minimum thermal capacitance rates (Kumar 2000; Kaushik and Tyagi 2002) and are given as below:

$$N_{\mathrm{H}} = \frac{(\mathrm{UA})_{\mathrm{H}}}{C_{\mathrm{H,min}}}, \quad N_{\mathrm{L}} = \frac{(\mathrm{UA})_{\mathrm{L}}}{C_{\mathrm{L,min}}}, \quad \text{and} \quad N_{\mathrm{R}} = \frac{(\mathrm{UA})_{\mathrm{R}}}{C_{\mathrm{wf}}} \tag{3.40}$$

Thus, defining the compressor and turbine efficiencies in terms of temperature as below:

$$\eta_{\mathrm{c}} = \frac{(T_{1\mathrm{s}} - T_4)}{(T_1 - T_4)} \quad \text{and} \quad \eta_{\mathrm{t}} = \frac{(T_2 - T_3)}{(T_2 - T_{3\mathrm{s}})} \tag{3.41}$$

Now from Eqs. (3.34), (3.35), (3.36), (3.37), (3.38), (3.39), (3.40), and (3.41) yields:

$$T_{3\mathrm{R}} = (1 - \varepsilon_{\mathrm{R}})T_3 + \varepsilon_{\mathrm{R}}T_1 \tag{3.42}$$

$$T_{1\mathrm{R}} = (1 - \varepsilon_{\mathrm{R}})T_1 + \varepsilon_{\mathrm{R}}T_3 \tag{3.43}$$

$$T_4 = (1 - y)T_{3\mathrm{R}} + yT_{\mathrm{L}1} \tag{3.44}$$

$$T_2 = (1 - x)T_{1\mathrm{R}} + xT_{\mathrm{H}1} \tag{3.45}$$

$$T_{1\mathrm{S}} = (1 - \eta_{\mathrm{c}})T_4 + T_1\eta_{\mathrm{c}} \tag{3.46}$$

$$T_{3\mathrm{S}} = \left(1 - \eta_{\mathrm{t}}^{-1}\right)T_2 + \eta_{\mathrm{t}}^{-1}T_3 \tag{3.47}$$

where $x = C_{\mathrm{H,\ min}}\varepsilon_{\mathrm{H}}/C_{\mathrm{w}}$, and $y = C_{\mathrm{L,\ min}}\varepsilon_{\mathrm{L}}/C_{\mathrm{w}}$. Again, using the second law of thermodynamics for a reversible Brayton cycle yields:

$$T_2/T_{1\mathrm{S}} = T_{3\mathrm{S}}/T_4 = \left(R_{\mathrm{p}}\right)^{(\gamma-1)/\gamma} \tag{3.48}$$

Substituting the values of T_1, T_2, T_3, and T_4 from Eqs. (3.42), (3.43), (3.44), (3.45), (3.46), and (3.47) in Eq. (3.48), we get a quadratic equation in T_3 (Kaushik and Tyagi 2002) as below:

$$AT_3^2 + BT_3 + C = 0 \tag{3.49}$$

which on solving for T_3, while treating T_1 as a constant, yields:

$$T_3 = \frac{1}{2A}\left[-B \pm \sqrt{B^2 - 4AC}\right] \tag{3.50}$$

where different parameters are given below:

$$\begin{aligned}
A &= a_2b_2 - \varepsilon_R(1-\varepsilon_R)(1-x)(1-y)\\
B &= b_4T_1 + b_5\\
C &= b_6T_1^2 + b_7T_1 + b_8\\
b_1 &= \eta_c + (1-\eta_c)(1-y)\varepsilon_R\\
b_2 &= (1-\eta_c)(1-\varepsilon_L)(1-\varepsilon_R)\\
b_3 &= y(1-\eta_c)T_{L1}\\
b_4 &= a_1b_2 + a_2b_1 - (1-x)(1-y)\left[(1-\varepsilon_R)^2 + \varepsilon_R^2\right]\\
b_5 &= a_2b_3 + a_3b_2 - y\varepsilon_R(1-x)T_{L1} - x(1-y)(1-\varepsilon_R)T_{H1}\\
b_6 &= a_1b_1 - \varepsilon_R(1-\varepsilon_R)(1-x)(1-y)\\
b_7 &= a_3b_1 + a_1b_3 - y(1-\varepsilon_R)(1-x)T_{L1} - x(1-y)\varepsilon_R T_{H1}\\
b_8 &= a_3b_3 - xyT_{H1}T_{L1}\\
a_1 &= \left(1-\eta_t^{-1}\right)(1-\varepsilon_R)(1-x)\\
a_2 &= \eta_t^{-1} + \left(1-\eta_t^{-1}\right)(1-x)\varepsilon_R\\
a_3 &= x\left(1-\eta_t^{-1}\right)T_{H1}
\end{aligned}$$

Now, using the first law of thermodynamics yields:

$$P = \dot{Q}_H - \dot{Q}_L = C_H\varepsilon_H(T_{H1} - T_{1R}) - C_L\varepsilon_L(T_{3R} - T_{L1}) \tag{3.51}$$

Substituting Eqs. (3.42) and (3.43) into Eq. (3.51) yields:

$$\begin{aligned}P = C_{wf}(xT_{H1} + yT_{L1}) - C_{wf}[x(1-\varepsilon_R) + \varepsilon_R y]T_1 + \frac{C_{wf}}{2A}\\
[y(1-\varepsilon_R) + \varepsilon_R x]\left[-B \pm \sqrt{B^2 - 4AC}\right]\end{aligned} \tag{3.52}$$

From Eq. (3.52), it is seen that P is a function of single variable T_1, because T_3 is also a function of T_1 as can be seen from Eq. (3.50). Thus maximizing P with respect to T_1, viz. $\frac{\partial^1 P}{\partial T_1} = 0$, yields

$$A_1T_1^2 + B_1T_1 + C_1 = 0 \tag{3.53}$$

Solving Eq. (3.53) for T_1, the optimum value of temperature T_1 is given as below:

$$T_{1,\mathrm{opt}} = \frac{1}{2A_1}\left[-B_1 \pm \sqrt{B_1^2 - 4A_1C_1}\right] \tag{3.54}$$

where the different parameters are given as below:

$$\begin{aligned}
A_1 &= b_4^2 - 4Ab_6 - a_5^2/a_4^2 \\
B_1 &= 2b_4b_5 - 4Ab_7 - 2a_6a_5/a_4^2 \\
C_1 &= b_5^2 - 4Ab_8 - a_6^2/a_4^2 \\
a_4 &= 2A[x(1-\varepsilon_R) + y\varepsilon_R] - b_4[y(1-\varepsilon_\mathrm{R}) + x\varepsilon_\mathrm{R}] \\
a_5 &= \left(b_4^2 - 4Ab_6\right)[y(1-\varepsilon_\mathrm{R}) + x\varepsilon_\mathrm{R}] \\
a_6 &= (b_4b_5 - 2Ab_7)[y(1-e_\mathrm{R}) + xe_\mathrm{R}]
\end{aligned}$$

Using Eqs. (3.52) and (3.54), one can find the optimal value of turbine exhaust temperature corresponding to the optimal power condition ($T_{3,\ \mathrm{opt}}$), while the values of other state point temperatures can be obtained from Eqs. (3.41), (3.42), (3.43), (3.44), (3.45), and (3.46) after substituting the values of $T_{1,\ \mathrm{opt}}$ and $T_{3,\ \mathrm{opt}}$. The maximum power output ($P_{\max}$), different heat transfer rates, and the corresponding efficiency can also be calculated as below:

$$Q_{\mathrm{H,opt}} = C_{\mathrm{H,min}}\varepsilon_\mathrm{H}\left(T_{\mathrm{H1}} - T_{\mathrm{1R,opt}}\right) \tag{3.55}$$

$$Q_{\mathrm{L,opt}} = C_{\mathrm{L,min}}\varepsilon_\mathrm{L}\left(T_{\mathrm{3R,opt}} - T_{\mathrm{L1}}\right) \tag{3.56}$$

$$P_{\max} = Q_{\mathrm{H,opt}} - Q_{\mathrm{L,opt}} \tag{3.57}$$

$$\eta_{\mathrm{opt}} = P_{\max}/Q_{\mathrm{H,opt}} \tag{3.58}$$

Although the thermal efficiency (η_{opt}) given by Eq. (3.58) is lesser than the Curzon–Ahlborn because they did not consider any kind of internal irreversibility which can be obtained by substituting values of different parameters. But in the present case, it is observed that the thermal efficiency of a real heat engine is always lesser than the Carnot and the Curzon and Ahlborn (1975) efficiencies as well as the efficiency given by Eq. (3.58) because there are more irreversibilities in the cycle which were not yet considered. The other causes of irreversibilities in a real cycle are due to various losses such as pressure loss, friction, heat loss, etc. in the turbine, compressor, and regenerator and in the connecting pipes and the direct heat leak loss from the source to the sink. Thus the thermal efficiency at maximum power output (η_m) of any real heat engine can be expressed (theoretically), in terms of source and sink temperatures along with the irreversibility parameter ($R_{\mathrm{K}\ -\ \mathrm{T}}$) defined by Kaushik and Tyagi (2002). Thus following Kaushik and Tyagi (2002), the bound for the efficiency of a real Brayton cycle having finite heat capacity of external reservoirs is given as below:

$$\eta_{\mathrm{m}} = 1 - \sqrt{\frac{R_{\mathrm{K-T}} T_{\mathrm{L1}}}{T_{\mathrm{H1}}}} \quad \text{i.e.,} \quad \eta_{\mathrm{actual}} < \eta_{\mathrm{opt}} < \eta_{\mathrm{C-A}} < \eta_{\mathrm{Carnot}} \tag{3.59}$$

where η_{Carnot} and $\eta_{\mathrm{C-A}}$ are, respectively, the Carnot and Curzon–Ahlborn efficiencies and $R_{\mathrm{K-T}}$ is the irreversibility parameter, always greater than unity for real heat engines.

It is important to note that the present cycle model presented here is more general, and the results obtained by earlier workers available in the literature (Ibrahim et al. 1991; Chen et al. 2001; Wu and Kiang 1992) can be drawn directly from this particular model as a special case. For example, in case of a simple endoreversible Brayton cycle without a regenerator given in the earlier section can be directly obtained from this particular model. In other words, the value of compressor and turbine efficiency is set to be unified and the effectiveness of the regenerator is set to zero, which means the cycle is internally reversible and there is no regenerator (viz. $\eta_{\mathrm{c}} = \eta_{\mathrm{t}} = 1.00$ and $\varepsilon_{\mathrm{R}} = 0$) for both finite and infinite heat capacity of external reservoirs. Similarly, if the external reservoirs are of infinite heat capacity, the expressions may reduce to:

$$\eta_{\mathrm{m}} = 1 - \sqrt{\frac{R_{\mathrm{K-T}} T_{\mathrm{L}}}{T_{\mathrm{H}}}} \quad \text{i.e.,} \quad \eta_{\mathrm{actual}} < \eta_{\mathrm{opt}} < \eta_{\mathrm{C-A}} < \eta_{\mathrm{Carnot}} \tag{3.60}$$

3.6 Discussion of Results

In order to have a numerical appreciation of the results, we consider the inlet external fluid temperature of heating source and cooling sink to be $T_{\mathrm{H1}} = 1250$ K and $T_{\mathrm{L1}} = 300$ K, respectively. Assuming the effectiveness of each heat exchanger, i.e. hot-side, cold-side, and the regenerative-side heat exchanger in the range of 0.60–1.00, compressor and turbine efficiencies in the range of 0.60–1.00 and the heat capacitance rates of external reservoirs and the heat engine working fluid in the range of 0.80–1.35 kW/K, respectively. The effects of each of these parameters on the maximum power output and the corresponding thermal efficiency were studied, while keeping the others as constant, and the discussion of results is given as below.

Effects of Effectiveness The variations of the maximum power output and the corresponding thermal efficiency against the effectiveness of different heat exchangers, viz. ε_{H}, ε_{L}, and ε_{R} are shown in Fig. 3.6 for a typical set of operating parameters mentioned above. It can be seen from Fig. 3.6 that the effect of sink-side heat exchanger effectiveness is more pronounced for the maximum power output, while the effect of the regenerator effectiveness is found to be more pronounced for the corresponding thermal efficiency. However, the effects of source-side heat exchanger effectiveness are found to be less pronounced not only for the maximum power output but also for the corresponding thermal efficiency. It is also seen that

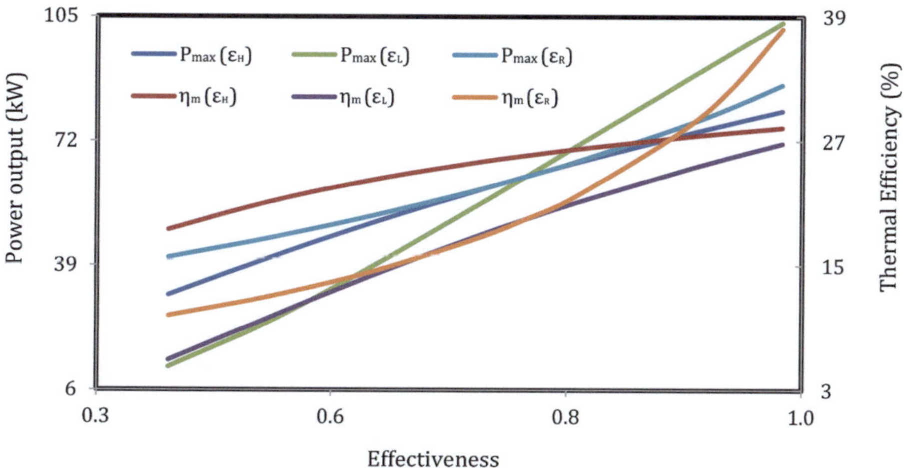

Fig. 3.6 Effects of effectiveness on the power output and thermal efficiency

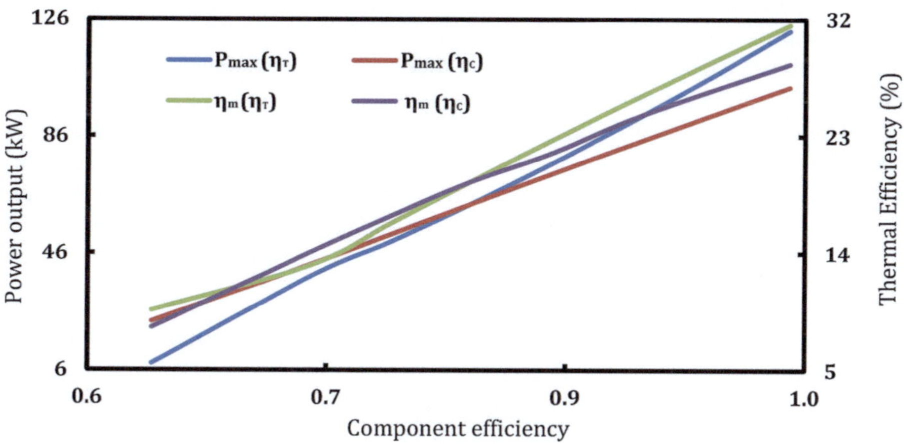

Fig. 3.7 Effects of component efficiency on the power output and thermal efficiency

the Brayton cycle with perfect heat exchangers ($\varepsilon_H = \varepsilon_L = \varepsilon_R = 1.00$) is much better than the Brayton cycle with real heat exchangers ($\varepsilon < 1.00$), but the former cycle is not practical because an ideal heat exchanger requires an infinite heat exchanger surface area or infinite longer heat addition/rejection time, which is neither economical nor favourable.

Effects of Component Efficiencies Figure 3.7 shows the effects of turbine and compressor efficiencies (η_t and η_c) on the maximum power output and the corresponding thermal efficiency of an irreversible regenerative Brayton cycle. It is seen from the figure that the effects of turbine efficiency are more pronounced for

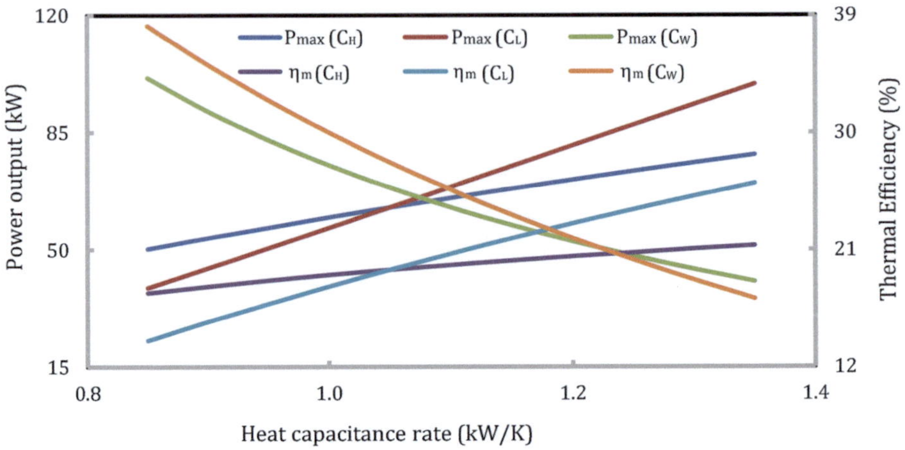

Fig. 3.8 Effects of heat capacitance rates on the power output and thermal efficiency

both the performance parameters, while it is found to be reversed in the case of compressor efficiency. This is mainly due to the fact that the Brayton cycle with higher turbine efficiency produces more power, while it consumes less power if the compressor efficiency is higher. As a result, the overall performance of the cycle enhances if the turbine efficiency is higher. Also the effect of turbine efficiency is more than that of the compressor efficiency on the overall performance of the cycle. However, the maximum power output and the corresponding thermal efficiency are found to be much better for ideal compression and ideal expansion processes, i.e. if compressor and turbine efficiencies are equal to 1.00, yet it is not practical because an ideal process cannot be achieved in real practice.

Effects of Heat Capacitance Rates The variation of the maximum power output and the corresponding thermal efficiency against the heat capacitance rates of different fluids (C_H, C_L and C_W) is plotted in Fig. 3.8 for a typical set of operating parameters, mentioned above. It is seen from Fig. 3.8 that the maximum power output and the corresponding thermal efficiency increase as the heat capacitance rates of the external reservoirs increase (C_H and C_L), while both the performance parameters (P_{max} and η_{opt}) decrease sharply and gradually as the heat capacitance rates of the working fluid increases. Also the variation of both the parameters P_{max} and η_{opt} is found to be more pronounced for the heat capacitance rates of the sink-side (C_L) fluid, while it is found to be reversed in the case of source-side heat capacitance rates (C_H) as can be seen from Fig. 3.8.

From Figs. 3.6, 3.7, and 3.8, it is observed that the heat capacitance rate and effectiveness on the sink side should be more than those of the source side (i.e. $C_L > C_H$ and $\varepsilon_L > \varepsilon_H$), and the turbine efficiency should be more than that of the compressor efficiency ($\eta_t > \eta_c$) for the better performance of an irreversible Brayton cycle. The optimal value of the heat capacitance rates (C_H) is always bounded with the heat capacitance rates of the heat engine working fluid (C_w) as well as the heat capacitance rate of the sink-side fluid (C_L), and the performance of Brayton cycle will be better for $C_H > C_L$ and $C_w < C_H$. The optimal value of the heat capacitance rates (C_L) is also bounded with the heat capacitance rates of the heat engine working fluid (C_w) as well as the heat capacitance rate of the source-side fluid (C_H). The output power and the corresponding efficiency of the Brayton cycle approach the maximum value when $C_L > C_H > C_W$, as shown in Fig. 3.8.

The results obtained in this study are useful to understand the deviation of the actual thermal efficiency from the ideal efficiency due to evolution of the external and internal irreversibilities associated with the system. The external irreversibility is due to the heat transfer rates to and from the cycle, and the internal irreversibility is due to non-isentropic compression and expansion process and other losses within the cycle, such as pressure loss, friction, heat loss in the regenerator, and so on. The heat capacitance rates (C_H, C_L, and C_w) also play important role for both the maximum power output and the corresponding thermal efficiency and found to be bounded as $C_L > C_H > C_W$ for the better performance of the cycle.

3.7 Conclusion

This chapter presents the basic background of the Brayton cycle, different versions of Brayton cycle, and modifications that took place in the Brayton cycle over time. Also the finite time thermodynamic analysis of an irreversible regenerative Brayton cycle having external and internal irreversibilities has been carried out for finite heat capacity of the external reservoirs. The external irreversibility is due to the finite temperature difference between the cycle and the external reservoirs, while the internal irreversibility is due to other losses such as friction, pressure drop and non-isentropic compression/pumping, and expansion processes in the compressor and the turbine, respectively. Effects of different parameters on the maximum power output and the corresponding thermal efficiency have been evaluated and presented graphically, besides some special cases that were also presented. The present model was found to be more general, because the cycle models discussed by different authors can be derived as the special case of this particular cycle model.

Chapter 4
Finite Time Thermodynamic Analysis of Modified Brayton Cycle

4.1 Introduction

It is well known that reheating in gas turbine engines limits the extent to which an isothermal heat addition is approached. With respect to a simple heat addition, when a compressible gas with subsonic velocity flows through a frictionless constant area duct with heat addition, the temperature of the gas increases along the duct. Also with respect to simple area change, when a compressible fluid/gas with subsonic velocity flows through a frictionless adiabatic duct with decreasing area, the temperature of the gas decreases along the duct. The idealized isothermal process consists of a compressible gas with subsonic velocity flowing through a frictionless converging duct, such that while heated all along the duct, any infinitesimal decrease in temperature due to simple area change is exactly compensated by the simple heat addition. It is noted that, since temperature of the gas is constant during the isothermal heat addition, the kinetic energy of the gas and hence the Mach number must increase in order to satisfy the conservation of energy. The appropriate application of the idealized isothermal process is to gas turbine engines operating with air. It is equally desirable that the Brayton cycle is modified by the isothermal heat addition process.

Based on the nature of these two flows, simple heat addition (Rayleigh flow) and simple area change (isentropic flow) may be combined in such a way so as to yield an isothermal heat addition process in a Brayton cycle (Vecchiarelli et al. 1997). Göektun and Yavuz (1999) studied the effect of isothermal heat addition using an open cycle regenerative Brayton heat engine with non-isentropic compression and expansion processes. Few studies (Kumar 2000; Tyagi et al. 2000; Kaushik et al.

The original version of this chapter was revised.
The book was inadvertently published with wrong 4.1 and 4.2 tables in chapter 4. It has been corrected and updated. A correction to this chapter is available at https://doi.org/10.1007/978-3-319-62812-7_13

S.C. Kaushik et al., *Finite Time Thermodynamics of Power and Refrigeration Cycles*,
https://doi.org/10.1007/978-3-319-62812-7_4

2003) have presented a more general analysis of an irreversible regenerative closed Brayton cycle with isothermal heat addition for the finite heat capacity of external heat sources and sink reservoirs and also showed that there is a significant improvement in the thermal efficiency of over 15–20% as compared to a conventional Brayton cycle discussed by earlier authors (Wu and Kiang 1990, 1991; Ibrahim et al. 1991; Kaushik and Tyagi 2002; Redcenco et al. 1998). Also there are other options to modify the traditional Brayton cycle using extra parts, such as intercooler, isothermal heat addition, regenerator, and their combinations. This chapter is an attempt to modify Brayton cycle with different combinations to enhance its performance for power generation applications.

4.2 Modified Brayton Cycles

The Brayton cycle can be modified in different ways by incorporating different components, such as regenerator, intercooler, reheat, isothermal heat addition, and/or the combination of these. However, the Brayton cycle with regeneration has been discussed in Chap. 3, while some of the other modifications are given as below.

4.2.1 Intercooled Brayton Cycle

An irreversible regenerative intercooled Brayton cycle coupled with the external reservoirs of finite heat capacities is shown in Fig. 4.1a along with its *T–s* diagram in Fig. 4.1b. The working fluid enters the compressor at state point 1 and is compressed up to state point 2 and then enters the intercooler and is cooled up to state point 3 by rejecting the heat to a finite heat capacity fluid whose temperature varies from T_{C1} to T_{C2}. The working fluid again enters the another compressor at state point 3 and is compressed up to state point 4 and then enters the regenerator where it is heated up to state point 4R by the turbine exhaust. In an ideal or real regenerator, the working fluid leaves the regenerator at the temperature equal to or less than the turbine exhaust (T_6), i.e. $T_{4R} \leq T_6$. The primary heat addition takes place between process 4−4R. The working fluid leaving the regenerator enters the hot-side heat exchanger and is heated up to state point 5 by a heat source of finite heat capacity whose temperature varies from T_{H1} to T_{H2}. The working fluid enters the turbine and expands up to state point 6, and the turbine exhaust enters the regenerator where it transfers the heat to the working fluid in the regenerator during the process 6–6R. Finally, the working fluid enters the cold-side heat exchanger and is cooled up to state point 1 by rejecting the heat to the heat sink of finite heat capacity whose temperature varies from T_{L1} to T_{L2}, thereby completing the cycle. The processes 2–2S and 3–4S are the isentropic, while the processes 1–2 and 3–4 are the non-isentropic of their respective-side compressors. Similarly processes 5–6S and 5–6 are the ideal and real expansion processes in the turbine, respectively. According to the cycle model shown in Fig. 4.1a, the different heat transfer rates, viz. $\dot{Q}_H$, $\dot{Q}_L$, $\dot{Q}_I$, and $\dot{Q}_R$, are given (Ibrahim et al. 1991; Kaushik and Tyagi 2002;

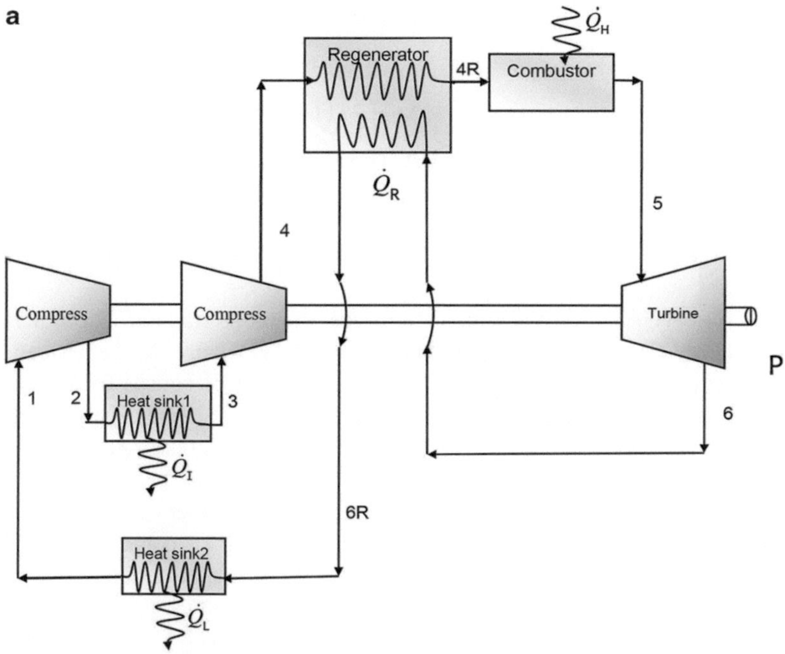

Fig. 4.1a Line diagram of regenerative intercooled Brayton cycle

Fig. 4.1b *T–s* diagram of regenerative intercooled Brayton cycle

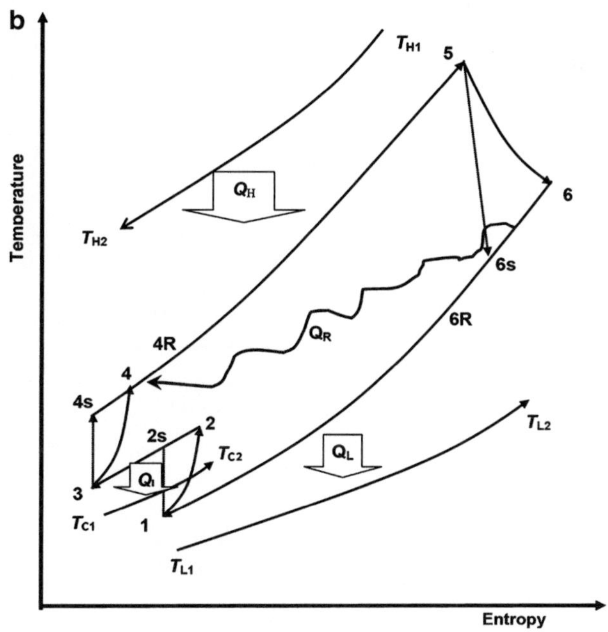

Redcenco et al. 1998; Tyagi 2009; Tyagi et al. 2000, 2007; Vecchiarelli et al. 1997; Göktun and Yavuz 1999) by:

$$\dot{Q}_{\rm H} = C_{\rm wf}(T_5 - T_{4\rm R}) = {\rm UA_H}({\rm LMTD})_{\rm H} = C_{\rm H}(T_{\rm H1} - T_{\rm H2}) \tag{4.1}$$

$$\dot{Q}_{\rm L} = C_{\rm wf}(T_{6\rm R} - T_1) = {\rm UA_L}({\rm LMTD})_{\rm L} = C_{\rm L}(T_{\rm L2} - T_{\rm L1}) \tag{4.2}$$

$$\dot{Q}_{\rm I} = C_{\rm wf}(T_2 - T_3) = {\rm UA_I}({\rm LMTD})_{\rm I} = C_{\rm I}(T_{\rm C2} - T_{\rm C1}) \tag{4.3}$$

$$\dot{Q}_{\rm R} = C_{\rm wf}(T_6 - T_{6\rm R}) = {\rm UA_R}({\rm LMTD})_{\rm R} = C_{\rm wf}(T_{4\rm R} - T_4) \tag{4.4}$$

where $({\rm UA})_{\rm H}$, $({\rm UA})_{\rm L}$, $({\rm UA})_{\rm I}$, and $({\rm UA})_{\rm R}$ are the heat transfer coefficient-area products; $({\rm LMTD})_{\rm H}$, $({\rm LMTD})_{\rm L}$, $({\rm LMTD})_{\rm I}$, and $({\rm LMTD})_{\rm R}$ are the Log Mean Temperature Differences; and $\varepsilon_{\rm H}$, $\varepsilon_{\rm L}$, $\varepsilon_{\rm I}$, and $\varepsilon_{\rm R}$ are the effectiveness for the source-, sink-, intercooler-, and regenerative-side heat exchangers, respectively, while, $C_{\rm H}$, $C_{\rm L}$, $C_{\rm I}$, and $C_{\rm wf}$ are, respectively, the heat capacitance rates of the fluids in the source-, sink-, and, intercooler-side reservoirs and within the cycle.

For counterflow heat exchangers, the different parameters are given (Kumar 2000; Tyagi et al. 2000, 2007; Vecchiarelli et al. 1997; Göktun and Yavuz 1999) as below:

$$\left.\begin{aligned} ({\rm LMTD})_{\rm H} &= \frac{(T_{\rm H1} - T_5) - (T_{\rm H2} - T_{4\rm R})}{\ln\left(\dfrac{(T_{\rm H1} - T_5)}{(T_{\rm H2} - T_{4\rm R})}\right)} \\ ({\rm LMTD})_{\rm L} &= \frac{(T_{6\rm R} - T_{\rm L2}) - (T_1 - T_{\rm L1})}{\ln\left(\dfrac{(T_{6\rm R} - T_{\rm L2})}{(T_1 - T_{\rm L1})}\right)} \end{aligned}\right\} \tag{4.5}$$

$$\left.\begin{aligned} ({\rm LMTD})_{\rm I} &= \frac{(T_2 - T_{\rm C2}) - (T_3 - T_{\rm C1})}{\ln\left(\dfrac{(T_2 - T_{\rm C2})}{(T_3 - T_{\rm C1})}\right)} \\ ({\rm LMTD})_{\rm R} &= \frac{(T_6 - T_{4\rm R}) - (T_{6\rm R} - T_4)}{\ln\left(\dfrac{(T_6 - T_{4\rm R})}{(T_{6\rm R} - T_4)}\right)} \end{aligned}\right\} \tag{4.6}$$

$$\varepsilon_{\rm H} = \frac{1 - {\rm e}^{-NTU_{\rm H}\left(1 - \frac{C_{\rm H,min}}{C_{\rm H,max}}\right)}}{1 - \frac{C_{\rm H,min}}{C_{\rm H,max}} {\rm e}^{-NTU_{\rm H}\left(1 - \frac{C_{\rm H,min}}{C_{\rm H,max}}\right)}} \tag{4.7}$$

$$\varepsilon_{\rm L} = \frac{1 - {\rm e}^{-NTU_{\rm L}\left(1 - \frac{C_{\rm L,min}}{C_{\rm L,max}}\right)}}{1 - \frac{C_{\rm L,min}}{C_{\rm L,max}} {\rm e}^{-NTU_{\rm L}\left(1 - \frac{C_{\rm L,min}}{C_{\rm L,max}}\right)}} \tag{4.8}$$

$$\varepsilon_{\mathrm{R}} = \frac{NTU_{\mathrm{R}}}{1 + NTU_{\mathrm{R}}} \text{ and } \varepsilon_{\mathrm{I}} \frac{1 - \mathrm{e}^{-NTU_{\mathrm{I}}\left(1 - \frac{C_{\mathrm{C},\min}}{C_{\mathrm{C},\max}}\right)}}{1 - \frac{C_{\mathrm{C},\min}}{C_{\mathrm{C},\max}} \mathrm{e}^{-NTU_{\mathrm{I}}\left(1 - \frac{C_{\mathrm{C},\min}}{C_{\mathrm{C},\max}}\right)}} \tag{4.9}$$

where *NTU*s are the number of transfer units on the respective-side heat exchangers. From Eqs. (4.1), (4.2), (4.3), (4.4), (4.5), (4.6), (4.7), (4.8), and (4.9), the heat transfer rates from the source to the heat engine ($\dot{Q}_{\mathrm{H}}$), from the heat engine to the sink ($\dot{Q}_{\mathrm{L}}$), and within the regenerator ($\dot{Q}_{\mathrm{R}}$), respectively, are given as below:

$$\dot{Q}_{\mathrm{H}} = C_{\mathrm{wf}}(T_5 - T_{4\mathrm{R}}) = C_{\mathrm{H,min}}\varepsilon_{\mathrm{H}}(T_{\mathrm{H1}} - T_{4\mathrm{R}}) \tag{4.10}$$

$$\dot{Q}_{\mathrm{L}} = C_{\mathrm{wf}}(T_{6\mathrm{R}} - T_1) = C_{\mathrm{L,min}}\varepsilon_{\mathrm{L}}(T_{6\mathrm{R}} - T_{\mathrm{L1}}) \tag{4.11}$$

$$\dot{Q}_{\mathrm{R}} = C_{\mathrm{wf}}(T_{4\mathrm{R}} - T_4) = C_{\mathrm{wf}}(T_6 - T_{6\mathrm{R}}) = C_{\mathrm{wf}}\varepsilon_{\mathrm{R}}(T_6 - T_4) \tag{4.12}$$

where $C_{\mathrm{H,\,min}}$ and $C_{\mathrm{H,\,max}}$ and $C_{\mathrm{L,\,min}}$ and $C_{\mathrm{L,\,max}}$ are, respectively, the minimum and the maximum capacitance rates of the external fluids on the source and sink sides, while ε_{H}, ε_{L}, and ε_{R} are, respectively, the effectiveness of hot-, cold-, and regenerative-side heat exchangers and, for counterflow heat exchanger, are defined as below:

$$\begin{pmatrix} C_{\mathrm{H,min}} = \min(C_{\mathrm{wf}}, C_{\mathrm{H}}) \\ C_{\mathrm{H,max}} = \max(C_{\mathrm{wf}}, C_{\mathrm{H}}) \end{pmatrix} \begin{pmatrix} C_{\mathrm{L,min}} = \min(C_{\mathrm{wf}}, C_{\mathrm{L}}) \\ C_{\mathrm{L,max}} = \max(C_{\mathrm{wf}}, C_{\mathrm{L}}) \end{pmatrix} \\ \text{and} \begin{pmatrix} C_{\mathrm{C,min}} = \min(C_{\mathrm{wf}}, C_{\mathrm{C}}) \\ C_{\mathrm{C,max}} = \max(C_{\mathrm{wf}}, C_{\mathrm{C}}) \end{pmatrix} \tag{4.13}$$

where NTU_{H}, NTU_{L}, and NTU_{R} are, respectively, the number of transfer units for different heat exchangers based on the minimum thermal capacitance rates (Ibrahim et al. 1991; Kaushik and Tyagi 2002; Redcenco et al. 1998; Tyagi et al. 2000, 2007; Tyagi 2009; Vecchiarelli et al. 1997; Göktun and Yavuz 1999) and are given as below:

$$NTU_{\mathrm{H}} = \frac{(\mathrm{UA})_{\mathrm{H}}}{C_{\mathrm{H,min}}},\ NTU_{\mathrm{L}} = \frac{(\mathrm{UA})_{\mathrm{L}}}{C_{\mathrm{L,min}}} \text{ and } NTU_{\mathrm{R}} = \frac{(\mathrm{UA})_{\mathrm{R}}}{C_{\mathrm{wf}}} \tag{4.14}$$

Thus, defining the compressor and turbine efficiencies in terms of temperature as below:

$$\eta_{\mathrm{c1}} = \frac{(T_{2\mathrm{s}} - T_1)}{(T_2 - T_1)};\ \eta_{\mathrm{c2}} = \frac{(T_{4\mathrm{s}} - T_3)}{(T_4 - T_3)} \text{ and } \eta_{\mathrm{t}} = \frac{(T_5 - T_6)}{(T_5 - T_{6\mathrm{s}})}, \tag{4.15}$$

From the cycle 1–2S–3–4S–5–6S–1 and the second law of thermodynamics for this case in an intercooled Brayton cycle (Tyagi et al. 2007) yields:

$$T_{2s}T_{4s}T_{6s} = T_1T_3T_5 \tag{4.16}$$

Using the equation of intercooling pressure ratio $R_{pi} = P_I/P_L = (T_{2S}/T_1)^{1-1/\gamma}$ and substituting the values of different state point temperatures into Eq. (4.16), one gets a quadratic equation in T_4 and T_6, and solving for T_4 yields:

$$T_4 = \frac{-B \pm \sqrt{B^2 - 4AC}}{2A} \tag{4.17}$$

where different parameters are given as below (Tyagi et al. 2007):

$$\begin{aligned}
&A = a_6b_1 - a_2a_8b_4, \quad B = b_7T_6 + b_8, \quad C = b_9T_6^2 \\
&\quad + b_{10}T_3 + b_{11}, a_1 = \eta_{c1}^{-1}\left(R_{pi}^{1-1/k} + \eta_{c1} - 1\right), a_2 = (1 - \eta_{c1} + a_1\eta_{c1}), a_3 = a_1(1 - x_3), \\
&a_4 = x_3T_{L1}, a_5 = a_3\varepsilon_R(1 - x_2), a_6 = a_3(1 - \varepsilon_R)(1 - x_2), a_7 = a_3(1 + a_4), a_8 = \eta_{c2} \\
&\quad + a_5(1 - \eta_{c2}), a_9 = a_6(1 - \eta_{c2}), a_{10} = a_7(1 - \eta_{c2}), b_1 = (1 - x_1)(1 - \varepsilon_R), \\
&\quad b_2 = (1 - x_1)\varepsilon_R, b_3 = x_1T_{H1}, b_4 = b_1\left(1 - \eta_T^{-1}\right), b_5 = \eta_T^{-1} \\
&\quad + b_2\left(1 - \eta_T^{-1}\right), b_6 = b_3\left(1 - \eta_T^{-1}\right), b_7 = a_5b_2 \\
&\quad + a_6b_1 - a_2(a_8b_5 + a_9b_4), b_8 = a_5b_3 + a_7b_1 \\
&\quad - a_2(a_8b_6 + a_{10}b_4), b_9 = a_6b_2 - a_2a_9b_5, b_{10} = a_6b_3 \\
&\quad + a_7b_2 - a_2(a_9b_6 + a_{10}b_5) \text{ and } b_{11} = a_7b_3 - a_2a_{10}b_6
\end{aligned}$$

The power output and the corresponding thermal efficiency in this case of an irreversible regenerative Brayton cycle with an intercooler are given as below (Tyagi 2009):

$$P = K_1 - \frac{a_{11}}{2A}\left(-(b_7T_6 + b_8) + \sqrt{(b_7T_6 + b_8)^2 - 4A\left(b_9T_6^2 + b_{10}T_6 + b_{11}\right)}\right) - a_{12}T_6 \tag{4.18}$$

$$\eta = \frac{K_1 - \frac{a_{11}}{2A}\left(-(b_7T_6 + b_8) + \sqrt{(b_7T_6 + b_8)^2 - 4A\left(b_9T_6^2 + b_{10}T_6 + b_{11}\right)}\right) - a_{12}T_6}{K_2 + \frac{K_3}{2A}\left(-(b_7T_6 + b_8) + \sqrt{(b_7T_6 + b_8)^2 - 4A\left(b_9T_6^2 + b_{10}T_6 + b_{11}\right)}\right) + K_4T_6} \tag{4.19}$$

where $K_1 = C_w(x_1T_{H1} + x_2(1 + a_1x_3)T_{L1} + x_3T_{c1})$, $K_2 = T_{H1}C_wx_1$, $K_3 = -C_wx_1(1 - \varepsilon_R)$, and $K_4 = -C_wx_1\varepsilon_R$. It is seen from Eq. (4.18) that P is a function of the single variable T_6 for a given set of operating conditions. Thus, using the extremal condition $\partial P/\partial T_6 = 0$ and Eq. (4.18), one can prove that when the power output attains its maximum value for a given set of operating parameters, the optimal temperature at state point 6 is determined by the following equation:

$$T_{6,\text{opt}} = \frac{-B_1 \pm \sqrt{B_1^2 - A_1C_1}}{A_1} \tag{4.20}$$

where $A_1 = a_{14} - (a_{14}/a_{13})^2$, $B_1 = 2(a_{15} - a_{14}a_{15}/a_{13}^2)$,

$$C_1 = b_8^2 - 4Ab_{11} - (a_{15}/a_{13})^2,$$

$$a_{11} = [(x_1 - k_0\varepsilon_H)(1 - \varepsilon_R) + \{x_2 - k_0\varepsilon_L + a_1(1 - x_2)x_3\}\varepsilon_R],$$
$$a_{14} = b_7^2 - 4Ab_9, a_{15} = b_7b_8 - 2Ab_{10},$$
$$a_{12} = [(x_1 - k_0\varepsilon_H)\varepsilon_R + \{x_2 - k_0\varepsilon_L + a_1(1 - x_2)x_3\}\ (1 - \varepsilon_R)] \text{ and}$$
$$a_{13} = (2Aa_{12} - a_{11}b_7)/a_{11}$$

Substituting the value of $T_{6,\ \text{opt}}$ from Eq. (4.20) into Eqs. (4.18) and (4.19), one can find the optimized power output and corresponding thermal efficiency, i.e. P_{max} and η_{m}. The optimum operating temperatures of other state points can also be calculated as listed in Table 4.1, where the typical operating parameters are chosen as $T_{\text{H1}} = 1200$ K, $T_{\text{L1}} = T_{\text{C1}} = 300$ K, $C_{\text{H}} = C_{\text{L}} = C_{\text{C}} = 1.0$ kW/K, $C_{\text{W}} = 1.05$ kW/K, $\varepsilon_{\text{H}} = \varepsilon_{\text{L}} = \varepsilon_{\text{I}} = \varepsilon_{\text{R}} = 0.85$, and $\eta_{\text{c1}} = \eta_{\text{c2}} = \eta_{\text{T}} = 0.90$. Using the cycle pressure ratio, one can also calculate the optimal value of cycle pressure (R_{P}) as below:

$$R_{\text{p}} = \frac{P_{\text{H}}}{P_{\text{L}}} = \frac{P_{\text{H}}}{P_{\text{I}}}\frac{P_{\text{I}}}{P_{\text{L}}} = (T_{4\text{S}}/T_3)^{\text{k}(\text{k}-1)}R_{\text{pi}} \tag{4.21}$$

Also, using the above results and a set of operating parameters which are the same as those given in Table 4.1, the graphs between the maximum power output and the intercooling pressure ratio, between the thermal efficiency at the maximum power output and the intercooling pressure ratio, and between the maximum power output and the corresponding thermal efficiency were plotted for different set of operating conditions, as shown in Figs. 4.4, 4.5, and 4.6, respectively.

4.2.2 *Isothermal Brayton Cycle*

The schematic and *T–s* diagrams of closed cycle regenerative Brayton heat engine with isothermal heat addition are shown in Figs. 4.2a and 4.2b, respectively. The basic components of this cycle are the compressor, regenerator, regular combustion chamber (RCC), converging combustion chamber (CCC), turbine, and the low-temperature heat exchanger (LTHE). The gas enters the compressor at state 1 and compressed up to state 2 (non-isentropically) and up to state 2 s (isentropically). At state 2, the cold gas leaving the compressor enters the regenerator, where it is heated up to 2R. In an ideal regenerator, the gas will leave the regenerator at the temperature of the turbine exhaust (state 5), i.e. $T_{2\text{R}} = T_5$. The isobaric heat addition process takes place between 2R–3 in RCC from a finite heat capacity source whose temperature varies from T_{H1} to T_{H2}. Further heat addition is accomplished in the CCC isothermally between states 3–4 from a finite heat capacity source whose temperature varies from T_{H3} to T_{H4}. When the gas leaves the CCC at state 4, it has lower pressure than at state 3, but the velocity and hence the kinetic energy of the gas has increased enormously due to the nature of CCC. The gas enters the turbine at state 4 and expands non-isentropically up to state 5 (ideally up to state 5S). The

Table 4.1 Effects of the intercooling pressure ratio on the optimum operating temperatures, cycle pressure ratio, maximum power output and the corresponding efficiency of an irreversible regenerative intercooled Brayton cycle

R_{pi}	$T_{1,opt}$ K	$T_{2S,opt}$ K	$T_{2,opt}$ K	$T_{3,opt}$ K	$T_{4S,opt}$ K	$T_{4,opt}$ K	$T_{4R,opt}$ K	$T_{5,opt}$ K	$T_{6S,opt}$ K	$T_{6,opt}$ K	$T_{6R,opt}$ K	R_{pm}	P^*_{max}	η_m
1.0	336.2	336.2	336.2	305.2	512.3	535.3	698.3	1128.3	670.6	716.4	553.4	6.18	0.577	0.392
1.2	334.2	350.9	352.8	307.5	499.7	521.1	687.2	1126.7	658.8	705.6	539.5	6.54	0.600	0.399
1.4	332.4	365.6	369.3	309.9	488.1	507.9	676.9	1125.3	648.0	695.7	526.7	6.90	0.618	0.404
1.6	330.7	380.3	385.8	312.3	477.3	495.6	667.4	1123.9	637.9	686.5	514.7	7.26	0.632	0.405
1.9	329.1	394.9	402.2	314.6	467.1	484.1	658.6	1122.7	628.6	678.0	503.5	7.61	0.642	0.407
2.2	327.6	409.5	418.6	316.9	457.6	473.3	650.3	1121.5	619.8	670.0	493.0	7.97	0.648	0.405
2.5	326.2	424.0	434.9	319.3	448.7	463.1	642.6	1120.4	611.7	662.6	483.1	8.32	0.652	0.402
2.9	324.8	438.5	451.1	321.6	440.3	453.5	635.4	1119.3	604.1	655.6	473.7	8.66	0.652	0.398
3.2	323.6	453.0	467.4	323.9	432.4	444.5	628.6	1118.4	596.9	649.1	464.9	9.00	0.651	0.393
3.7	322.4	467.4	483.6	326.2	424.9	435.9	622.2	1117.5	590.2	642.9	456.6	9.34	0.647	0.386
4.1	321.2	481.9	499.7	328.5	417.8	427.8	616.2	1116.6	583.9	637.1	448.7	9.67	0.640	0.379
4.6	320.2	496.3	515.8	330.8	411.1	420.0	610.5	1115.8	577.9	631.7	441.2	10.00	0.633	0.371
5.2	319.1	510.6	531.9	333.1	404.7	412.6	605.1	1115.0	572.3	626.5	434.0	10.30	0.623	0.362
5.8	318.2	525.0	548.0	335.4	398.6	405.6	600.0	1114.3	566.9	621.7	427.2	10.60	0.612	0.353
6.4	317.2	539.3	564.0	337.7	392.8	398.9	595.2	1113.6	561.9	617.0	420.7	11.00	0.599	0.343
7.1	316.4	553.6	580.0	340.0	387.2	392.5	590.6	1112.9	557.0	612.6	414.5	11.30	0.585	0.333
7.8	315.5	567.9	595.9	342.3	381.9	386.3	586.3	1112.3	552.5	608.5	408.5	11.60	0.569	0.322
8.6	314.7	582.2	611.9	344.6	376.8	380.4	582.1	1111.7	548.2	604.5	402.8	11.90	0.553	0.311

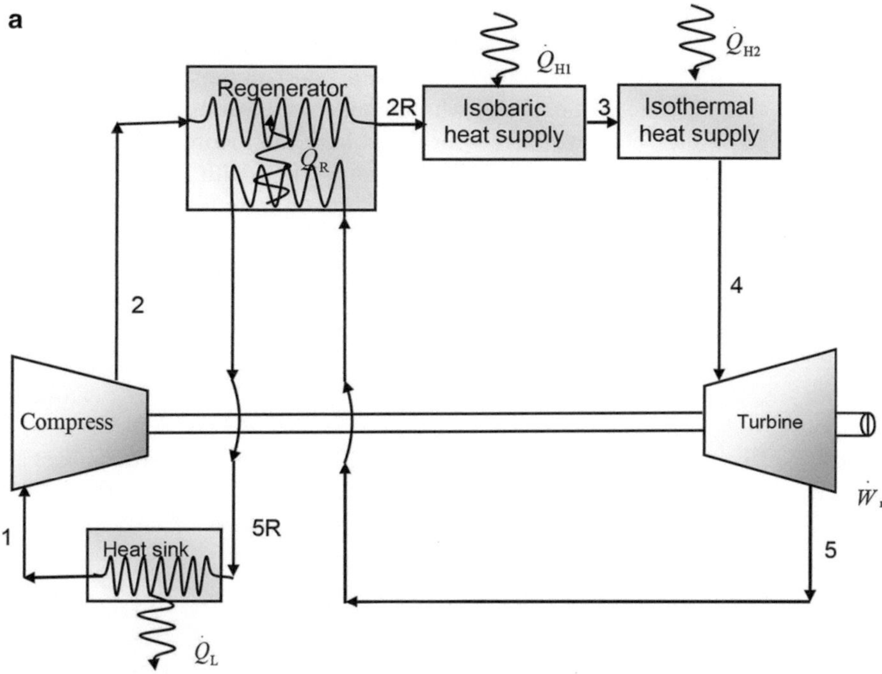

Fig. 4.2a Line diagram of regenerative isothermal Brayton cycle

Fig. 4.2b *T–s* diagram of regenerative isothermal Brayton cycle

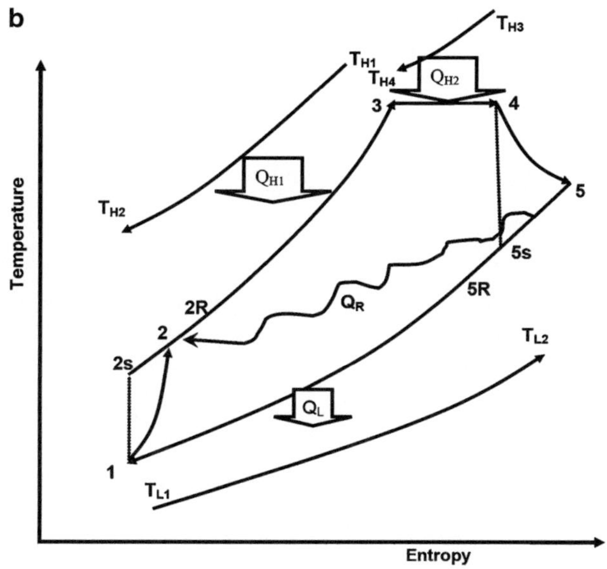

hot gas leaves the turbine at state 5; enters regenerator where it is cooled up to state 5R isobarically, by supplying the heat to the compressor outlet gas; and finally enters LTHE at state 5R and cooled up to state 1 at constant pressure rejecting the heat to heat sink of finite heat capacity whose temperature varies from T_{L1} to T_{L2}, completing the cycle. Thus we have considered here the theoretical model of a modified regenerative Brayton cycle 1–2–2R–3–4–5–5R–1 with real processes.

If $\dot{Q}_H$, $\dot{Q}_L$, and $\dot{Q}_R$ are the heat transfer rates to and from the heat engine and within the regenerator, respectively, then:

$$\dot{Q}_H = \dot{Q}_{H1} + \dot{Q}_{H2} = \dot{Q}_{2R-3} + \dot{Q}_{3-4} \tag{4.22}$$

$$\dot{Q}_L = \dot{Q}_{5R-1} \tag{4.23}$$

$$\dot{Q}_R = \dot{Q}_{2-2R} = \dot{Q}_{5-5R} \tag{4.24}$$

where the different heat transfer rates are given by (Tyagi et al. 2002; Kaushik et al. 2003) Eqs. (4.1), (4.2), (4.3), (4.4), (4.5), (4.6), (4.7), (4.8), (4.9), (4.10), (4.11), (4.12), (4.13), (4.14), and (4.15), while $\dot{Q}_{3-4}$ is given as below:

$$\dot{Q}_{3-4} = \frac{1}{2}\dot{m}_{wf}\left(V_4^2 - V_3^2\right) = C_{H1}(T_{H3} - T_{H4}) = \mathrm{UA}_{H1}(\mathrm{LMTD})_{H1} \tag{4.25}$$

where $(UA)_{H1}$, $(LMTD)_{H1}$, and C_{H1}, are, respectively, overall heat transfer coefficient-area product, Log Mean Temperature Difference, and the heat capacitance rates in the external fluids on conversing combustion chamber (CCC) and $\dot{m}_{wf}$ is the mass flow rate of the working fluid. For counterflow, heat exchanger can be defined similar to those mentioned above. Also the isothermal heat addition during state 3–4 can also be defined in terms of Mach number (M) as below:

$$M = V/V_s \tag{4.26a}$$

where V_s is the speed of sound and for a perfect gas yields:

$$V_s = (kRT)^{1/2} \tag{4.26b}$$

where k is the specific heat ratio and R is the universal gas constant of the working fluid, using Eqs. (4.26a) and (4.26b) yields:

$$\dot{Q}_{3-4} = \frac{1}{2}C_{wf}(k-1)T_3\left(M_4^2 - M_3^2\right) \tag{4.27}$$

Using Eqs. (4.10), (4.11), (4.12), and (4.27) and Fig. 4.2b, yields:

$$\dot{Q}_{H1} = \dot{Q}_{2R-3} = C_{wf}(T_3 - T_{2R}) = C_{H,min}\varepsilon_H(T_{H1} - T_{2R}) \tag{4.28}$$

$$\dot{Q}_{H2} = \dot{Q}_{3-4} = \frac{1}{2}C_{wf}(k-1)T_3\left(M_4^2 - M_3^2\right) = C_{H1,min}\varepsilon_{H1}(T_{H3} - T_3) \tag{4.29}$$

$$\dot{Q}_{\mathrm{L}} = \dot{Q}_{5\mathrm{R}-1} C_{\mathrm{wf}}(T_{5\mathrm{R}} - T_1) = C_{\mathrm{L,min}} \varepsilon_{\mathrm{L}}(T_{5\mathrm{R}} - T_{\mathrm{L1}}) \tag{4.30}$$

$$\dot{Q}_{\mathrm{R}} = C_{\mathrm{wf}}(T_{2\mathrm{R}} - T_2) = C_{\mathrm{wf}}(T_5 - T_{5\mathrm{R}}) = C_{\mathrm{wf}} \varepsilon_{\mathrm{R}}(T_5 - T_2) \tag{4.31}$$

where ε's are the effectiveness of the various heat exchangers, and for counterflow heat exchangers, the LMTDs and ε's are defined in Eqs. (4.5), (4.6), (4.7), (4.8), and (4.9), while for isothermal heat exchanger, they are defined as:

$$(\mathrm{LMTD})_{\mathrm{H1}} = \frac{(T_{\mathrm{H3}} - T_4) - (T_{\mathrm{H4}} - T_3)}{\ln\left(\frac{(T_{\mathrm{H3}} - T_4)}{(T_{\mathrm{H4}} - T_3)}\right)} (\text{while } T_3 = T_4) \text{ and } \varepsilon_{\mathrm{H1}} = \frac{NTU_{\mathrm{H1}}}{1 + NTU_{\mathrm{H1}}} \tag{4.32}$$

where $NTU_{\mathrm{H1}} = (\mathrm{UA})_{\mathrm{H1}}/C_{\mathrm{H1,\,min}}$ and other parameters are similar to those defined in Eqs. (4.13), (4.14,), and (4.15).

As there is some pressure drop during the isothermal heat addition, using the second law of thermodynamics following Vecchiarelli et al. (1997), Göktun and Yavuz (1999) and Kaushik et al. (2003) yields:

$$\alpha \frac{T_{2\mathrm{s}}}{T_1} = \frac{T_4}{T_{5\mathrm{S}}} \Rightarrow T_{2\mathrm{S}} T_{5\mathrm{S}} = \alpha^{-1} T_3 T_1 \tag{4.33}$$

where $\alpha = (r_{\mathrm{t}})^{(\gamma - 1)/\gamma}$ and $r_{\mathrm{t.}}$ is the isothermal pressure drop ratio. Substituting the values of different temperatures into Eq. (4.33), following earlier authors (Tyagi 2009; Tyagi et al. 2000, 2007; Vecchiarelli et al. 1997; Göktun and Yavuz 1999) yields:

$$A_3 T_2^2 + B_3 T_2 + C_3 = 0 \tag{4.34}$$

where A_3, B_3, and C_3 are constants and are given as below:

$$\begin{aligned}
&x = C_{\mathrm{H}} \varepsilon_{\mathrm{H}} / C_{\mathrm{W}} \quad y = C_L \varepsilon_L / C_{\mathrm{W}} \quad x_1 = C_{\mathrm{H1}} \varepsilon_{\mathrm{H1}} / C_{\mathrm{W}} \\
&a_1 = \eta_{\mathrm{c}} + (1 - y)\varepsilon_{\mathrm{R}}(1 - \eta_{\mathrm{c}}) \quad b_1 = (1 - y)\,(1 - \varepsilon_{\mathrm{R}})\,(1 - \eta_{\mathrm{c}}) \\
&c_1 = y T_{\mathrm{L1}}(1 - \eta_{\mathrm{c}}) \\
&a_2 = (1 - \varepsilon_{\mathrm{R}})(1 - x)\left(1 - \eta_{\mathrm{T}}^{-1}\right) \quad b_2 = \eta_{\mathrm{T}}^{-1} + \varepsilon_{\mathrm{R}}(1 - x)\left(1 - \eta_{\mathrm{T}}^{-1}\right) \\
&c_2 = x T_{\mathrm{H1}}\left(1 - \eta_{\mathrm{T}}^{-1}\right) \\
&a_3 = \varepsilon_{\mathrm{R}}(1 - y) \quad b_3 = (1 - \varepsilon_{\mathrm{R}})\,(1 - y) \quad c_3 = y T_{\mathrm{L1}} \\
&a_4 = (1 - x)\,(1 - \varepsilon_{\mathrm{R}}) \quad b_4 = (1 - x)\varepsilon_{\mathrm{R}} \quad c_4 = x T_{\mathrm{H1}} \\
&a_5 = a_1 b_2 + a_2 b_1 - \alpha^{-1}(a_3 b_4 + a_4 b_3) \quad b_5 = b_1 b_2 - \alpha a^{-1} b_3 b_4 \\
&c_5 = c_1 c_2 - \alpha^{-1} c_3 c_4 \\
&a_6 = a_1 c_2 + a_2 c_1 - \alpha^{-1}(a_3 c_4 + a_4 c_3) \quad b_6 = b_1 c_2 + b_2 c_1 - \alpha^{-1}(b_3 c_4 + b_4 c_3) \\
&A_3 = a_1 a_2 - \alpha^{-1} a_3 a_4 \quad B_3 = a_5 T_5 + a_6 \quad \text{and } C_3 = b_5 T_5^2 + b_6 T_5 + c_5
\end{aligned}$$

Again, using the first law of thermodynamics for this cycle model yields:

$$P = k_{01} - a_7T_2 - b_7T_5 \tag{4.35}$$

$$\eta = \frac{k_{01} - a_7T_2 - b_7T_5}{k_{02} - b_8T_2 - b_9T_5} \tag{4.36}$$

From Eq. (4.35), it can be seen that P is a function of single variable T_5 (as T_2 is also function of T_5), thus using Eq. (4.35) and its extremal condition, viz.$\frac{\partial P}{\partial T_5} = 0$, yields:

$$T_{5,\text{opt}} = \frac{1}{2A_4}\left(-B_4 \pm \sqrt{B_4^2 - 4A_4C_4}\right) \tag{4.37}$$

where the various parameters are given as below:

$$\begin{aligned}
&k_{01} = C_{\rm W}\{x(1-x_1)T_{\rm H1} + x_1T_{\rm H3} + yT_{\rm L1}\}; k_{02} = C_{\rm W}\{x(1-x_1)T_{\rm H1} + x_1T_{\rm H3}\};\\
&a_7 = C_{\rm W}\{x(1-\varepsilon_{\rm R}) + y\varepsilon_{\rm R} + x_1(1-x)(1-e_{\rm R})\};\\
&k_{02} = C_{\rm W}\{x(1-x_1)T_{\rm H1} + x_1T_{\rm H3} + yT_{\rm L1}\};\\
&b_7 = C_{\rm W}\{y(1-\varepsilon_{\rm R}) + x\varepsilon_{\rm R} + \varepsilon_{\rm R}(1-x)x_1\};\\
&a_8 = (a_5a_7 - 2Ab_7)/a_7;\;\; a_9 = a_5{}^2 - 4Ab_5;\\
&b_9 = C_{\rm W}\{x\varepsilon_{\rm R} + \varepsilon_{\rm R}(1-x)x_1\};\\
&b_8 = C_{\rm W}\{x(1-\varepsilon_R) + x_1(1-x)(1-\varepsilon_{\rm R})\}; a_{10} = a_5a_6 - 2Ab_6;\\
&A_4 = a_9 - (a_9/a_8)^2; B_4 = a_{10} - a_9a_{10}/a_8{}^2 \text{ and } C_4 = a_6{}^2 - 4A_3c_5 - (a_{10}/a_8)^2
\end{aligned}$$

Substituting the value of $T_{5,\text{ opt}}$ into Eq. (4.34), one can find $T_{2,\text{ opt}}$, and from Eqs. (4.35) and (4.36), one can obtain the maximum power output ($P_{\rm max}$) and the corresponding thermal efficiency ($\eta_{\rm m}$) for a given set of operating parameters, given above.

4.2.3 *Intercooled Isothermal Brayton Cycle*

The *T*–*s* diagram of a closed-cycle regenerative intercooled isothermal Brayton cycle heat engine is shown in Figs. 4.3a and 4.3b. The basic components of this cycle are the intercooler, two compressors, regenerator, regular combustion chamber (RCC), converging combustion chamber (CCC), turbine, and the low- and high-temperature heat exchangers. The gas enters the compressor at state point 1 and is compressed up to state point 2 (non-isentropically) or up to state point 2S (isentropically). The gas then enters the intercooler with a constant pressure $P_{\rm I}$ where it rejects the heat to the heat sink of finite heat capacity whose temperature varied from $T_{\rm C1}$ to $T_{\rm C2}$ and is cooled down up to state point 3. It again enters the second compressor and is compressed up to state point 4 (ideally up to state point 4S). At state point 4, the cold gas leaving the second

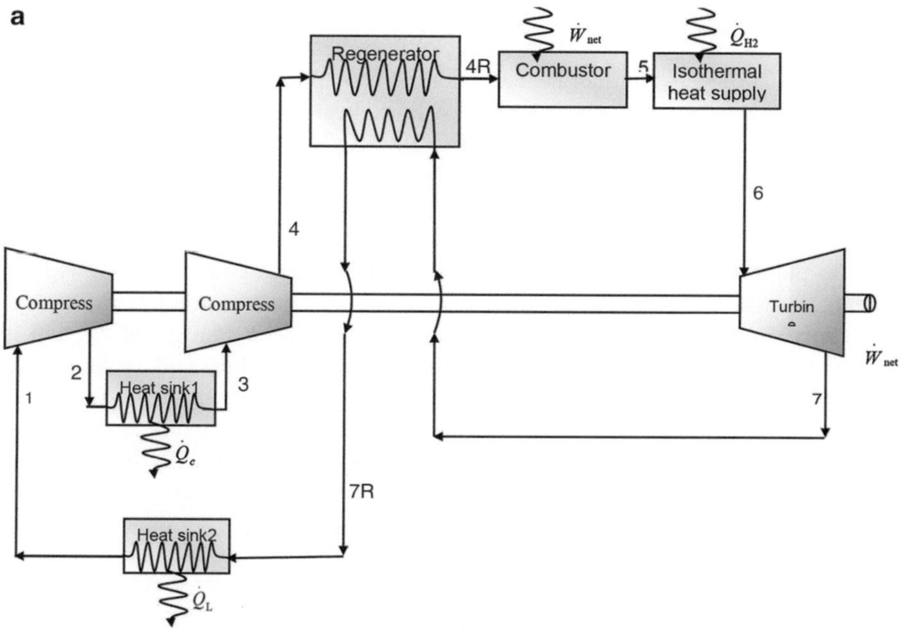

Fig. 4.3a Line diagram of regenerative intercooled isothermal Brayton cycle

Fig. 4.3b *T–s* diagram of regenerative isothermal intercooled Brayton cycle

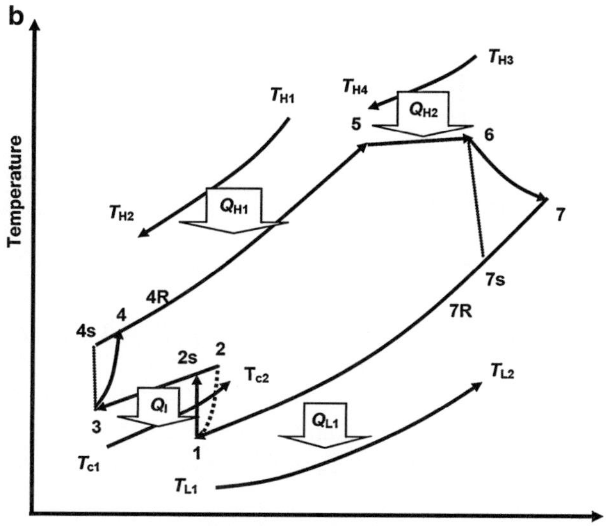

compressor enters the regenerator at constant pressure P_H, where it is heated up to state point 4R by the turbine exhaust. In a real or an ideal regenerator, the gas will leave the regenerator at the temperature less than or equal to that of the turbine exhaust (state point 7), i.e. $T_{4R} \leq T_7$.

The isobaric heat addition process with a same pressure P_H takes place during process 4R–5 in an RCC from a heat source of finite heat capacity whose temperature varies from T_{H1} to T_{H2}. Further heat addition is accomplished in a CCC isothermally during process 5–6, from a heat source of finite heat capacity whose temperature varies from T_{H3} to T_{H4}. When the gas leaves the CCC at state point 6, it has a lower pressure than that at state point 5, but the velocity and hence the kinetic energy of the gas has been increased enormously due to the nature of a CCC. The gas enters the turbine at state point 6 and expands non-isentropically up to state point 7 (ideally up to state point 7S). The hot gas leaves the turbine at state point 7 and enters the regenerator where it is cooled up to state point 7R at a constant pressure P_L.

Finally, it enters the low-temperature heat exchanger at state point 7R and is cooled down up to state point 1 by rejecting the heat to the heat sink of finite heat capacity whose temperature varied from T_{L1} to T_{L2}, thereby completing the cycle. Thus, we have considered here the theoretical model of an irreversible regenerative intercooled modified Brayton cycle 1–2–3–4–4R–5–6–7–7R–1 with the real compression and expansion processes. Following earlier sections, the different heat transfer rates to and from the cycle are written as below:

$$\dot{Q}_{in} = \dot{Q}_{H1} + \dot{Q}_{H2} = \dot{Q}_{4R-5} + \dot{Q}_{5-6} \tag{4.38}$$

$$\dot{Q}_{out} = \dot{Q}_{L1} + \dot{Q}_{C} = \dot{Q}_{5R-1} + \dot{Q}_{2-3} \tag{4.39a}$$

$$\dot{Q}_{R} = \dot{Q}_{4-4R} = \dot{Q}_{7-7R} \tag{4.39b}$$

where

$$\dot{Q}_{H1} = (\mathrm{UA})_H(\mathrm{LMTD})_H = C_w(T_5 - T_{4R}) = C_H(T_{H1} - T_{H2}) \tag{4.40}$$

$$\dot{Q}_{H2} = (\mathrm{UA})_{H1}(\mathrm{LMTD})_{H1} = C_{H1}(T_{H3} - T_{H4}) = C_{H1}\varepsilon_{H1}(T_{H3} - T_5) \tag{4.41}$$

$$\begin{aligned}\dot{Q}_{L1} &= (\mathrm{UA})_L(\mathrm{LMTD})_L = C_w(T_{7R} - T_1) = C_L(T_{L2} - T_{L2}) \\ &= C_L\varepsilon_L(T_{7R} - T_{L1})\end{aligned} \tag{4.42}$$

$$\begin{aligned}\dot{Q}_{C} &= (\mathrm{UA})_I(\mathrm{LMTD})_I = C_W(T_2 - T_3) = C_I(T_{C2} - T_{C1}) \\ &= C_I\varepsilon_I(T_2 - T_{C1})\end{aligned} \tag{4.43}$$

$$\begin{aligned}\dot{Q}_{R} &= (\mathrm{UA})_R(\mathrm{LMTD})_R = C_W(T_7 - T_{7R}) = C_W(T_{4R} - T_4) \\ &= C_W\varepsilon_R(T_7 - T_4)\end{aligned} \tag{4.44}$$

where C_H, C_{H1}, C_L, C_I, and C_W are the heat capacitance rates in the external fluids on RCC, CCC, cold-side reservoir, intercooler-side reservoir, and the working fluid, respectively, and $(\mathrm{UA})_H$, $(\mathrm{UA})_{H1}$, $(\mathrm{UA})_L$, $(\mathrm{UA})_I$, and $(\mathrm{UA})_R$ are the overall heat transfer coefficient-area products between the external reservoirs and RCC, CCC, low-temperature heat exchanger, intercooling heat exchanger, and the regenerative heat exchanger, respectively. ε's are the effectiveness of the various heat exchangers, and (LMTD)'s are the Log Mean Temperature Differences between the external reservoirs and the heat engine, as defined in the earlier sections (Kaushik 1999; Tyagi et al. 2007).

With the help of Fig. 4.3b, the compressors and turbine efficiencies are given as below:

$$\eta_I = (T_{2S} - T_1)/(T_2 - T_1), \eta_C = (T_{4S} - T_3)/(T_4 - T_3) \\ \text{and } \eta_T = (T_5 - T_7)/(T_5 - T_{7S}) \tag{4.45}$$

Using Eqs. (4.40), (4.41), (4.42), (4.43), (4.44), and (4.45) yields:

$$T_3 = a_3T_7 + a_4T_4 + a_5 \tag{4.46a}$$

$$T_5 = b_1T_7 + b_2T_4 + b_3 \tag{4.46b}$$

$$T_1 = c_1T_7 + c_2T_4 + c_3 \tag{4.46c}$$

$$T_{2S} = (1 - \eta_I)T_1 + \eta_I T_2 \tag{4.46d}$$

$$T_{7S} = b_4T_4 + b_5T_7 + b_6 \tag{4.46e}$$

$$T_{4S} = a_6T_4 + a_7T_7 + a_8 \tag{4.46f}$$

where the different parameters are given as below:

$$x_1 = C_H\varepsilon_H/C_{H,\min}, x_2 = C_{L,\min}\varepsilon_L/C_W, \ x_3 = C_{I,\min}\varepsilon_I/C_W, a_2 = \{1 + \eta_I(a_1 - 1)\},$$
$$a_1 = \left\{1 + \eta_I^{-1}\left(R_{pi}^{\gamma} - 1\right)\right\}, R_{Pi} = P_I/P_L \text{ and } k = 1$$
$$- 1/\gamma, c_1 = (1 - \varepsilon_R)(1 - x_2), c_2 = \varepsilon_R(1 - x_2),$$
$$c_3 = x_2T_{L1}, b_1 = (1 - \varepsilon_R)(1 - x_1), b_2 = \varepsilon_R(1 - x_1), b_3 = x_1T_{H1}, b_4 = \left(1 - \eta_T^{-1}\right)b_2,$$
$$b_5 = \left(1 - \eta_T^{-1}\right)b_1 + \eta_T^{-1}, \ b_6 = \left(1 - \eta_T^{-1}\right)b_3, a_3 = a_1c_1(1 - x_3), a_4 = a_1c_2(1 - x_3),$$
$$a_5 = a_1c_3(1 - x_3) + x_3T_{C1}, \ a_6 = (1 - \eta_C)a_1$$
$$+ \eta_C, \ a_7 = (1 - \eta_C)a_3 \text{ and} a_8 = (1 - \eta_C)a_5$$

In this case of an intercooled modified Brayton cycle (Tyagi 2009; Tyagi and kaushik 2005; Wang et al. 2003; Tyagi et al. 2000, 2007) yields:

$$\frac{T_6}{T_{7S}} = \left(\frac{P_6P_5}{P_5P_7}\right)^{\gamma} = \left(\frac{P_6P_H}{P_HP_L}\right)^{\gamma} = \alpha\left(\frac{P_H}{P_L}\right)^{\gamma} = \alpha\frac{T_{2S}}{T_1}\frac{T_{4S}}{T_3} \tag{4.47}$$

where $\alpha = (P_6/P_H)^{\gamma} < 1$ is the isothermal pressure drop. Using Eqs. 4.46 (a–f) and (4.47), one can get a quadratic equation in T_4 or T_7 which on solving for T_7 yields:

$$T_7 = \frac{-B_4 \pm \sqrt{B_4^2 - 4A_4C_4}}{2A_4} \tag{4.48}$$

where the different parameters are given as below:

$$A = b_1b_3 - b_7b_5a_7, B = b_8T_4 + b_9, C = a_9T_4^2$$
$$+ a_{10}T_4 + a_{11}, b_7 = \alpha_{a_2}, b_8 = a_1b_4 + a_2b_3 - b_7(a_7b_4 + a_6b_5),$$
$$b_9 = b_1a_5 + b_3a_3 - b_7(a_8b_5 + a_7b_6), a_9 = a_4b_2$$
$$- b_7b_4a_6, a_{10} = a_5b_2 + a_4b_3 - b_7(a_8b_4 + a_6b_6) \text{ and } a_{11} = a_5b_1$$
$$- b_7b_6a_8$$

From the first law of thermodynamics, the power output and the thermal efficiency are defined as:

$$P = K - b_{10}T_7 - b_{11}T_4 \tag{4.49}$$

$$\eta = \frac{P}{Q_{\mathrm{H}}} = \frac{K - b_{10}T_7 - b_{11}T_4}{x_1T_{H1} + x_4T_{H3} - (x_1T_{4R} + x_4T_5)} \tag{4.50}$$

where the different parameters are as below:

$$K = C_{\mathrm{W}}\{(1 - x_4)x_1T_{\mathrm{H1}} + x_3T_{\mathrm{L1}}(1 - x_2) + x_3T_{\mathrm{C1}} + x_4T_{\mathrm{H3}}\},$$
$$b_{10} = C_{\mathrm{W}}\{x_1(1 - \varepsilon_{\mathrm{R}}) + x_2\varepsilon_{\mathrm{R}} + x_3a_1c_2 + x_4b_2\},$$
$$b_{11} = C_{\mathrm{W}}\{x_1\varepsilon_{\mathrm{R}} + x_2(1 - \varepsilon_{\mathrm{R}}) + x_3a_1c_1 + x_4b_1\} \text{ and } x_4 = C_{\mathrm{H1,min}}\varepsilon_{\mathrm{H1}}/C_{\mathrm{W}}.$$

Using Eq. (4.49) and its extremal condition, viz. $\frac{\partial P}{\partial T_4} = 0$, one can prove that the power output attains its maximum at the optimal value of T_4 given by:

$$T_{4,\mathrm{opt}} = \left(B_5 \pm \sqrt{B_5^2 - A_5C_5}\right)/A_5 \tag{4.51}$$

where the different parameters are as below:

$$A_5 = b_8^2 - 4A_4b_9 - a_{13}^2/a_{12}^2, B_1 = b_8b_9 - 2A_4a_{10} - a_{13}a_{14}/a_{12}^2,$$
$$C_1 = b_9^2 - 4A_4a_{11} - a_{14}^2/a_{12}^2, a_{12} = (2A_4b_9 - b_8b_{11})/b_{11},$$
$$a_{13} = \left(b_8^2 - 4A_4a_9\right) \text{ and } a_{14} = b_8b_9 - 2A_4a_{10}.$$

Using the above equations, one can calculate the maximum power output, the corresponding thermal efficiency, and the other cycle parameters of an irreversible regenerative intercooled isothermal Brayton cycle for different intercooling pressure ratios, which are listed in Table 4.2, where the operating parameters are similar to those selected in the previous sections. It is also important to note that this is a more general case, and the relations obtained in other sections are the special case of this cycle model. For example, if the intercooling pressure ratio is unity, i.e. there is no intercooler in the cycle ($R_{\mathrm{pi}} = 1.0$), the relation is similar to be obtained in Sect. 4.2.2, and if there is no isothermal heat addition, viz. $\dot{Q}_{\mathrm{H2}} = 0.0$, this cycle model reduces to that of given in Sect. 4.2.1. Finally, if there is no intercooler and no isothermal heat addition ($R_{\mathrm{pi}} = 1.0$, $\dot{Q}_{\mathrm{H2}} = 0.0$), this cycle model will reduce to that of an irreversible regenerative Brayton cycle studied and discussed in previous chapter.

Table 4.2 Effects of the intercooling pressure ratio on the maximum power output, the corresponding thermal efficiency and the other cycle parameters of an irreversible regenerative intercooled modified Brayton cycle

R_{pi}	$T_{1,opt}$	$T_{2,opt}$	$T_{2S,opt}$	$T_{3,opt}$	$T_{4S,opt}$	$T_{4,opt}$	$T_{4R,opt}$	$T_{5,opt}$	$T_{7,opt}$	T_{7Sopt}	$T_{7R,opt}$	$R_{p,opt}$	P^*_{max}	η_m
1.00	362.4	362.4	362.4	314.7	537.4	561.7	583.7	769.9	587.6	567.3	565.5	2.74	0.721	0.449
1.09	357.8	389.3	365.7	321.2	518.8	540.7	571.0	766.5	576.4	555.3	546.1	3.11	0.774	0.463
1.13	355.7	403.2	367.4	324.3	510.1	530.8	565.1	764.9	571.2	549.6	536.9	3.29	0.796	0.466
1.23	351.5	432.0	370.9	331.1	493.8	511.9	553.9	761.9	561.3	539.3	519.3	3.69	0.828	0.466
1.28	349.6	446.9	372.6	334.5	486.2	502.9	548.5	760.5	556.6	534.2	510.9	3.91	0.839	0.464
1.38	345.8	477.8	376.3	341.8	471.3	485.7	538.6	757.9	547.9	524.5	495.1	4.37	0.853	0.454
1.43	344.1	493.7	378.1	345.5	464.3	477.5	533.9	756.6	543.8	520.1	487.5	4.61	0.850	0.448
1.54	340.7	526.7	381.9	353.3	451.3	461.9	525.0	754.2	536.1	511.8	473.3	5.13	0.842	0.431
1.61	339.1	543.7	383.8	357.3	444.7	454.4	520.7	753.1	532.4	507.9	466.1	5.41	0.833	0.421
1.72	335.9	578.9	387.7	365.5	432.7	440.2	512.8	751.2	525.6	500.5	453.2	5.97	0.805	0.396
1.78	334.5	597.1	389.7	369.8	427.1	433.3	509.0	750.3	522.4	497.1	446.7	6.27	0.787	0.382
1.92	331.7	634.6	393.7	378.6	416.1	420.3	502.0	748.2	516.4	490.7	434.7	6.92	0.741	0.352
2.12	327.7	693.9	399.8	392.6	401.1	402.2	492.5	745.6	508.4	482.2	417.9	7.92	0.649	0.332
2.21	326.5	714.5	401.9	397.4	396.4	396.2	489.4	744.9	505.9	479.4	412.7	8.28	0.613	0.282
2.35	324.1	757.0	406.1	407.4	387.4	385.2	483.9	743.4	501.3	474.5	402.6	9.02	0.532	0.239
2.42	323.2	778.9	408.2	412.5	383.1	379.9	481.3	742.7	499.2	472.2	397.8	9.41	0.487	0.217
2.59	320.8	824.0	412.5	423.1	375.2	369.7	476.5	741.4	495.3	467.9	388.5	10.21	0.388	0.173
2.86	317.8	895.2	419.1	439.9	363.8	355.3	470.0	739.7	490.2	462.3	375.5	11.50	0.218	0.094

4.3 Discussion of Results

The different models of a modified irreversible Brayton cycle have been presented so far including the combination of intercooler, regenerator, and isothermal heat addition. In order to have numerical appreciation of the results, the variation of performance parameters such as the maximum power output and the corresponding thermal efficiency against the cycle parameters was plotted for different cycle models, and the discussion of results is given for each model separately, as below:

4.3.1 Intercooled Brayton Cycle

Since the intercooling pressure ratio is one of the important parameters in the investigation of an intercooled regenerative Brayton cycle, one can discuss the effects of the various operating parameters on the performance of an irreversible regenerative intercooled Brayton cycle, as below.

Effect of Pressure Ratios

It is seen from Table 4.1 that the outlet temperatures of the intercooling compressor and the inlet temperature of the second compressor as well as the cycle compression ratio increase, while the temperatures of all other state points decrease as the intercooling pressure ratio increases. However, the maximum power output and the corresponding thermal efficiency first increase and then decrease as the either pressure ratio increases, as shown in Fig. 4.4. It shows clearly that there exists an optimal value of the intercooling pressure ratio for other given parameters at which the optimized power output attains its maximum value, i.e. $(P_{max})_{max}$. The conclusion may provide an important theoretical instruction for the optimal design of an intercooler of a regenerative Brayton cycle.

It is important to note the fact that the intercooling pressure ratio is always larger than unity and less than the cycle pressure ratio, i.e.:

$$1 < R_{Pi} < R_p \tag{4.52}$$

When $R_{Pi} = 1$ or $R_{Pi} = R_p$, the cycle becomes an irreversible Brayton cycle without intercooler, which has been discussed by a number of authors (Wu and Kiang 1990, 1991; Ibrahim et al. 1991; Kaushik and Tyagi 2002; Redcenco et al. 1998; Kaushik 1999). Obviously, the values of the intercooling pressure ratio are different for different operating states. It is important how to choose the optimal value of the intercooling pressure ratio in the investigation of a regenerative intercooled Brayton heat engine.

According to Table 4.1 or Fig. 4.4, in principle, one can determine the rational range of the optimal values of the intercooling pressure ratio as:

$$\left(R_{pi,opt}\right)_{\eta} \le R_{pi,opt} \le \left(R_{pi,opt}\right)_{m} \tag{4.53}$$

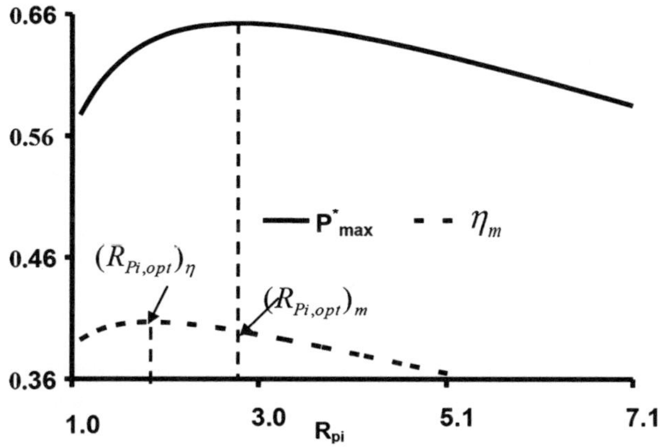

Fig. 4.4 Performance parameters versus intercooling pressure ratio

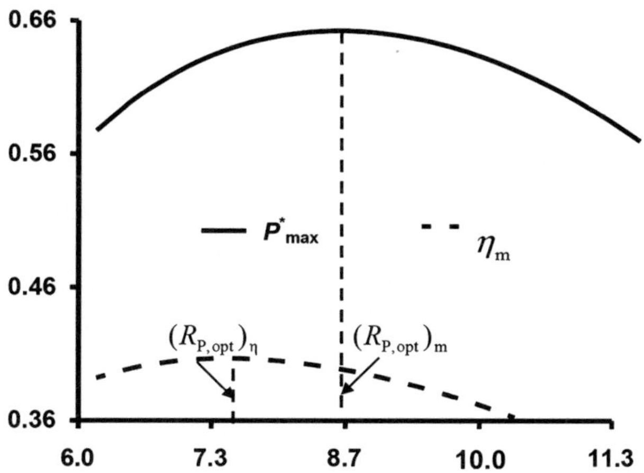

Fig. 4.5 Performance parameters versus cycle pressure ratio

where $(R_{pi\,,\,opt})_m$ and $(R_{pi\,,\,opt})_\eta$ are, respectively, the intercooling pressure ratios at points of the maximum power output and maximum efficiency, as indicated in Fig. 4.4. Thus using Eq. (4.52), one can give another optimum criterion for the cycle pressure ratio, as below:

$$\left(R_{p,opt}\right)_\eta \le R_{p,opt} \le \left(R_{p,opt}\right)_m \tag{4.54}$$

where $(R_{p,\ opt})_m$ and $(R_{p,\ opt})_\eta$ are, respectively, the optimal cycle pressure ratios at the points of the maximum power and maximum efficiency, as indicated in Fig. 4.5. According to Eq. (4.54) and Table 4.1, one can also give the optimum criteria of the temperatures of the various state points. The criteria will be helpful for engineers to optimally design the irreversible regenerative intercooled Brayton cycle.

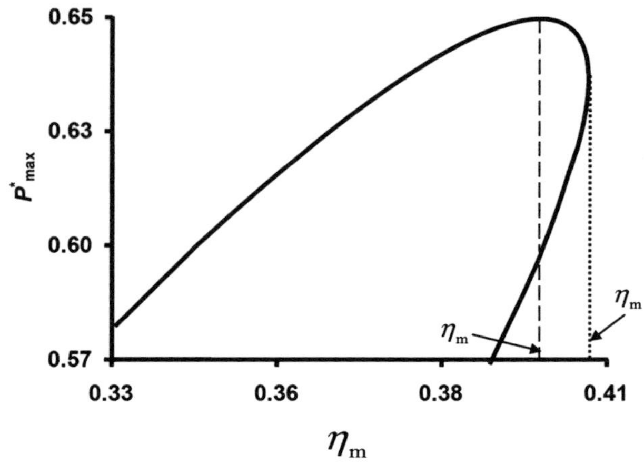

Fig. 4.6 Maximum power output vs. thermal efficiency

Maximum Power Output vs. Thermal Efficiency

Figure. 4.6 shows the variation of the dimensionless maximum power output against the thermal efficiency. It is seen from the figure that the maximum power output is not a monotonic function of the thermal efficiency. When the maximum power output is situated in the regions of the $P^*_{\max}\tilde{}\eta_m$ curve with a positive slope, it will decrease as the thermal efficiency is decreased. Obviously, these regions are not optimal. It is thus clear that the maximum power output should be situated in the region of the $P^*_{\max}\tilde{}\eta_m$ curve with a negative slope. When the maximum power output is in the region, it will increase as the thermal efficiency is decreased, and vice versa. It is shown once again that the optimal region of the intercooling pressure ratio should be determined by Eq. (4.53).

4.3.2 Isothermal Brayton Cycle

In order to have a numerical appreciation of the results, different cycle parameters such as the heat sources/sink inlet temperatures, the effectiveness of various heat exchangers, the turbine and compressor efficiencies, the heat capacitance rates of fluids, etc. were considered in the practical range, mentioned above. The effect of each of these parameters constant on the various state point temperatures, heat transfer rates, maximum power output, and the thermal efficiency of an irreversible regenerative isothermal Brayton cycle was studied, and the discussion of results is given below.

Effect of Effectiveness

The effect of effectiveness on various heat exchangers on the power output and the corresponding thermal efficiency of an irreversible regenerative Brayton cycle heat engine are shown in Fig. 4.7. It is seen from these figures that the power output and the corresponding thermal efficiency increase as the effectiveness on sink-side,

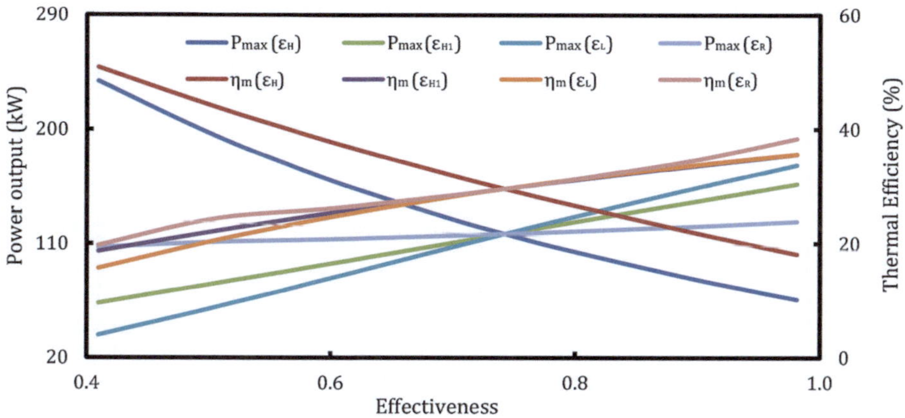

Fig. 4.7 Effects of effectiveness on the power output and thermal efficiency

hot-side (isothermal), and the regenerative-side heat exchangers increases, while both the parameters decrease as the effectiveness on the isobaric side heat exchanger increases. The effect of isobaric side effectiveness is more pronounced for both the parameters.

The higher values of εH, εH1, and εL decrease the external irreversibilities on their respective-side heat exchanger by forcing the working fluid to transfer heat to and from the sink/sources reservoirs at faster pace by decreasing the heat transfer times and the temperature differences. But the higher value of εL increases the power more than that of the higher value of εH1, while the higher value of εH decreases not only the power output but also the corresponding thermal efficiency. So it is desirable to have the relation ε L > εH1 > εH for better performance of the cycle. Again, the Brayton cycle with ideal regenerator is more efficient than that of the Brayton cycle with real regenerator, but it requires infinite regenerative time or area, which is not the case in practice. Hence, it would be impossible to conclude new results in the analysis of Brayton cycle, if the regenerative losses were not considered.

Effect of Reservoir Temperatures

The effects of heat sources reservoir temperatures on the power output and the corresponding thermal efficiency of an irreversible regenerative Brayton cycle are shown in Fig. 4.8. It is seen from these figures that as the temperature of the isobaric side and the sink-side heat exchangers increases, the power output as well as the corresponding thermal decreases, while both the parameters increase as the temperature of isothermal side heat exchanger increases. Thus it can be concluded that the inlet temperature for the isobaric heat addition should be less than that of the inlet temperature of isothermal heat addition for better performance. This is due to the fact that the heat addition in the former increases the quantity of energy, while in the latter it increases the quality of energy, as there is isothermal heat addition, and the heat supplied by the latter only increases the kinetic energy rather than thermal

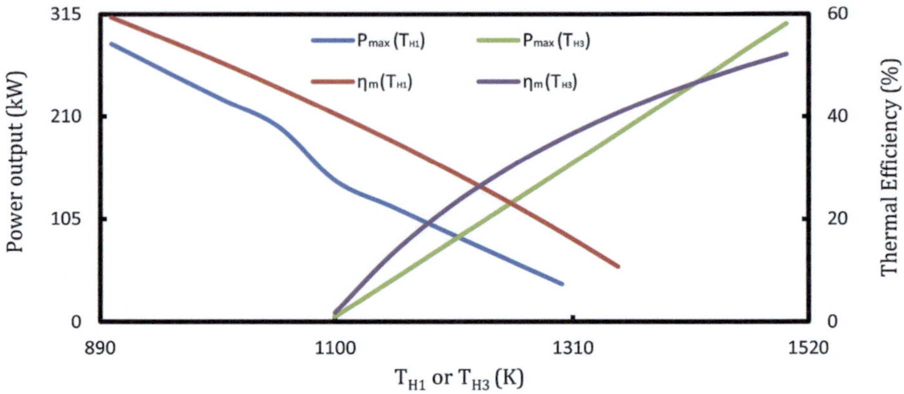

Fig. 4.8 Effects of source temperatures on the power output and thermal efficiency

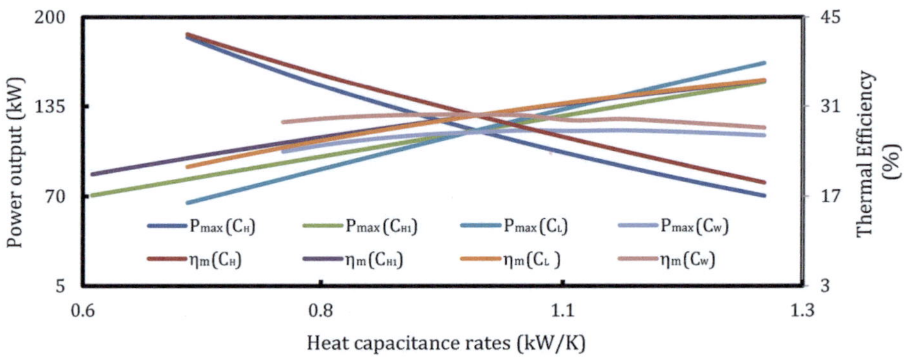

Fig. 4.9 Effects of heat capacitance rates on the power output and thermal efficiency

energy unlike the former source. As a result, the performance of the cycle enhancement is higher in the latter case.

Effect of Heat Capacitance Rates

The effects of various heat capacitance rates on the power output and the corresponding thermal are shown in Fig. 4.9. It is seen that the power output and the corresponding thermal efficiency increase as the heat capacitance rates on the isothermal side, sink side, and the cycle fluid increase, while both the parameters decrease as the heat capacitance rate on the isobaric side heat exchanger increases. The effect of isobaric side heat capacitance rate is more pronounced than the other heat capacitance rates not only on the power output but also on the corresponding thermal efficiency.

The higher values of heat capacitance rates decrease the external irreversibilities on their respective-side heat exchanger by forcing the working fluid to transfer heat to and from the reservoirs by decreasing the heat transfer times and/or the temperature differences. But higher value of C_L increases more power than higher value of other heat capacitance rates (viz. C_{H1}, C_H, or C_W), while lower value of C_H increases not only the power output but also the corresponding thermal efficiency. Thus for the better performance of these cycles, it is found these heat capacitance rates should be in the order of $C_L > C_{H1} > C_W > C_H$.

4.3.3 Intercooled Isothermal Brayton Cycle

Using the mathematical formulation given in Eqs. (4.49), (4.50), and (4.51) and a set of operating parameters which are similar to those as given in Table 4.2, the graphs between the maximum power output and the intercooling and cycle pressure ratios, between the thermal efficiency at the maximum power output and the intercooling and cycle pressure ratios, between the maximum power output and the corresponding thermal efficiency, between the power output and the turbine outlet temperature, and between the thermal efficiency and turbine outlet temperature for a typical set of operating condition have been plotted, as shown in Figs. 4.10a, 4.10b, 4.11a, 4.11b 4.12, 4.13a, 4.13b, respectively.

Effects of Pressure Ratios

On one hand, it is seen from Table 4.2 that the outlet temperatures of the intercooling compressor and the inlet temperature of the second compressor as well as the cycle pressure ratio increase, while the temperatures of all other state points decrease as the intercooling pressure ratio increases. On the other hand, the maximum power output and the thermal efficiency at the maximum power output first increase and then decrease as the pressure ratio increases, as shown in Figs. 4.10a and 4.10b. It shows clearly that there exists an optimal value of the

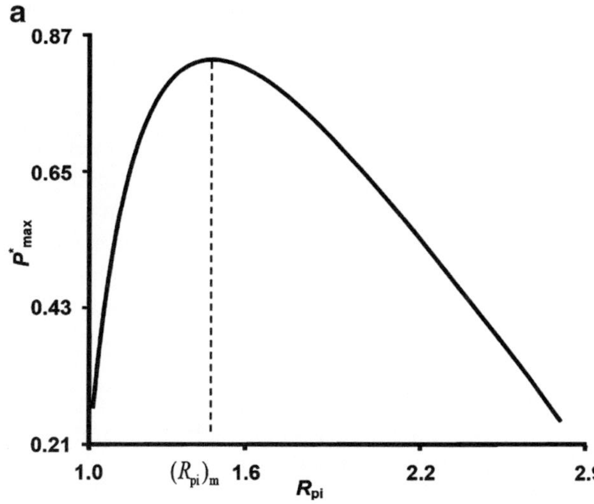

Fig. 4.10a The variation of power output with respect to intercooling pressure ratio

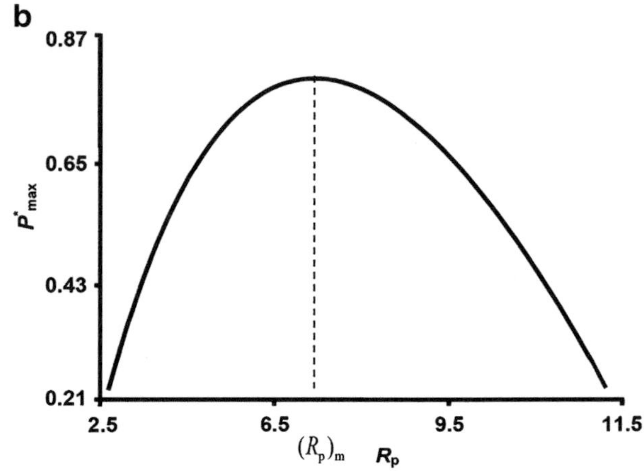

Fig. 4.10b The variation of power output with respect to intercooling pressure ratio

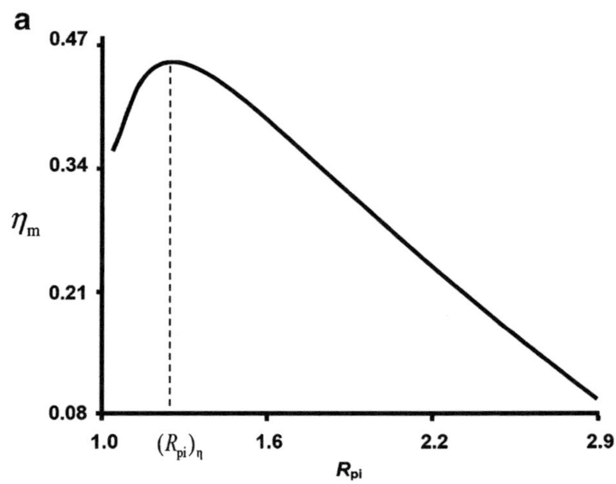

Fig. 4.11a The variation of the thermal efficiency at the maximum power output with respect to the intercooling pressure ratio

intercooling pressure ratio for other given parameters at which the optimized power output and the corresponding thermal efficiency attain their maximum values, i.e. $(P_{max})_{max}$ and $(\eta_m)_{max}$. This conclusion may provide an important theoretical instruction for the optimal design of the intercooler of this cycle model.

Obviously, the values of the intercooling pressure ratio are different for different operating states. It is important how to choose the optimal value of the intercooling pressure ratio in the investigation of an irreversible regenerative intercooled isothermal Brayton cycle. According to Table 4.2 and Figs. 4.10a and 4.10b, the rational range of the optimal values of the intercooling pressure ratio from the point of view of the power output as well as from the point of view of the thermal efficiency can be determined by the relation:

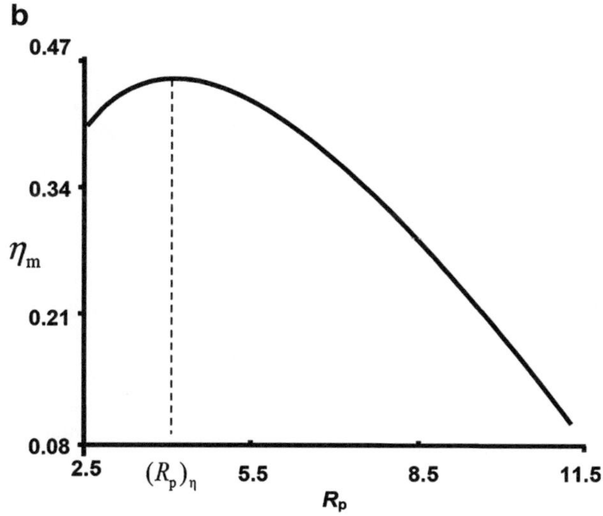

Fig. 4.11b The variation of the thermal efficiency at the maximum power output with respect to the cycle pressure ratio

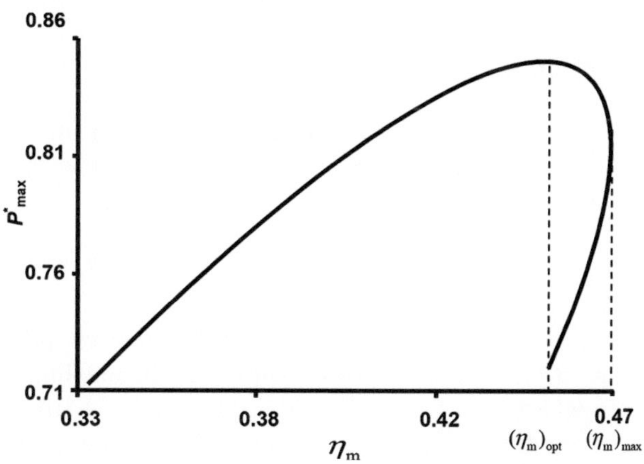

Fig. 4.12 The maximum power output versus thermal efficiency

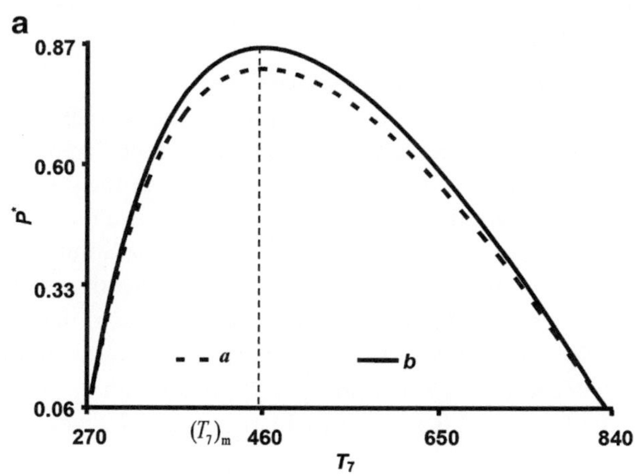

Fig. 4.13a The dimensionless power output vs. turbine outlet temperature (a, $R_{pi} = 1.17$ and b, $R_{pi} = 1.38$)

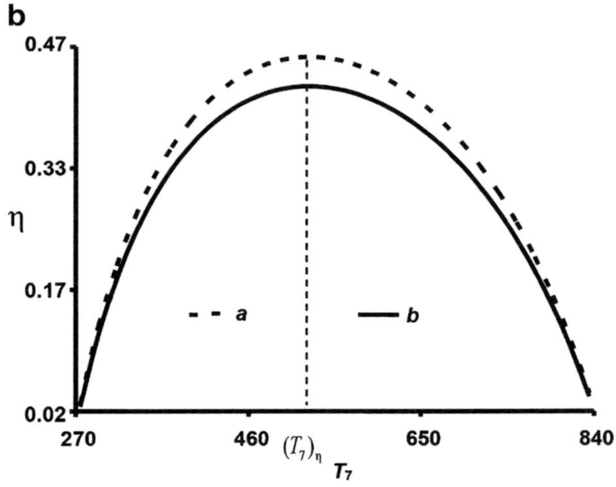

Fig. 4.13b Thermal efficiency vs. turbine outlet temperature (a, $R_{pi} = 1.17$ and b, $R_{pi} = 1.38$)

$$\left(R_{pi}\right)_\eta \leq R_{pi} \leq \left(R_{pi}\right)_m \tag{4.55}$$

where $(R_{pi})_m$ and $(R_{pi})_\eta$ are, respectively, the intercooling pressure ratios at the maximum point of the optimized power output and the maximum value of the corresponding thermal efficiency, as indicated in Figs. 4.10a, 4.10b, 4.11a, and 4.11b. Thus, using Eq. (4.55), one can give another optimum criterion for the cycle pressure ratio, as below:

$$\left(R_p\right)_\eta \leq R_p \leq \left(R_p\right)_m \tag{4.56}$$

where $(R_p)_m$ and $(R_p)_\eta$ are, respectively, the cycle pressure ratios at the maximum point of the optimized power output and the maximum value of the corresponding thermal efficiency, as indicated in Figs. 4.10a, 4.10b, 4.11a, and 4.11b.

According to Eq. (4.56) and Table 4.2, one can also give the optimum criteria of the temperatures of the various state points. The criteria will be helpful for engineers to optimally design the irreversible regenerative intercooled isothermal Brayton cycle.

Power Output* vs. *Thermal Efficiency

Figure 4.12 shows the variation of the maximum power output against the corresponding thermal efficiency. It is seen from Fig. 4.12 that the maximum power output is not a monotonic function of the corresponding thermal efficiency. The maximum power output first increases and then decreases as the corresponding thermal efficiency is increased. When the maximum power output is situated in the regions of the $P^*_{max} \sim \eta_m$ curve with a positive slope, it will decrease as the corresponding thermal efficiency is decreased, and hence these regions are not the optimal regions. It is thus clear that the maximum power output should be situated in the region of the $P^*_{max} \sim \eta_m$ curve with the negative slope. When the power output falls in this region, it will increase as the thermal efficiency is decreased and vice versa. It is shown once again that the optimal region of the intercooling pressure ratio should be determined by Eq. (4.55).

Effects of Turbine Outlet Temperature

The variations of the power output and the thermal efficiency of an irreversible regenerative intercooled modified Brayton cycle against the turbine outlet temperature for two different values of the intercooling pressure ratio are shown in Figs. 4.13a and 4.13b. It is seen from these figures that both the parameters, viz. the power output and the thermal efficiency, first increase and then decrease as the turbine outlet temperature is increased. It is clear from these figures that there exist different optimal values of the turbine outlet temperature for a given value of the intercooling pressure ratio at which the power output and the thermal efficiency attain their maximum values, from the point of view of maximum power output as well as from the point of view of the thermal efficiency.

The optimal criteria of the turbine outlet temperature for a given value of the intercooling pressure ratio along with other cycle parameters can be given by:

$$(T_7)_\eta \geq T_7 \geq (T_7)_{\mathrm{m}} \tag{4.57}$$

where $(T_7)_{\mathrm{m}}$ and $(T_7)_\eta$ are, respectively, the turbine outlet temperatures at the maximum power output point and the maximum efficiency point for a typical set of operating parameters, mentioned above.

So far we have given some optimal criteria of an irreversible regenerative intercooled isothermal Brayton cycle heat engine with finite heat capacities of external reservoirs. The present cycle model is more general and the results obtained by earlier workers (Wu and Kiang 1990, 1991; Ibrahim et al. 1991; Kaushik and Tyagi 2002; Redcenco et al. 1998; Kaushik 1999; Kumar 2000; Wang et al. 2003; Tyagi 2009; Tyagi et al. 2000, 2007; Vecchiarelli et al. 1997 Göktun and Yavuz 1999) can be directly obtained from the present cycle model. For example, if there is no superheater, viz. $\alpha = 1$, the present cycle model will reduce to that given in Sect. 4.2.1, and if there is no intercooler, the cycle model will reduce to that given in Sect. 4.2.2. Again, if $\alpha = 1$ and the reservoirs are of infinite heat capacities, this cycle model will reduce to that used in Wang et al. (2003); and if there is no intercooler and $\alpha = 1$, the present cycle model will reduce to those adopted in Tyagi et al. (2000, 2007), Vecchiarelli et al. (1997), and Göktun and Yavuz (1999) for finite as well as infinite heat capacities of the external reservoirs along with other cycle parameters for a typical set of operating condition.

4.4 Conclusion

In this chapter, realistic models of Brayton cycle including regenerative intercooled, regenerative isothermal, and regenerative intercooled with isothermal heat addition for the finite heat capacities of external reservoirs have been studied in detail. The analysis of regenerative intercooled and intercooled isothermal Brayton cycles exhibit that there exists an optimal value of the intercooling pressure ratio at which these cycle models attain their maximum performance for a given set of operating parameters.

Also, from the analysis of an irreversible regenerative isothermal Brayton cycle, it is found that the inlet temperature for the isobaric heat addition should be less than the inlet temperature of isothermal heat addition for better performance because the heat addition in the former increases the quantity of energy, while in the latter it increases the quality of energy because there is isothermal heat addition and the heat supplied by the latter only increases the kinetic energy rather than thermal energy unlike the former source. Again there is a significant improvement in the performance of Brayton cycle (above 15%) with isothermal heat addition over the conventional Brayton cycle.

Chapter 5
Finite Time Thermodynamic Analysis of Complex Brayton Cycle

5.1 Introduction

In the present chapter, the effects of some important cycle parameters on the thermodynamic performance of complex Brayton cycles are studied by means of analytical relations using the concept of finite time thermodynamics (Curzon and Alhborn 1975). The analysis presented in this paper demonstrates the main parameters governing changes in performance variables and provides simple relations amongst the different cycle parameters. Thus, the use of analytical relations allows an understanding of the effects of different cycle parameters, such as, the turbine outlet temperature, reheat, intercooling, isothermal heat addition, cycle pressure ratios, etc. and provides a means for a quick estimation of such effects in a wide range.

5.2 Complex Brayton Cycle

The regenerative Brayton cycle having the combination of more than two additional components such as intercooler and reheat is sometime called complex Brayton cycle for both finite and infinite heat capacity of external reservoirs. The complex Brayton cycle can also be modified by incorporating isothermal heat addition apart from the combination of different components mentioned above. Thus, the present chapter will deal with such combination as regenerator, intercooler, and reheat with and without isothermal heat addition, as given below.

S.C. Kaushik et al., *Finite Time Thermodynamics of Power and Refrigeration Cycles*,
https://doi.org/10.1007/978-3-319-62812-7_5

5.2.1 Intercooled–Reheat Brayton Cycle

The schematic and *T–s* diagram of an irreversible regenerative–intercooled–reheat (complex) Brayton cycle heat engine is shown in Figs. 5.1a and 5.1b. As can be seen from Fig. 5.1a, the working fluid enters the first compressor at state point 1 and is compressed up to state point 2 (ideally up to state point 2S) and then enters the first heat sink where it is cooled down up to state point 3 by rejecting the heat to the finite heat capacity heat sink whose temperature varies from T_{C1} to T_{C2}. The working fluid again enters the second compressor and is compressed up to state point 4 (ideally up to state point 4S) and then enters the regenerator where it is heated up to state point 4R by the exhaust of a low-pressure turbine. The working fluid leaving the regenerator enters the first heat source (hot reservoir) and is heated up to state point 5, while the temperature of the hot reservoir varies from T_{H1} to T_{H2}. Then it enters the high-pressure turbine and expands up to state point 6 (ideally up to state point 6S) and then enters the second heat source where it is heated up to state point 7 by the finite heat capacity heat source whose temperature varies from T_{H3} to T_{H4}.

The working fluid again enters the low-pressure turbine and expands up to state point 8 (ideally up to state point 8S) and enters the regenerator where it partly transfers its heat to the outlet of the second compressor and is cooled down up to state point 8R. Finally, the working fluid enters the second heat sink and is cooled down up to state point 1 by rejecting the heat to the finite heat capacity heat sink whose temperature varies from T_{L1} to T_{L2}, thereby, completing the cycle. Thus, the study is performed on a closed version 1–2–3–4–4R–5–6–7–8–8R–1 of an irreversible intercooled–regenerative–reheat Brayton cycle with the non-isentropic

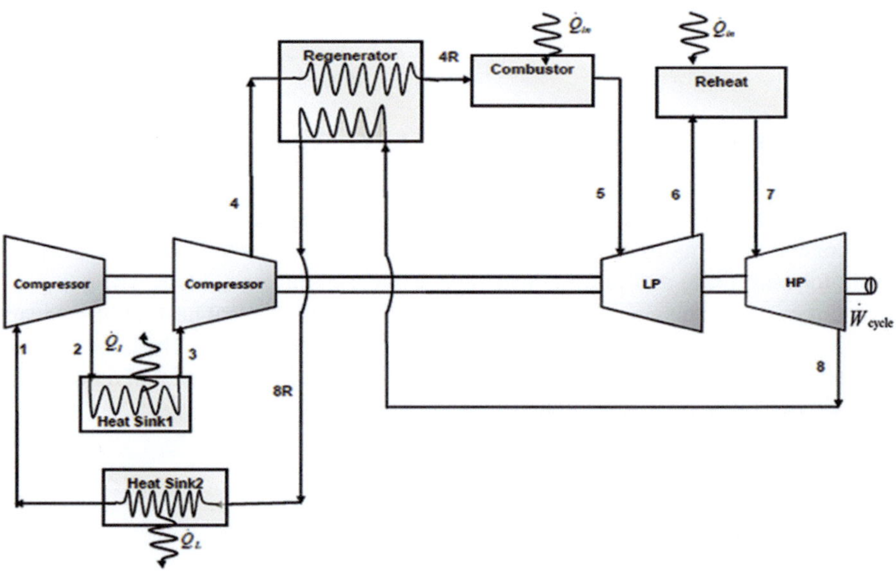

Fig. 5.1a Schematic diagram of an irreversible complex Brayton cycle

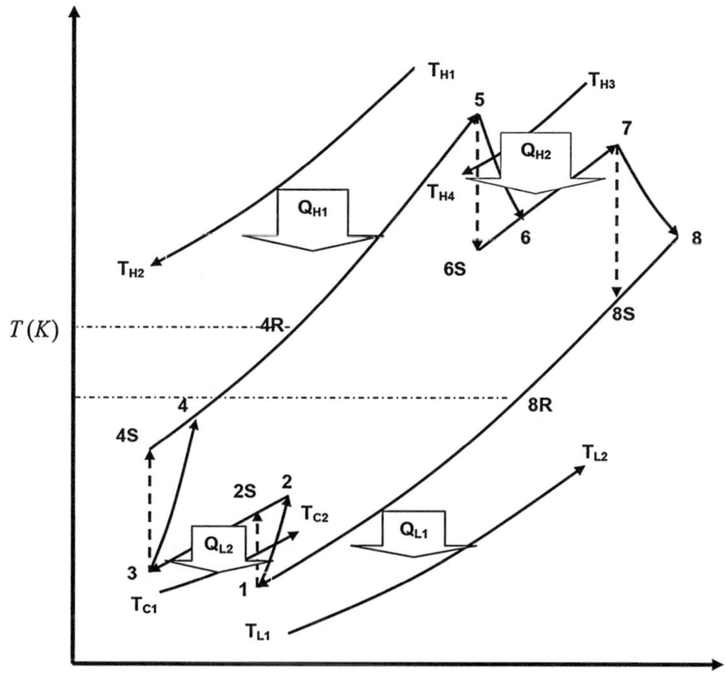

Fig. 5.1b T–s diagram of an irreversible complex Brayton

compression and expansion processes for the finite heat capacity of external reservoirs. According to the cycle model shown in Fig. 5.1b, the various heat transfer rates to and from the cycle (Q_{H1}, Q_{H2}, Q_{L1}, and Q_{L2}) are given by Tyagi et al. (2006a, b, 2008):

$$\begin{aligned} Q_{H1} &= (UA)_{H1}(LMTD)_{H1} = C_W(T_5 - T_{4R}) = C_{H1}(T_{H1} - T_{H2}) \\ &= C_{H1,\min}\varepsilon_{H1}(T_{H1} - T_{4R}) \end{aligned} \tag{5.1}$$

$$\begin{aligned} Q_{H2} &= (UA)_{H2}(LMTD)_{H2} = C_W(T_7 - T_6) = C_{H2}(T_{H3} - T_{H4}) \\ &= C_{H2,\min}\varepsilon_{H2}(T_{H3} - T_6) \end{aligned} \tag{5.2}$$

$$\begin{aligned} Q_{L1} &= (UA)_{L1}(LMTD)_{L1} = C_W(T_{8R} - T_1) = C_{L1}(T_{L2} - T_{L1}) \\ &= C_{L1,\min}\varepsilon_{L1}(T_{8R} - T_{L1}) \end{aligned} \tag{5.3}$$

$$\begin{aligned} Q_{L2} &= (UA)_{L2}(LMTD)_{L2} = C_W(T_2 - T_3) = C_{L2}(T_{C2} - T_{C1}) \\ &= C_{L2,\min}\varepsilon_{L2}(T_2 - T_{C1}) \end{aligned} \tag{5.4}$$

and the regenerative heat transfer rate is given by:

$$\begin{aligned} Q_R &= (UA)_R(LMTD)_R = C_W(T_8 - T_{8R}) = C_W(T_{4R} - T_4) \\ &= C_W\varepsilon_R(T_8 - T_4) \end{aligned} \tag{5.5}$$

where $(UA)_J$'s (J = H1, H2, L1, L2, and R) are the overall heat transfer coefficient-area products, while ε_J's are, respectively, the effectiveness on the source-, reheat-, sink-, intercooler-, and the regenerative-side heat exchangers, defined as (Wang et al. 2003; Tyagi and Kaushik 2005; Tyagi 2009; Tyagi et al. 2006a, b, 2008; Negri-di et al. 1995; Bhargava and Peretto 2002):

$$\varepsilon_k = \frac{\{1 - e^{-NTU_k(1-C_{km})}\}}{\{1 - C_{km}e^{-NTU_k(1-C_{km})}\}} \quad (C_{km} = C_{k,\min}/C_{k,\max}) \quad \text{and} \quad \varepsilon_R = \frac{NTU_R}{1 + NTU_R} \tag{5.6}$$

where C_k's and C_W are, respectively, the heat capacitance rates (mass flow rate multiplied by the specific heat) of the external reservoirs and within the cycle.

While $C_{k,\min} = \min(C_k, C_W)$, $C_{k,\max} = \max(C_k, C_W)$ (here k = H1, H2, L1, and L2) and NTU_J's$\{=(UA)_J/C_W\}$ are the number of transfer units, based on the minimum thermal capacitance, and (LMTD)s$_J$ are the Log Mean Temperature Differences on their respective side heat exchangers, defined as (Wang et al. 2003; Tyagi and Kaushik 2005; Tyagi 2009; Tyagi et al. 2006a, b, 2008; Negri-di et al. 1995; Bhargava and Peretto 2002):

$$(LMTD)_{H1} = [(T_{H2} - T_4) - (T_{H1} - T_5)]\left(\ln\frac{(T_{H2} - T_4)}{(T_{H1} - T_5)}\right)^{-1} \tag{5.7a}$$

$$(LMTD)_{H2} = [(T_{H4} - T_6) - (T_{H3} - T_7)]\left(\ln\frac{(T_{H4} - T_6)}{(T_{H3} - T_7)}\right)^{-1} \tag{5.7b}$$

$$(LMTD)_{L1} = [(T_{8R} - T_{L2}) - (T_1 - T_{L1})]\left(\ln\frac{(T_{8R} - T_{L2})}{(T_1 - T_{L1})}\right)^{-1} \tag{5.7c}$$

$$(LMTD)_{L2} = [(T_2 - T_{C2}) - (T_3 - T_{C1})]\left(\ln\frac{(T_2 - T_{C2})}{(T_3 - T_{C1})}\right)^{-1} \tag{5.7d}$$

$$(LMTD)_{R} = [(T_8 - T_{4R}) - (T_{8R} - T_4)]\left(\ln\frac{(T_8 - T_{4R})}{(T_{8R} - T_4)}\right)^{-1} \tag{5.7e}$$

With the help of Fig. 5.1b, the efficiencies of the compression and expansion devices can be written as (Tyagi et al. 2006a, b, 2008):

$$\eta_{C1} = \frac{(T_{2S} - T_1)}{(T_2 - T_1)}, \quad \eta_{C2}\frac{(T_{4S} - T_3)}{(T_4 - T_3)} \tag{5.8a}$$

$$\eta_{T1} = \frac{(T_5 - T_{6S})}{(T_5 - T_6)} \text{and} \eta_{T2}\frac{(T_7 - T_{8S})}{(T_7 - T_8)} \leq 1.0 \tag{5.8b}$$

Using Eqs. (5.1), (5.2), (5.3), (5.4), (5.5,), and (5.8a, b), we may get the expressions for temperature at different state points, following Tyagi et al. (2006a, b) as below:

$$T_{8R} = (1 - \varepsilon_R)T_8 + \varepsilon_R T_4 \tag{5.9a}$$

$$T_{4R} = (1 - \varepsilon_R)T_4 + \varepsilon_R T_8 \tag{5.9b}$$

$$T_5 = (1 - x_1)T_{4R} + x_1 T_{H1} \tag{5.9c}$$

$$T_7 = (1 - x_2)T_6 + x_2 T_{H3} \tag{5.9d}$$

$$T_1 = (1 - y_1)T_{8R} + y_1 T_{L1} \tag{5.9e}$$

$$T_3 = (1 - y_2)T_2 + y_2 T_{C1} \tag{5.9f}$$

$$T_{2S} = (1 - \eta_{C1})T_1 + \eta_{C1} T_2 \tag{5.9g}$$

$$T_{4S} = (1 - \eta_{C2})T_3 + \eta_{C2} T_4 \tag{5.9h}$$

$$T_{6S} = \left(1 - \eta_{T1}^{-1}\right)T_5 + \eta_{T1}^{-1} T_6 \tag{5.9i}$$

$$T_{8S} = \left(1 - \eta_{T2}^{-1}\right)T_7 + \eta_{T2}^{-1} T_8 \tag{5.9j}$$

where $x_i = C_{H,\,min}\varepsilon_H/C_w$ and $y_i = C_{L,\,min}\varepsilon_L/C_w$ ($i = 1, 2$, $H =$ H1, H2, and $L =$ L1, L2). Rearranging Eqs. (5.9a), (5.9b), (5.9c), (5.9d), (5.9e), (5.9f), (5.9g), (5.9h), (5.9i), and (5.9j), yields:

$$T_1 = a_1 T_4 + b_1 T_8 + c_1 \quad \text{and} \quad T_2 = aT_1 \tag{5.10a}$$

$$T_3 = a_2 T_4 + b_2 T_8 + c_2 \tag{5.10b}$$

$$T_{4S} = a_3 T_4 + b_3 T_8 + c_3 \tag{5.10c}$$

$$T_5 = a_4 T_4 + b_4 T_8 + c_4 \quad \text{and} \quad T_6 = bT_5 \tag{5.10d}$$

$$T_7 = a_5 T_4 + b_5 T_8 + c_5 \tag{5.10e}$$

$$T_{8S} = a_6 T_4 + b_6 T_8 + c_6 \tag{5.10f}$$

where the different parameters are as below:

$$\begin{aligned}
&a_1 = (1 - \varepsilon_R)(1 - y_1), b_1 = \varepsilon_R(1 - y_1), c_1 = y_1 T_{L1}, a_2 = a(1 - y_2)b_1,\\
&b_2 = a(1 - y_2)a_1, c_2 = \{a(1 - y_2)y_1 + y_1\}T_{L1}, a_3 = \eta_{C2} + (1 - \eta_{C2})a_2,\\
&b_3 = (1 - \eta_{C2})b_2, c_3 = (1 - \eta_{C2})c_2, a_4 = (1 - \varepsilon_R)(1 - x_1), b_4 = \varepsilon_R(1 - x_1),\\
&c_4 = x_1 T_{H1}, a_5 = b(1 - x_2)b_4, b_5 = b(1 - x_2)a_4, c_5 = b(1 - x_2)c_4 + x_2 T_{H3},\\
&a_6 = \left(1 - \eta_{T2}^{-1}\right)a_5, b_6 = \eta_{T2}^{-1} + \left(1 - \eta_{T2}^{-1}\right)b_5, c_6 = \left(1 - \eta_{T2}^{-1}\right)c_5,\\
&\quad a = \left(R_{pi}^{1-1/\gamma} + \eta_{C1} - 1\right)\eta_{C1}^{-1} \text{ and } b = \left(1/R_{ph}^{1-1/\gamma} + \eta_{T1}^{-1} - 1\right)\eta_{T1}
\end{aligned}$$

Using the second law of thermodynamics in this case of a closed cycle 1–2S–3–4S–5–6S–7–8S–1 intercooled–reheat Brayton heat engine (Tyagi et al. 2006a, b, 2008) yields:

$$T_{2S}T_{4S}T_{6S}T_{8S} = T_1 T_3 T_5 T_7 \tag{5.11}$$

Substituting the values of intercooling and reheat pressure ratios, i.e. $R_{\mathrm{pi}}^{1-1/\gamma} = T_{2S}/T_1$, $R_{\mathrm{ph}}^{1-1/\gamma} = T_{6S}/T_5$, and different temperatures from Eqs. (5.10a), (5.10b), (5.10c), (5.10d), (5.10e) and (5.10f) into Eq. (5.11), one can get a quadratic equation in T_4 and T_8, which on solving for T_4 yields (Tyagi et al. 2006a, b, 2008; Kaushik and Tyagi 2002):

$$T_4 = \frac{-B \pm \sqrt{B^2 - 4AC}}{2A} \tag{5.12}$$

where the different parameters are as below:

$$\begin{aligned} &A = a_2a_5 - \alpha a_3a_6, B = a_7T_8 + a_8, C = a_9T_8^2 + a_{10}T_8 + a_{11}, \alpha = \left(R_{\mathrm{pi}}/R_{\mathrm{ph}}\right)^{1-1/\gamma},\\ &a_7 = a_2b_5 + a_5b_2 - \alpha(a_3b_6 + a_6b_3), a_8 = a_2c_5 + a_5c_2 - \alpha(a_3c_6 + a_6c_3),\\ &a_9 = b_2b_5 - \alpha b_3b_6,\\ &a_{10} = b_2c_5 + b_5c_2 - \alpha(b_3c_6 + b_6c_3)\,\text{and}\,a_{11} = c_2c_5 - \alpha c_3c_6 \end{aligned}$$

Using the first law of thermodynamics, the power output and the corresponding thermodynamic efficiency in this case of an irreversible regenerative–intercooled–reheat Brayton cycle heat engine are given by Tyagi et al. (2006a, b, 2008):

$$P = K_1 - \frac{a_{12}}{2A}\left(\begin{array}{c} -(a_7T_6 + a_8) \\ \pm\sqrt{(a_7T_8 + a_8)^2 - 4A\left(a_9T_8^2 + a_{10}T_8 + a_{11}\right)} \end{array}\right) - a_{13}T_8 \tag{5.13}$$

$$\eta = \frac{K_1 - \frac{a_{12}}{2A}\left(\begin{array}{c} -(a_7T_8 + a_8) \\ \pm\sqrt{(a_7T_8 + a_8)^2 - 4A\left(a_9T_8^2 + a_{10}T_8 + a_{11}\right)} \end{array}\right) - a_{13}T_8}{K_2 - \frac{b_7}{2A}\left(\begin{array}{c} -(a_7T_8 + a_8) \\ \pm\sqrt{(a_7T_8 + a_8)^2 - 4A\left(a_9T_8^2 + a_{10}T_8 + a_{11}\right)} \end{array}\right) - b_8T_8} \tag{5.14}$$

where the different parameters are given as below:

$$\begin{aligned} K_1 &= C_W\{(1 - ax_2)x_1T_{H1} + x_2T_{H3} + (1 - by_2)y_1T_{L1} + y_2T_{C1}\},\\ K_2 &= C_W\{x_1T_{H1} + x_2T_{H3} + x_2c_4\}, K_3 = a_{12}K_3 - b_7K_2, K_4 = b_8K_2 - a_{13}K_3,\\ a_{12} &= C_W\{x_1(1 - \varepsilon_R) + bx_2(1 - x_1)(1 - \varepsilon_R) + y_1\varepsilon_R + ay_2\varepsilon_R(1 - y_1)\}, \quad ,\\ b_8 &= C_W\{x_1\varepsilon_R + bx_2b_4\}, b_7 = C_W\{x_1(1 - \varepsilon_R) + bx_2a_4\}\,\text{and}\\ a_{13} &= C_W\{x_1\varepsilon_R + bx_2(1 - x_1)\varepsilon_R + y_1(1 - \varepsilon_R) + ay_2(1 - \varepsilon_R)(1 - y_1)\} \end{aligned}$$

5.2.2 *Isothermal Intercooled–Reheat Brayton Cycle*

The *T*–*s* diagram of a closed cycle regenerative modified complex Brayton cycle is shown in Figs. 5.2a and 5.2b. The basic components of this cycle are the intercooler, two compressors, regenerator, regular combustion chamber (RCC), converging combustion chamber (CCC), reheater, two turbines, and the low- and

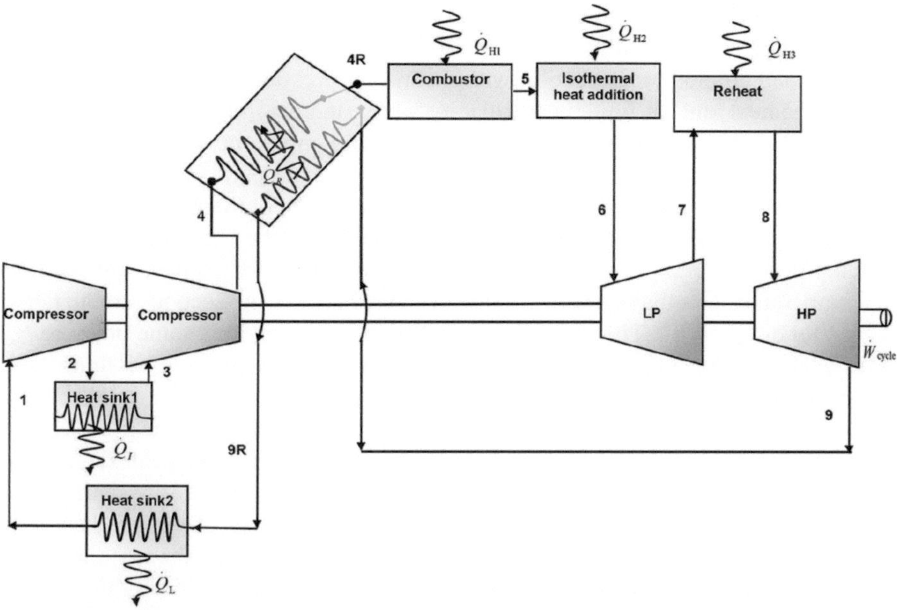

Fig. 5.2a Schematic diagram of an irreversible modified complex Brayton cycle

Fig. 5.2b *T–s* diagram of an irreversible isothermal complex Brayton

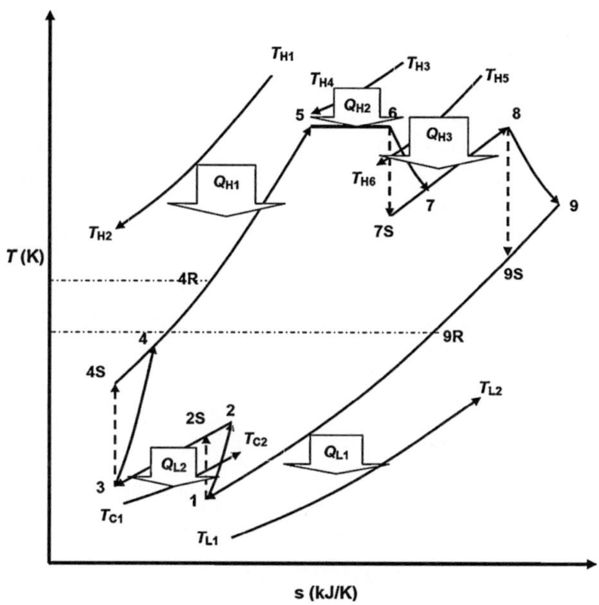

high-temperature heat exchangers. The working fluid enters the compressor at state point 1 and is compressed up to state point 2 or isentropically up to state point 2S. The gas then enters the intercooler with a constant pressure P_I where it rejects the heat to the heat sink of finite heat capacity whose temperature varied from T_{C1} to T_{C2} and is cooled down up to state point 3. It again enters the second compressor and is compressed up to state point 4 (ideally up to state point 4S). At state point 4, the working fluid leaving the second compressor enters the regenerator with a constant pressure P_H and heated up to state point 4R by the exhaust of the second turbine. In a real or an ideal regenerator, the working fluid leaves the regenerator at the temperature less than or equal to that of the exhaust of the second turbine (state point 9), i.e. $T_{4R} \leq T_9$. The isobaric heat addition process with a same pressure P_H takes place during process 4R–5 in an RCC from a heat source of finite heat capacity whose temperature varies from T_{H1} to T_{H2}.

Further heat addition is accomplished in a CCC isothermally during process 5–6, from a heat source of finite heat capacity whose temperature varies from T_{H3} to T_{H4}. When the gas leaves the CCC at state point 6, it has a lower pressure than that at state point 5, but the velocity and hence, the kinetic energy of the gas has been increased enormously due to the nature of a CCC. The gas enters the first turbine at state point 6 and expands non-isentropically up to state point 7 (ideally up to state point 7S) and again enters the reheater and heated up to state point 8. The exhaust of the second turbine at state point regenerator where it is cooled up to state point 7R at a constant pressure P_L and finally enters the low-temperature heat exchanger at state point 9R and is cooled down up to state point 1 by rejecting the heat to the heat sink of finite heat capacity whose temperature varied from T_{L1} to T_{L2}, thereby completing the cycle. Thus, the theoretical model of an irreversible modified complex Brayton cycle 1–2–3–4–4R–5–6–7–8–9–9R–1 with the real compression and expansion processes is considered in this study.

Let Q_H and Q_L be the heat transfer rates to and from the heat engine, respectively, and Q_R the regenerative heat transfer rate; then (Tyagi et al. 2006a, b, 2008):

$$\begin{aligned} Q_{H1} &= (UA)_{H1}(LMTD)_{H1} = C_W(T_5 - T_{4R}) = C_{H1}(T_{H1} - T_{H2}) \\ &= C_{H1,min}\varepsilon_{H1}(T_{H1} - T_{4R}) \end{aligned} \quad (5.15)$$

$$\begin{aligned} Q_{H2} &= (UA)_{H2}(LMTD)_{H2} = \dot{m}\left(V_6^2 - V_5^2\right)/2 = C_{H2}(T_{H3} - T_{H4}) \\ &= C_{H2,min}\varepsilon_{H2}(T_{H3} - T_5) \end{aligned} \quad (5.16)$$

$$\begin{aligned} Q_{H3} &= (UA)_{H3}(LMTD)_{H3} = C_W(T_8 - T_7) = C_{H1}(T_{H5} - T_{H6}) \\ &= C_{H3,min}\varepsilon_{H3}(T_{H5} - T_7) \end{aligned} \quad (5.17)$$

$$\begin{aligned} Q_{L1} &= (UA)_{L1}(LMTD)_{L1} = C_W(T_{9R} - T_1) = C_{L1}(T_{L2} - T_{L1}) \\ &= C_{L1,min}\varepsilon_{L1}(T_{9R} - T_{L1}) \end{aligned} \quad (5.18)$$

$$\begin{aligned} Q_{L2} &= (UA)_{L2}(LMTD)_{L2} = C_W(T_2 - T_3) = C_{L2}(T_{C2} - T_{C1}) \\ &= C_{L2,min}\varepsilon_{L2}(T_2 - T_{C1}) \end{aligned} \quad (5.19)$$

$$\begin{aligned} Q_R &= (UA)_{LR}(LMTD)_R = C_W(T_9 - T_{9R}) = C_W(T_{4R} - T_4) \\ &= C_W\varepsilon_{R1}(T_9 - T_4) \end{aligned} \quad (5.20)$$

where C_k ($k = \mathrm{H1, H2, H3, L1, L2}$) and C_W are, respectively, the heat capacitance rates (mass flow rate times specific heat) of the external fluids and the working fluid, respectively. $(UA)_J$ and $(UA)_R$ are, respectively, the overall heat transfer coefficient-area products on the different reservoirs and cycle and the regenerative heat exchanger. V_5 and V_6 are the velocities of the working fluid at state points 5 and 6, respectively, $\dot{m}$ is the mass flow rate of the working fluid, $(\mathrm{LMTD})_J$'s are the Log Mean Temperature Differences between the external reservoirs and the cycle and ε_J's ($J = k, R$) are the effectiveness of the various heat exchangers, while $C_{k,\min}$ and $C_{k,\max}$ are defined as (Tyagi 2009; Tyagi et al. 2006a, b, 2008; Negri-di et al. 1995; Bhargava and Peretto 2002):

$$C_{k,\min} = \min(C_k, C_W) \quad \text{and} \quad C_{k,\max} = \max(C_k, C_W) \tag{5.21}$$

With the help of Fig. 5.2b, the compression and expansion efficiencies are given as (Tyagi et al. 2006a, b, 2008):

$$\eta_{C1} = \frac{(T_{2S} - T_1)}{(T_2 - T_1)} \text{ and } \eta_{C2} = \frac{(T_{4S} - T_3)}{(T_4 - T_3)}, \tag{5.22a}$$

$$\eta_{T1} = \frac{(T_6 - T_7)}{(T_6 - T_{7S})}, \quad \eta_{T2} = \frac{(T_8 - T_9)}{(T_8 - T_{9S})} \le 1.0 \tag{5.22b}$$

Now from Eqs. (5.15), (5.16), (5.17), (5.18), (5.19), (5.20) and (5.22a, b), it yields:

$$T_1 = a_1 T_4 + b_1 T_9 + c_1 \tag{5.23a}$$

$$T_3 = a_2 T_4 + b_2 T_9 + c_2 \tag{5.23b}$$

$$T_{4S} = a_3 T_4 + b_3 T_7 + c_3 \tag{5.23c}$$

$$T_5 = a_4 T_7 + b_4 T_4 + c_4 \tag{5.23d}$$

$$T_8 = a_5 T_4 + b_5 T_7 + c_5 \tag{5.23e}$$

$$T_{9S} = a_6 T_4 + b_6 T_7 + c_6 \tag{5.23f}$$

$$T_2 = a' T_1, T_{7S} = b' T_6 \, \text{and} \, T_5 = T_6 \tag{5.23g}$$

where the different parameters are as below:

$$x_i = C_{Hi,\min}\varepsilon_{Hi}/C_W (i = 1, 2, 3), y_j = C_{Lj,\min}\varepsilon_{Lj}/C_W (j = 1, 2),$$
$$a' = \left\{1 + \left(R_{pi}^{k} - 1\right)\eta_{C1}^{-1}\right\}, b' = \left\{1 - \eta_{T1}\left(1 - R_{ph}^{-k}\right)\right\}, R_{Pi} = P_2/P_L,$$
$$R_{Ph} = P_H/P_7, R_t = P_H/P_6, k = 1 - 1/\gamma, a_1 = \varepsilon_R(1 - y_1),$$
$$b_1 = (1 - \varepsilon_R)(1 - y_1), c_1 = y_1 T_{L1}, a_2 = a'(1 - y_1)a_1, b_2 = a'(1 - y_1)b_1,$$
$$c_2 = a'(1 - y_1)c_1 + y_2 T_{C1}, a_3 = (1 - \eta_{C2})a_2 + \eta_{C2}, b_3 = (1 - \eta_{C2})b_2,$$
$$c_3 = (1 - \eta_{C2})c_2, a_4 = (1 - \varepsilon_R)(1 - x_1), b_4 = \varepsilon_R(1 - x_1), c_4 = x_1 T_{H1},$$
$$a_5 = b'(1 - x_3)a_4, b_5 = b'(1 - x_3)b_4, c_5 = b'(1 - x_3)c_4 + x_3 T_{H5},$$
$$a_6 = \left(1 - \eta_{T2}^{-1}\right)a_5, b_6 = \left(1 - \eta_{T2}^{-1}\right)b_5, c_6 = \left(1 - \eta_{T2}^{-1}\right)c_5$$

In this case of a modified complex Brayton cycle (Tyagi et al. 2006a, b, 2008; Kaushik and Tyagi 2002) yields:

$$\left(\frac{P_{\mathrm{H}}}{P_{\mathrm{L}}}\right)^{\mathrm{k}}=\left(\frac{P_{\mathrm{H}}P_{2\mathrm{S}}}{P_{2\mathrm{S}}P_{\mathrm{L}}}\right)^{\mathrm{k}}=\left(R_{\mathrm{pi}}\frac{P_{\mathrm{H}}}{P_{2\mathrm{S}}}\right)^{\mathrm{k}}=R_{\mathrm{pi}}^{\mathrm{k}}\frac{T_{4\mathrm{S}}}{T_{3}} \tag{5.24a}$$

Similarly,

$$\left(\frac{P_{\mathrm{H}}}{P_{\mathrm{L}}}\right)^{\mathrm{k}}=\left(\frac{P_{\mathrm{H}}P_{6}P_{7\mathrm{S}}}{P_{6}P_{7\mathrm{S}}P_{\mathrm{L}}}\right)^{\mathrm{k}}=\left(R_{\mathrm{pt}}R_{\mathrm{ph}}\frac{P_{7\mathrm{S}}}{P_{\mathrm{L}}}\right)^{\mathrm{k}}=\left(R_{\mathrm{pt}}R_{\mathrm{ph}}\right)^{\mathrm{k}}\frac{T_{8}}{T_{9\mathrm{S}}} \tag{5.24b}$$

where R_{pi} and R_{ph} are, respectively, the intercooling, reheat pressure ratios, while R_{pt} is the isothermal pressure drop during process 5–6 and γ is the specific heat ratio and. Combining Eqs. (5.23a), (5.23b), (5.23c), (5.23d), (5.23e), (5.23f), (5.23g), and (5.24a), (5.24b), and solving for T_4 (treating T_9 as independent along with the other cycle parameters) yields:

$$T_{4}=\frac{-B_{3}+\sqrt{B_{3}^{2}-4A_{3}C_{3}}}{2A_{3}} \tag{5.25}$$

where the parameters are as below:

$$\begin{aligned}
&A_{3}=a_{3}a_{6}-\alpha a_{2}a_{5}, B_{3}=a_{7}T_{4}+b_{7}, C_{3}=a_{8}T_{4}^{2}+b_{8}T_{4}+c_{7},\\
&a_{7}=a_{3}b_{6}+a_{6}b_{3}-\alpha(a_{2}b_{5}+a_{5}b_{2}), b_{7}=a_{3}c_{6}+a_{6}c_{3}-\alpha(a_{2}c_{5}+a_{5}c_{2}),\\
&a_{8}=b_{3}b_{6}-\alpha b_{2}b_{5}, b_{8}=b_{3}c_{6}+b_{6}c_{3}-\alpha(b_{2}c_{5}+b_{5}c_{2}),\\
&c_{7}=c_{3}c_{6}-\alpha c_{2}c_{5}\text{ and }\alpha=\left(R_{ph}R_{pt}/R_{pi}\right)^{k}
\end{aligned}$$

The power output and the thermal efficiency are defined as (Tyagi et al. 2006a, b, 2008; Kaushik and Tyagi 2002):

$$P=Q_{\mathrm{H1}}+Q_{\mathrm{H2}}+Q_{\mathrm{H3}}-Q_{\mathrm{L1}}-Q_{\mathrm{L2}}=K_{1}-a_{9}T_{4}-b_{9}T_{9} \tag{5.26}$$

$$\eta=\frac{P}{Q_{\mathrm{H1}}+Q_{\mathrm{H2}}+Q_{\mathrm{H3}}}=\frac{K_{1}-a_{9}T_{4}-b_{9}T_{9}}{K_{2}-c_{8}T_{4}-c_{9}T_{9}} \tag{5.27}$$

while the different parameters are as below:

$$\begin{aligned}
&K_{1}=C_{\mathrm{W}}\{x_{1}T_{\mathrm{H1}}+x_{2}T_{\mathrm{H3}}+x_{3}T_{\mathrm{H5}}+y_{1}T_{\mathrm{L1}}+y_{2}T_{\mathrm{C1}}-(x_{2}+b'x_{3})c_{4}-a'y_{2}c_{1}\},\\
&a_{9}=C_{\mathrm{W}}\{x_{1}(1-\varepsilon_{\mathrm{R}})+y_{1}\varepsilon_{\mathrm{R}}+y_{2}a'a_{1}+(x_{2}+x_{3}b')a_{3}\},\\
&c_{9}=C_{\mathrm{W}}\{x_{1}\varepsilon_{\mathrm{R}}+(x_{2}+x_{3}b')b_{4}\}, c_{8}=C_{\mathrm{W}}\{x_{1}(1-\varepsilon_{\mathrm{R}})+(x_{2}+x_{3}b')a_{4}\},\\
&b_{9}=C_{\mathrm{W}}\{x_{1}\varepsilon_{\mathrm{R}}+y_{1}(1-\varepsilon_{\mathrm{R}})+y_{2}a'b_{1}+(x_{2}+x_{3}b')b_{3}\}\text{ and}\\
&K_{2}=C_{\mathrm{W}}\{x_{1}T_{\mathrm{H1}}+x_{2}T_{\mathrm{H3}}+x_{3}T_{\mathrm{H5}}-(x_{2}+b'x_{3})c_{4}\}
\end{aligned}$$

It is seen from Eqs. (5.25), (5.26) and (5.27), that the power and efficiency are functions of a single variable for a given set of cycle parameters. Thus using Eqs. (5.26) and (5.27), their most extreme conditions, i.e. $\frac{\partial P}{\partial T_9} = 0$ and $\frac{\partial \eta}{\partial T_9} = 0$, along with Eq. (5.25), one can obtain the optimal values of T_9 as:

$$\left(T_{9,\mathrm{opt}}\right)_{P_{\mathrm{opt}}} = \frac{-B_4 \pm \sqrt{B_4^2 - A_4 C_4}}{A_4} \tag{5.28}$$

$$\left(T_{9,\mathrm{opt}}\right)_{\eta_{\mathrm{opt}}} = \frac{-B_5 \pm \sqrt{B_5^2 - A_5 C_5}}{A_5} \tag{5.29}$$

where $\left(T_{9,opt}\right)_{\mathrm{P_{opt}}}$ and $\left(T_{9,\mathrm{opt}}\right)_{\tilde{C}\,\mathrm{opt}}$ are, respectively, the values corresponding to the maximum power and maximum efficiency at a given set of cycle parameters, while the other parameters are:

$$\begin{aligned}
A_4 &= a_{10} - a_{10}^2/c_{10}^2, B_4 = b_{10} - a_{10}b_{10}/c_{10}^2, C_4 = a_{11} - b_{10}^2/c_{10}^2,\\
A_5 &= a_{10} - b_{12}^2/a_{12}^2, B_5 = b_{10} - b_{12}c_{13}/a_{12}^2, C_5 = a_{11} - c_{13}^2/a_{12}^2,\\
a_{10} &= a_7^2 - 4Aa_8, a_{11} = b_7^2 - 4Ac_7, a_{12} = 2Ac_{12} + a_7c_7 - b_7b_{11},\\
b_{10} &= a_7b_7 - 2Ab_8, b_{11} = b_9c_8 - a_9c_9, b_{12} = a_{10}c_{11} - b_{10}b_{11},\\
c_{10} &= 2Ab_9/a_9 - a_7, c_{11} = a_9K_2 - c_8K_1, c_{12} = c_9K_1 - b_9K_2 \text{ and}\\
c_{13} &= b_{10}c_{11} - b_{11}a_{11}
\end{aligned}$$

It is also important to note that the case presented here is of finite heat capacity of external reservoirs, and similar case may also be studied for infinite heat capacity of the external reservoirs, and the readers may take it as an exercise for both the cases.

5.3 Discussion of Results

Using the above equations, one can calculate the maximum power output, the corresponding thermal efficiency as well as the maximum efficiency and the corresponding power, and the other cycle parameters for a given set of operating conditions. The graphs are plotted for the optimized power and the corresponding efficiency as well as the optimized efficiency and the corresponding power against the intercooling, reheat, and cycle pressure ratios along with other cycle parameters, while the discussion of results for the two different cases are given separately, as below.

5.3.1 Intercooled–Reheat Brayton Cycle

It is seen from Eqs. (5.13) and (5.14) that P and η are the functions of a single variable T_8, for a given set of operating conditions. Thus using Eqs. (5.13) and (5.14) and a typical set of operating parameters such as $R_{\text{pi}}=5$, $T_{\text{H1}}=T_{\text{H3}}=1500K$, $\varepsilon_{\text{H1}}=\varepsilon_{\text{H2}}=\varepsilon_{\text{L1}}=\varepsilon_{\text{L2}}=\varepsilon_{\text{R}}=\eta_{\text{C1}}=\eta_{\text{C2}}=\eta_{\text{T1}}=\eta_{\text{T2}}=0.90$, $T_{\text{L1}}=T_{\text{C1}}=300\text{K}$, $\text{R}_{\text{ph}}=8$, and $C_{\text{W}}=0.95\,\text{kWK}^{-1}$, some characteristic curves were plotted for the power output and the corresponding thermodynamic efficiency against the turbine outlet temperature (T_8) and between the power output and efficiency and obtained the results, as below.

Performance Versus Turbine Outlet Temperature The variations of the dimensionless power output and the corresponding thermodynamic efficiency against the outlet temperature of the low-pressure turbine (T_8) are shown in Figs. 5.3a and 5.3b for a given set of operating parameters mentioned above. It is seen from these figures that the power output and the thermodynamic efficiency first increase and then decrease as T_8 is increased. Thus, there are optimal values of T_8 at which the power output and the corresponding thermodynamic efficiency attain their maximum values for a typical set of operating conditions. Also the optimal value of T_8 corresponding to the maximum power point is different and lower than that of the optimal value of T_8 corresponding to the maximum efficiency point, i.e. $\left(T_{8,\text{opt}}\right)_{\text{p}*} < \left(T_{8,\text{opt}}\right)_{\eta}$, as can be seen from these figures. Again the optimal value of T_8 will change, if any of these operating parameters mentioned above is changed.

Thus, using Figs. 5.3a and 5.3b and a typical set of operating parameters mentioned above, one can give some optimal criteria for the optimal values of T_8 corresponding to the point of the maximum power output and the maximum efficiency as:

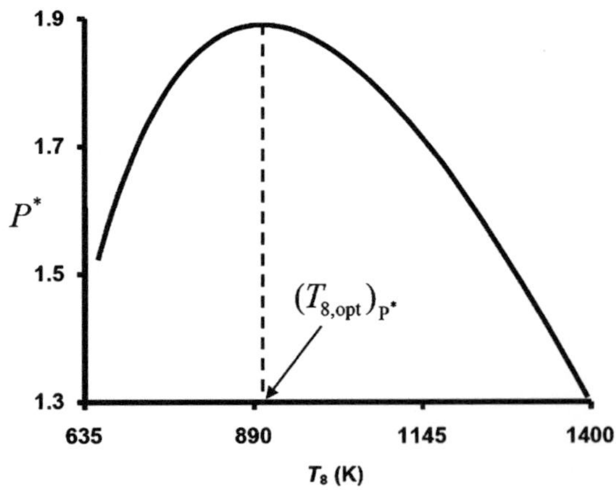

Fig. 5.3a Dimensionless power vs turbine outlet temperature

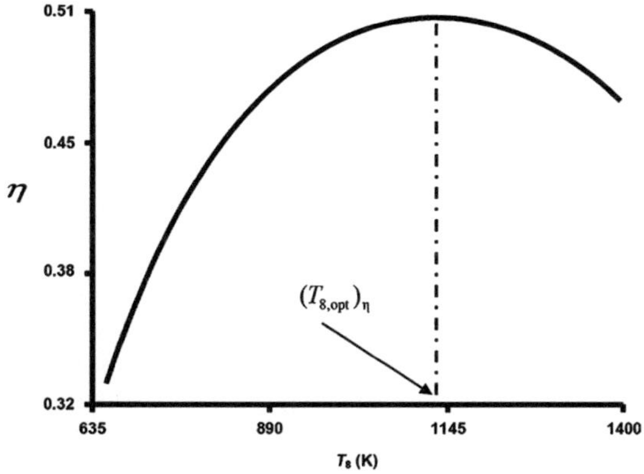

Fig. 5.3b Thermodynamic efficiency vs turbine outlet temperature

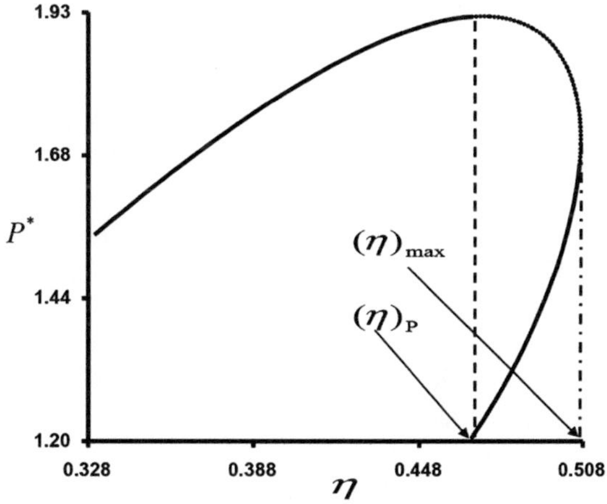

Fig. 5.3c Dimensionless power output vs thermodynamic efficiency

$$\left(T_{8,\text{opt}}\right)_{p*} < T_{8,\text{opt}} < \left(T_{8,\text{opt}}\right)_{\eta} \tag{5.30}$$

where $\left(T_{8,\text{opt}}\right)_{p*}$ and $(T_{8,\text{opt}})_\eta$ are, respectively, the optimal values of T_8 corresponding the maximum point of power output and thermodynamic efficiency, while $T_{8,\text{opt}}$ is the optimum obtainable value of T_8 which exists somewhere between those two extreme points, mentioned above. Thus, Eq. (5.30) gives the optimal criterion for the outlet temperature of the low-pressure turbine of an irreversible regenerative–intercooled–reheat Brayton cycle for a given set of operating parameters.

Power Output vs. Thermodynamic Efficiency The variation of the dimensionless power output against the thermodynamic efficiency is shown in Fig. 5.3c. It is seen from Fig. 5.3c that the power output is not a monotonic function of the thermodynamic efficiency. The power output first increases and then decreases as the thermodynamic efficiency is increased. Thus, there is an optimal value of thermodynamic efficiency at which the power output attains its maximum value for a typical set of operating parameters. Also the maximum power point and the maximum efficiency point exist, but the power output at the point of the maximum efficiency is lower than that can be attained, as seen from Fig. 5.3c.

When the power output is situated in the region of the $(P^*)_{\max} \sim (\eta)_{\max}$ curve with a positive slope, it will decrease as the corresponding thermodynamic efficiency is decreased and vice versa. Obviously, these regions are not optimal. It is thus clear that the maximum power output should be situated in the region of the $(P^*)_{\max} \sim (\eta)_{\max}$ curve with the negative slope. When the power output falls in this region, it will increase as the thermodynamic efficiency is decreased, and vice versa. Thus, it is shown that the optimal region of the thermodynamic efficiency should be determined by the relation:

$$(\eta)_{\max} \leq (\eta)_{\mathrm{opt}} \leq (\eta)_{\mathrm{P}^*} \tag{5.31}$$

where $(\eta)_{\max}$, $(\eta)_{\mathrm{opt}}$ and $(\eta)_{\mathrm{P}}$ are, respectively, the maximum point of the thermodynamic efficiency, the optimal obtainable efficiency, and the efficiency corresponding to the maximum point of power output. Hence, the optimum operating region falls somewhere between two extreme points mentioned above and represented in Fig. 5.3c with the dotted portion

The physical meaning of the above-mentioned results can be explained in different ways, based on the irreversibility associated with the cycle, as well as on the concept of the energy saving. For example, if the temperature at state point 8 (T_8) is too low (lower than T_{L2}) then the regenerative effect is negative. In other words, the heat will be in the reverse direction, i.e. from the outlet of the second compressor to the outlet of the low-pressure turbine, and hence, the utility of the regenerator will vanish. Also, if the temperature at state point 8 (T_8) is too high (higher than T_{H3}) then the heat will flow directly from the source to the sink; again, the utility of the reheat will vanish.

Thus, between these two limits mentioned above, there are optimal values of T_8 at which the cycle attains its maximum performance for a given set of operating parameters. Again, if the low-pressure turbine produces more work, then the power output of the cycle will be higher and the outlet temperature (T_8) will be lower, resulting a less energy transfer by the regenerator and, hence, a lower value of the cycle efficiency. Also, if the low-pressure turbine produces less work, then the power output of the cycle will be lower, and the outlet temperature (T_8) will be higher, resulting a more energy transfer by the regenerator and, hence, a higher value of the cycle efficiency. As a result, one may get the curves shown in Figs. 5.3a, 5.3b, and 5.3c.

Optimum Performance Parameters Based on Figs. 5.3a, 5.3b, and 5.3c and Eq. (5.30), it is observed that there is an optimal value of T_8 at which the cycle attains its maximum power output for a given set of cycle parameters. Thus, using Eq. (5.13) and its extremal condition $\frac{\partial P}{\partial T_8} = 0$, one can prove that the optimal value of T_8 corresponding to the maximum point of the optimized power output is determined by the following equation from Eqs. (5.14) and (5.30):

$$\left(T_{8,\mathrm{opt}}\right)_{P_{\max}} = \frac{-B_1 \pm \sqrt{B_1^2 - A_1C_1}}{A_1} \tag{5.32}$$

where the different parameters arc given as:

$A_1 = a_{14} - a_{14}^2/a_{16}^2, B_1 = a_{15} - a_{14}a_{15}/a_{16}^2, C_1 = a_{17} - a_{15}^2/a_{16}^2,$
$a_{14} = a_7^2 - 4Aa_9, a_{15} = a_7a_8 - 2Aa_{10}, a_{16} = (a_7a_{12} - 4Aa_{13})/a_{12}$ and
$a_{17} = a_8^2 - 4Aa_{11}$

Substituting the value of $\left(T_{8,\mathrm{opt}}\right)_{\mathrm{P_{max}}}$ and a typical set of operating parameters into Eqs. (5.12), (5.13), (5.14) and (5.9a), (5.9b), (5.9c), (5.9d), (5.9e), (5.9f), (5.9g), (5.9h), (5.9i), (5.9j), one can calculate the maximum value of the optimized power output and corresponding thermodynamic efficiency, i.e. $P_{\max}$ and η_{m}, and the optimum operating temperatures of other state points, which are listed in Tables 5.1 and 5.2 for different values of intercooling and reheat pressure ratios. Using the intercooling and reheat pressure ratios:

$$\left(R_{\mathrm{p,opt}}\right)_{\mathrm{P_{max}}} = \frac{P_{\mathrm{H}}}{P_{\mathrm{L}}} = \frac{P_{\mathrm{H}}}{P_2}\frac{P_2}{P_{\mathrm{L}}} = (T_{4\mathrm{S}}/T_3)^{\mathrm{k}/(\mathrm{k}-1)}\left(R_{\mathrm{pi,opt}}\right)_{\mathrm{P_{max}}} \tag{5.33}$$

$$\left(R_{\mathrm{p,opt}}\right)_{\mathrm{P_{max}}} = \frac{P_{\mathrm{H}}}{P_{\mathrm{L}}} = \frac{P_{\mathrm{H}}}{P_6}\frac{P_6}{P_{\mathrm{L}}} = (T_7/T_{8\mathrm{S}})^{\mathrm{k}/(\mathrm{k}-1)}\left(R_{\mathrm{ph,opt}}\right)_{\mathrm{P_{max}}} \tag{5.34}$$

One can also calculate the cycle pressure ratio corresponding to the maximum value of the optimized power output $\left(R_{\mathrm{p,opt}}\right)_{\mathrm{P_{max}}}$, for different values of the intercooling and reheat pressure ratios, as listed in Tables 5.1 and 5.2. Again from Fig. 5.3b, it is observed that there exists an optimal value of T_8 at which the cycle attains its maximum efficiency. Thus using Eq. (5.14) and its most extreme condition $\frac{\partial \eta}{\partial T_8} = 0$, one can further give the optimal value of T_8 corresponding to the maximum point of the optimized thermodynamic efficiency, as:

$$\left(T_{8,\mathrm{opt}}\right)_{\eta_{\max}} = \frac{-B_2 \pm \sqrt{B_2^2 - A_2C_2}}{A_2} \tag{5.35}$$

where the parameters are given as below:

$A_2 = a_{14} - b_{10}^2/b_{12}^2, B_2 = a_{15} - b_{10}b_{11}/b_{12}^2, C_2 = a_{17} - b_{11}^2/b_{12}^2,$
$b_9 = b_7a_{13} - b_8a_{12}, b_{10} = b_9a_{15} - K_4a_{14}, b_{11} = (b_9a_{17} - K_4a_{15})$, and
$\mathrm{b}_{12} = K_4a_7 + 2AK_5 - b_8b_9$

Table 5.1 Effects of R_{pi} on the maximum power output, the corresponding efficiency, state point temperatures, and the cycle pressure ratio for a given set of cycle parameters as mentioned in Figs. 5.2a, b

R_{pi}	$T_{1,opt}$	$T_{2,opt}$	$T_{2S,opt}$	$T_{3,opt}$	$T_{4S,opt}$	$T_{4,opt}$	$T_{4R,opt}$	$T_{5,opt}$	$T_{6S,opt}$	$T_{6,opt}$	$T_{7,opt}$	$T_{8,opt}$	$T_{8S,opt}$	$T_{8R,opt}$	P^*_{max}	η_m	$R_{p,opt}$
1.0	363	363	363	309	662	701	1058	1437	744	813	1402	1097	1064	741	1.79	0.450	14.4
1.5	357	405	400	315	621	655	1019	1431	741	810	1401	1060	1022	696	1.91	0.476	16.2
2.5	349	465	454	324	574	602	975	1425	738	807	1401	1016	974	643	2.03	0.493	18.6
3.5	344	509	493	330	545	569	948	1421	736	805	1401	990	945	611	2.08	0.495	20.3
4.5	341	545	524	335	525	546	929	1418	735	803	1400	972	924	589	2.10	0.492	21.7
5.5	339	575	551	339	509	528	915	1416	734	802	1400	958	909	571	2.11	0.488	22.8
6.5	337	601	575	343	496	514	903	1415	733	801	1400	947	896	557	2.11	0.482	23.7
7.5	335	625	596	346	486	501	894	1413	732	800	1400	937	886	545	2.10	0.476	24.5
8.5	334	646	615	349	477	491	886	1412	731	800	1400	929	877	535	2.10	0.470	25.2
9.5	332	666	632	352	469	482	879	1411	731	799	1400	923	870	526	2.09	0.464	25.9
10.5	331	684	648	355	462	474	872	1410	730	798	1400	917	863	518	2.08	0.458	26.5
11.5	330	701	663	357	456	467	867	1410	730	798	1400	911	857	511	2.07	0.452	27.0
12.5	329	716	678	359	450	460	862	1409	730	798	1400	907	852	505	2.05	0.446	27.5
13.5	328	731	691	362	445	455	858	1408	729	797	1400	902	847	499	2.04	0.441	28.0
14.5	328	745	704	364	441	449	853	1408	729	797	1400	898	843	494	2.03	0.435	28.4
15.5	327	759	716	366	436	444	850	1407	729	797	1400	895	839	489	2.02	0.430	28.8
16.5	326	772	727	367	432	440	846	1407	729	796	1399	891	835	485	2.00	0.425	29.2
17.5	326	784	738	369	429	435	843	1406	728	796	1399	888	832	481	1.99	0.420	29.5
18.5	325	796	749	371	425	431	840	1406	728	796	1399	885	828	477	1.98	0.415	29.9
19.5	325	807	759	372	422	428	837	1405	728	796	1399	883	825	473	1.96	0.410	30.2

Table 5.2 Effects of R_{pi} on the maximum efficiency, the corresponding power output, state point temperatures, and the cycle pressure ratio for a given set of cycle parameters as mentioned in Figs. 5.2a, b

R_{ph}	$T_{1,opt}$	$T_{2,opt}$	$T_{2S,opt}$	$T_{3,opt}$	$T_{4S,opt}$	$T_{4,opt}$	$T_{4R,opt}$	$T_{5,opt}$	$T_{6S,opt}$	$T_{6,opt}$	$T_{7,opt}$	$T_{8,opt}$	$T_{8S,opt}$	$T_{8R,opt}$	P^*_{max}	η_m	$R_{p,opt}$
1.0	316	522	501	332	374	379	700	1386	1386	1386	1484	736	653	414	1.65	0.524	7.6
1.5	320	527	507	332	396	403	733	1390	1238	1254	1465	770	693	439	1.85	0.544	9.2
2.5	325	535	514	334	425	435	778	1397	1075	1107	1444	816	746	473	2.02	0.549	11.7
3.5	328	541	520	334	445	458	810	1401	980	1022	1432	849	784	497	2.09	0.543	13.6
4.5	331	545	524	335	461	476	835	1405	914	963	1423	875	814	515	2.12	0.535	15.3
5.5	333	549	527	336	475	490	855	1408	865	919	1417	896	838	531	2.13	0.526	16.8
6.5	335	552	530	336	486	503	873	1410	826	885	1412	914	859	544	2.13	0.518	18.2
7.5	336	555	533	336	496	514	889	1413	794	856	1408	931	878	555	2.13	0.510	19.5
8.5	338	557	535	337	505	523	903	1415	768	832	1405	945	894	566	2.12	0.502	20.6
9.5	339	559	537	337	513	532	916	1417	745	812	1402	958	909	575	2.11	0.494	21.7
10.5	340	561	539	337	520	540	927	1418	724	794	1399	970	923	583	2.10	0.486	22.8
11.5	342	563	541	338	527	548	938	1420	707	778	1397	981	935	591	2.09	0.479	23.7
12.5	343	565	543	338	533	555	948	1421	691	764	1395	992	947	598	2.07	0.473	24.7
13.5	344	566	544	338	539	561	957	1422	676	751	1393	1001	958	605	2.06	0.466	25.6
14.5	345	568	546	338	544	567	966	1424	663	739	1391	1010	968	612	2.04	0.460	26.4
15.5	345	569	547	338	549	573	974	1425	651	729	1390	1019	978	618	2.03	0.454	27.3
16.5	346	571	548	339	554	578	982	1426	640	719	1388	1027	987	623	2.01	0.448	28.1
17.5	347	572	549	339	559	583	990	1427	630	710	1387	1035	996	629	2.00	0.442	28.8
18.5	348	573	551	339	563	588	997	1428	620	701	1386	1042	1004	634	1.99	0.437	29.6
19.5	348	574	552	339	568	593	1004	1429	612	693	1385	1049	1012	639	1.97	0.431	30.3

Substituting the value of $(T_{8,\mathrm{opt}})_{\eta_{\max}}$ and a typical set of operating parameters into Eqs. (5.12), (5.13), (5.14) and (5.9a), (5.9b), (5.9c), (5.9d), (5.9e), (5.9f), (5.9g), (5.9h), (5.9i), (5.9j), one can find the maximum value of the optimized thermodynamic efficiency and corresponding power output, i.e. $\eta_{\max}$ and P_{m}, and the optimum operating temperatures of other state points, which are also listed in Tables 5.3 and 5.4 for different values of intercooling and reheat pressure ratios. Again, using the intercooling and reheat pressure ratios:

$$(R_{\mathrm{p,opt}})_{\eta_{\max}} = \frac{P_{\mathrm{H}}}{P_{\mathrm{L}}} = \frac{P_{\mathrm{H}}}{P_2}\frac{P_2}{P_{\mathrm{L}}} = (T_{4\mathrm{S}}/T_3)^{\mathrm{k}/(\mathrm{k}-1)}(R_{\mathrm{pi,opt}})_{\eta_{\max}} \tag{5.36}$$

$$(R_{\mathrm{p,opt}})_{\eta_{\max}} = \frac{P_{\mathrm{H}}}{P_{\mathrm{L}}} = \frac{P_{\mathrm{H}}}{P_6}\frac{P_6}{P_{\mathrm{L}}} = (T_7/T_{8\mathrm{S}})^{\mathrm{k}/(\mathrm{k}-1)}(R_{\mathrm{ph,opt}})_{\eta_{\max}} \tag{5.37}$$

One can further calculate the optimal value of cycle pressure ratio corresponding to the maximum point of the optimized thermodynamic efficiency $(R_{\mathrm{p,opt}})_{\eta_{\max}}$, for different values of the intercooling and reheat pressure ratios, as given in Tables 5.3 and 5.4. Since the intercooling and reheat pressure ratios are amongst the important parameters of an intercooled–reheat Brayton cycle, it is worth to discuss the effects of these pressure ratios on the performance of an irreversible regenerative–intercooled–reheat Brayton cycle heat engine to further reveal the general characteristics of this cycle. Using the above results and a set of operating parameters which are similar to those given in Tables 5.1, 5.2, 5.3, and 5.4, some performance characteristic curves are plotted amongst different cycle parameters, as below.

Effects of Pressure Ratios It is seen in Tables 5.1, 5.2, 5.3, and 5.4 that the outlet temperatures of the intercooling compressor and the inlet temperature of the second compressor as well as the cycle compression ratio increase, while the temperatures of all other state points decrease as the intercooling pressure ratio increases. On one hand, the outlet temperatures of the high-pressure turbine and the inlet temperature of the low-pressure turbine decrease, while the temperatures of all other state points as well as the cycle pressure ratio increase with increasing the reheat pressure ratio, as can be seen from Tables 5.2 and 5.4. On the other hand, the maximum power output and the corresponding thermodynamic efficiency, as well as the maximum thermodynamic efficiency and the corresponding power output, first increase and then decrease as the either pressure ratio is increased.

Thus it is seen clearly from these figures and Tables 5.1, 5.2, 5.3, and 5.4 that there are the optimal values of the intercooling and reheat pressure ratios at which the optimized power output and the corresponding thermodynamic efficiency, as well as the maximum thermodynamic efficiency and the corresponding power output attain their maximum values, i.e. $(P^*_{\max})_{\max}$, $(\eta_{\mathrm{m}})_{\max}$, $(\eta_{\max})_{\max}$, and $(P^*_{\mathrm{m}})_{\max}$ for a given cycle parameters.

The maximum value of the optimized power output is found to be more than that of the maximum value of the optimal power output. Similarly, the maximum value of the optimized efficiency is found to be more than that of the maximum value of

Table 5.3 Effects of R_{ph} on the maximum power output, the corresponding efficiency, state point temperatures, and the cycle pressure ratio for a given set of cycle parameters as mentioned in Figs. 5.2a, b

R_{pi}	$T_{1,opt}$	$T_{2,opt}$	$T_{2S,opt}$	$T_{3,opt}$	$T_{4S,opt}$	$T_{4,opt}$	$T_{4R,opt}$	$T_{5,opt}$	$T_{6S,opt}$	$T_{6,opt}$	$T_{7,opt}$	$T_{8,opt}$	$T_{8S,opt}$	$T_{8R,opt}$	P^*_m	η_{max}	$R_{p,opt}$
1.0	362	362	362	309	629	664	1272	1467	760	831	1404	1339	1332	732	1.10	0.582	12.1
1.5	355	403	398	314	584	614	1242	1463	758	828	1404	1312	1302	684	1.21	0.601	13.1
2.5	347	463	451	322	538	562	1198	1457	755	825	1404	1268	1253	632	1.31	0.612	15.1
3.5	343	508	491	327	513	534	1165	1452	752	822	1403	1235	1217	604	1.36	0.612	16.9
4.5	341	544	524	332	497	515	1139	1448	750	820	1403	1209	1187	584	1.39	0.608	18.5
5.5	339	575	551	336	485	502	1118	1445	749	818	1403	1187	1163	570	1.41	0.603	19.9
6.5	337	602	575	339	476	491	1100	1443	747	817	1402	1168	1142	559	1.42	0.598	21.3
7.5	336	626	597	342	469	483	1085	1441	746	816	1402	1152	1124	550	1.42	0.592	22.6
8.5	335	648	617	345	463	476	1071	1439	745	815	1402	1137	1108	542	1.43	0.586	23.8
9.5	334	668	635	348	458	471	1059	1437	744	814	1402	1124	1093	536	1.42	0.580	24.9
10.5	333	687	652	350	454	466	1048	1435	743	813	1402	1113	1080	530	1.42	0.574	26.1
11.5	332	705	667	352	450	461	1038	1434	743	812	1402	1102	1069	525	1.42	0.569	27.1
12.5	332	721	682	355	447	457	1029	1433	742	811	1402	1092	1058	521	1.41	0.563	28.1
13.5	331	737	696	357	444	454	1020	1431	741	810	1401	1083	1048	517	1.40	0.558	29.1
14.5	330	752	709	358	442	451	1013	1430	741	810	1401	1075	1039	513	1.39	0.552	30.1
15.5	330	766	722	360	439	448	1005	1429	740	809	1401	1067	1030	510	1.39	0.547	31.0
16.5	330	779	734	362	437	445	999	1428	740	809	1401	1060	1022	507	1.38	0.542	31.9
17.5	329	792	746	364	435	443	992	1427	739	808	1401	1053	1015	504	1.37	0.537	32.8
18.5	329	804	757	365	433	441	986	1427	739	808	1401	1047	1008	501	1.36	0.532	33.6
19.5	328	816	767	367	432	439	981	1426	739	807	1401	1041	1001	499	1.35	0.528	34.4

Table 5.4 Effects of R_{ph} on the maximum efficiency, the corresponding power output, state point temperatures, and the cycle pressure ratio for a given set of cycle parameters as mentioned in Figs. 5.2a, b

R_{ph}	$T_{1,opt}$	$T_{2,opt}$	$T_{2S,opt}$	$T_{3,opt}$	$T_{4S,opt}$	$T_{4,opt}$	$T_{4R,opt}$	$T_{5,opt}$	$T_{6S,opt}$	$T_{6,opt}$	$T_{7,opt}$	$T_{8,opt}$	$T_{8S,opt}$	$T_{8R,opt}$	P^*_m	η_{max}	$R_{p,opt}$
1.0	317	523	502	329	361	365	869	1410	1410	1410	1487	924	862	421	1.03	0.664	6.9
1.5	320	527	506	329	373	377	922	1417	1262	1278	1468	982	928	438	1.19	0.671	7.9
2.5	324	534	513	330	394	401	983	1426	1098	1130	1447	1047	1003	466	1.33	0.665	9.3
3.5	327	539	518	331	413	422	1020	1431	1001	1044	1435	1087	1048	488	1.38	0.657	10.8
4.5	330	543	522	332	429	439	1047	1435	934	984	1426	1115	1080	507	1.41	0.648	12.3
5.5	332	547	526	332	442	455	1068	1438	884	939	1420	1136	1105	523	1.42	0.639	13.6
6.5	334	550	529	333	455	468	1085	1441	844	904	1415	1154	1125	537	1.42	0.631	14.9
7.5	336	553	532	333	466	481	1100	1443	811	874	1411	1169	1142	550	1.42	0.623	16.2
8.5	337	556	534	333	476	492	1112	1445	784	850	1407	1181	1156	561	1.42	0.616	17.4
9.5	339	559	537	334	486	503	1123	1446	760	829	1404	1192	1169	572	1.41	0.609	18.6
10.5	340	561	539	334	495	513	1133	1448	739	810	1401	1202	1180	582	1.40	0.603	19.8
11.5	342	563	541	334	503	522	1142	1449	721	794	1399	1211	1190	591	1.39	0.597	20.9
12.5	343	565	543	335	511	531	1150	1450	705	779	1397	1219	1199	600	1.38	0.591	22.0
13.5	344	567	545	335	519	539	1157	1451	690	766	1395	1226	1207	608	1.37	0.585	23.1
14.5	345	569	547	335	526	547	1164	1452	676	754	1393	1233	1215	616	1.35	0.580	24.2
15.5	346	571	548	335	533	555	1171	1453	664	743	1392	1239	1222	623	1.34	0.575	25.2
16.5	347	572	550	336	539	562	1176	1454	653	733	1390	1245	1229	630	1.33	0.571	26.2
17.5	348	574	551	336	545	569	1182	1455	642	723	1389	1250	1235	637	1.32	0.565	27.3
18.5	349	575	553	336	551	575	1187	1455	632	715	1388	1255	1241	643	1.30	0.561	28.3
19.5	350	577	554	336	557	582	1192	1456	623	706	1387	1260	1246	650	1.29	0.556	29.3

the optimal efficiency for the same of operating parameters, as can be seen from Figs. 5.4a and 5.4b and Tables 5.1, 5.2, 5.3, and 5.4. Also, the optimum values of intercooling and reheat pressure ratios are different for different parameters. In other words, the optimum values of the intercooling and reheat pressure ratios at the maximum point of the optimized power output are different and higher than those of the optimum values of the intercooling and reheat pressure ratios at the maximum point of the optimized thermodynamic efficiency. This conclusion will provide some important theoretical instruction for the optimal design and operation of an irreversible regenerative–intercooled–reheat Brayton cycle heat engine.

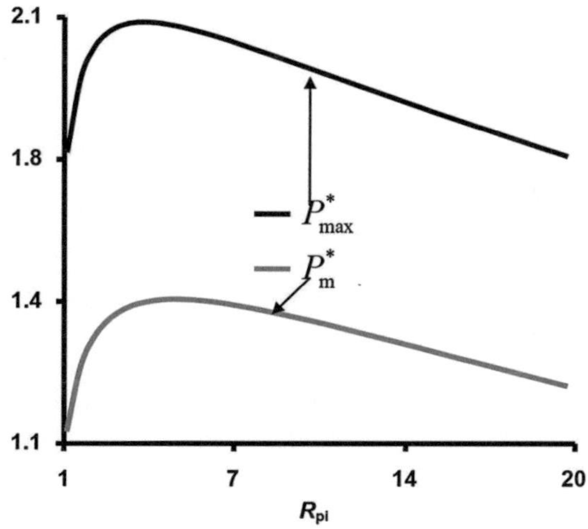

Fig. 5.4a Maximum and optimal powers vs intercooling pressure ratio

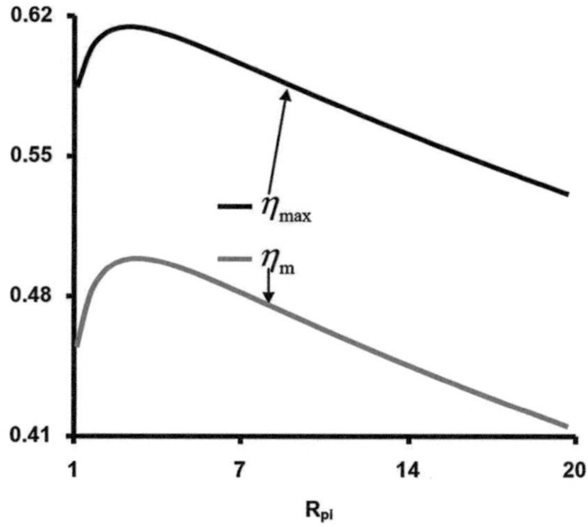

Fig. 5.4b Maximum and optimal efficiencies vs intercooling pressure ratio

In fact, the physical significance of Figs. 5.4a and 5.4b and Tables 5.1, 5.2, 5.3, and 5.4 may clearly be explained from Fig. 5.4b in different ways. For example, if cycle pressure ratio tends to unity (as the intercooling and reheat pressure ratios are always less than the cycle pressure ratio), the heat will flow directly from the source to the sink, so no real cycle is performed. Again, if cycle pressure ratio tends to the reservoirs temperature ratio $(T_{H1}/T_{L1})^{\gamma/(\gamma-1)}$, the heat transfer rates to and from the cycle tend to zero; again, no real cycle is performed. Thus at the above-mentioned limits, the significant cycle cannot be formed, so the power output and corresponding thermodynamic efficiency tend to zero. Hence, there are necessarily some optimal values of the intercooling, reheat, and cycle pressure ratios between unity and $(T_{H1}/T_{L1})^{\gamma/(\gamma-1)}$, at which the optimized power and the optimal efficiency as well as the optimized efficiency and the optimal power attain their maximum values, i.e. $\left(P^*_{\max}\right)_{\max}$, $(\eta_m)_{\max}$, $(\eta_{\max})_{\max}$, and $\left(P^*_{\max}\right)_{\max}$ for a typical set of operating parameters.

It is also important to note the fact that the intercooling and reheat pressure ratios are always more than unity and less than the cycle pressure ratio, i.e.

$$1 < R_{pi} < R_p \tag{5.38}$$

$$1 < R_{ph} < R_p \tag{5.39}$$

When $R_{pi} = 1$ or $R_{pi} = R_p$ and $R_{ph} = 1$ or $R_{ph} = R_p$, the cycle becomes an irreversible regenerative Brayton cycle, which has been discussed by Curzon and Alhborn (1975). Similarly, when $R_{ph} = 1$ or $R_{ph} = R_p$, the cycle becomes an irreversible intercooled Brayton cycle without reheat which also has been discussed by other workers (Wang et al. 2003; Tyagi and Kaushik 2005; Tyagi 2009; Tyagi et al. 2007; Negri-di et al. 1995), and if $R_{pi} = 1$ or $R_{pi} = R_p$ the cycle becomes an irreversible reheat Brayton cycle (Bhargava and Peretto 2002).

Obviously, the optimal values of the intercooling and reheat pressure ratios are different for different operating states. It is important how to choose the optimal values of these pressure ratios in the investigation of an irreversible regenerative–intercooled–reheat Brayton cycle heat engine. Hence, according to Tables 5.1, 5.2, 5.3, and 5.4, one can, in principle, determine the rational range of the optimal values of the intercooling and reheat pressure ratios, as:

$$\left(R_{pi,opt}\right)_{P^*_{\max}} > R_{pi,opt} > \left(R_{pi,opt}\right)_{\eta_{\max}} \tag{5.40}$$

$$\left(R_{ph,opt}\right)_{P^*_{\max}} > R_{ph,opt} > \left(R_{ph,opt}\right)_{\eta_{\max}} \tag{5.41}$$

where $\left(R_{pi,opt}\right)_{P^*_{\max}}$, $\left(R_{pi,opt}\right)_{\eta_{\max}}$ and $\left(R_{ph,opt}\right)_{P^*_{\max}}$, $\left(R_{ph,opt}\right)_{\eta_{\max}}$ are, respectively, the intercooling and reheat pressure ratios at the maximum points of the optimized power output and thermodynamic efficiency, while $R_{pi\,,\,opt}$ and $R_{pi\,,\,opt}$ are the optimal obtainable values of the intercooling and reheat pressure ratios, respectively. Using Eqs. (5.40) and (5.41) and Figs. 5.5a and 5.5b, one can give another optimum criterion for the cycle pressure ratio, as:

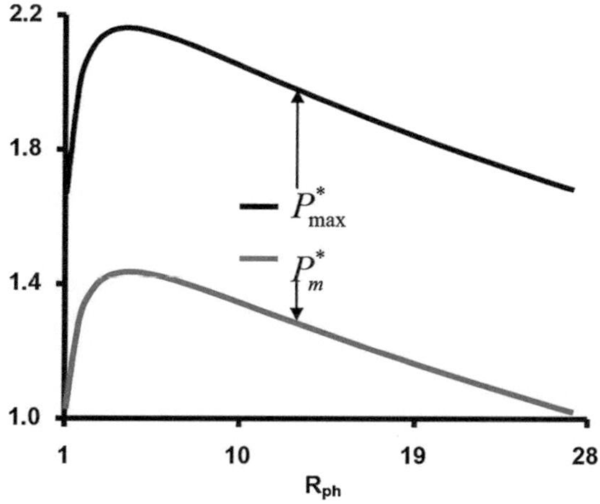

Fig. 5.5a Maximum and optimal powers vs intercooling pressure ratio

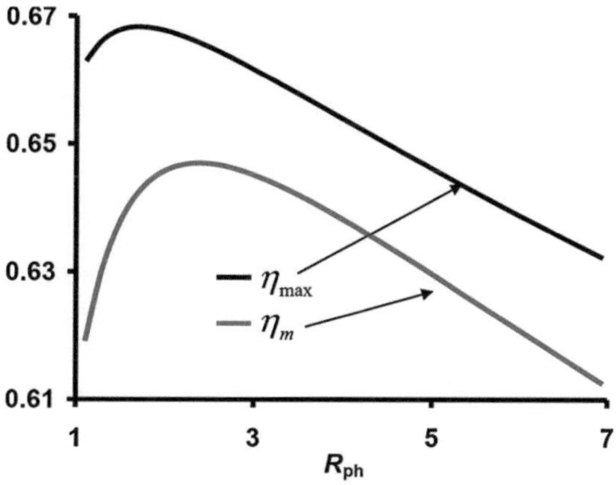

Fig. 5.5b Maximum and optimal efficiencies vs reheat pressure ratio

$$\left(R_{p,opt}\right)_{P^*_{max}} > R_{p,opt} > \left(R_{p,opt}\right)_{\eta_{max}} \tag{5.42}$$

where $\left(R_{p,opt}\right)_{P^*_{max}}$ and $\left(R_{p,opt}\right)_{\eta_{max}}$ are, respectively, the optimal values of the cycle pressure ratios at the maximum points of the optimized power and thermodynamic efficiency, while $R_{p,opt}$ is the optimal obtainable value of the cycle pressure ratio. According to Eqs. (5.40) and (5.41) and Tables 5.1, 5.2, 5.3 and 5.4, one can also give the optimum criteria for the temperatures of the various state points. The criteria will be helpful for engineers to optimally design and operate an irreversible regenerative–intercooled–reheat Brayton cycle.

Thus, according to Figs. 5.5a and 5.5b and Tables 5.1, 5.2, 5.3 and 5.4, one can also determine the rational range of the optimal value of cycle pressure ratio of a complex Brayton cycle, as given by Eq. (5.42) and Figs. 5.5a and 5.5b for a typical set of operating parameters.

5.3.2 Isothermal Intercooled–Reheat Brayton Cycle

Using the equations given in Sect. 5.2.2, one can calculate the maximum power output, the corresponding thermal efficiency as well as the maximum efficiency and the corresponding power, and the other cycle parameters for a given set of operating condition such as $\varepsilon_J = \eta' s = 0.90$, $T_{L1} = T_{C1} = 300\text{K}$, $T_{\text{Hi}} = 1500\text{K}$, $C_{\text{k}} = 1.05\text{kW/K}$, $C_{\text{W}} = 1.05\text{kW/K}$, $R_{\text{pt}} = 0.85$, and $R_{\text{pt}} = R_{\text{ph}} = 5.0$. The graphs are plotted for the optimized power and the corresponding efficiency as well as the optimized efficiency and the corresponding power against the intercooling, reheat, and cycle pressure ratios along with other cycle parameters as shown in Figs. (5.6)–(5.12), and the discussion of results is as given below.

Figures (5.6a, 5.6b, and 5.7) show the variations of the optimized power and the corresponding efficiency as well as the optimized efficiency and the corresponding power against the intercooling and reheat pressure ratios for a typical set of operating conditions mentioned above. It is seen from these figures that all the parameters first increase and then decrease as the intercooling and reheat pressure ratios are increased. It shows clearly that there exists an optimal value of either pressure ratio at which these parameters attain their maximum values, also the optimal values of the intercooling and reheat pressure ratios corresponding to the maximum power are much higher than those of the optimal values of these pressure ratios corresponding to the maximum efficiency for both the cases. Thus according to Figs. 5.6a, 5.6b and 5.7, the rational range of the optimal values of the intercooling and reheat pressure ratios from the point of view of the power as

Fig. 5.6a Optimized and optimal power vs intercooling pressure ratio

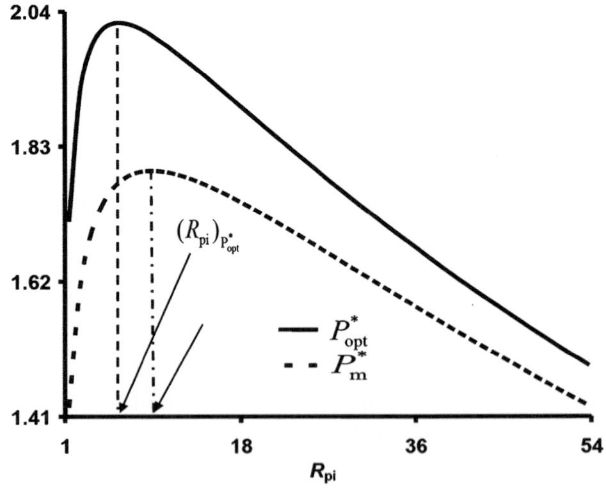

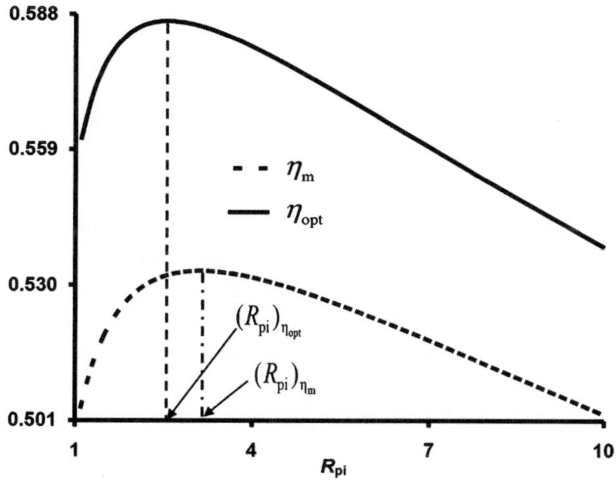

Fig. 5.6b Optimized and optimal efficiency vs intercooling pressure ratio

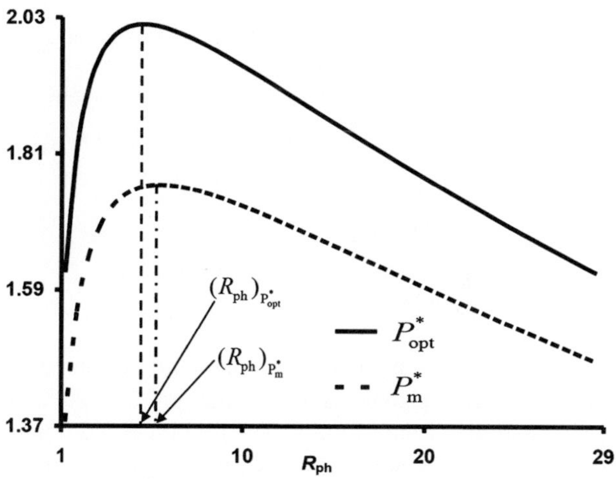

Fig. 5.7 Optimized and optimal power vs reheat pressure ratio

well as from the point of view of the efficiency can be determined by the following relations:

$$\begin{aligned} &\left(R_{pi,opt}\right)_{P_{opt}} \geq \left(R_{pi,opt}\right) \geq \left(R_{pi,opt}\right)_{\eta_{opt}} \text{ and} \\ &\left(R_{pi,opt}\right)_{\eta_m} \leq \left(R_{pi,opt}\right) \leq \left(R_{pi,opt}\right)_{P_m} \end{aligned} \tag{5.43}$$

$$\begin{aligned} &\left(R_{ph,opt}\right)_{P_{opt}} \geq \left(R_{ph,opt}\right) \geq \left(R_{ph,opt}\right)_{\eta_{opt}} \text{ and} \\ &\left(R_{ph,opt}\right)_{\eta_m} \leq \left(R_{ph,opt}\right) \leq \left(R_{ph,opt}\right)_{P_m} \end{aligned} \tag{5.44}$$

where $\left(R_{pi,opt}\right)_{P_{opt}}$, $\left(R_{pi,opt}\right)_{\eta_{opt}}$, $\left(R_{pi,opt}\right)_{\eta_m}$, and $\left(R_{pi,opt}\right)_{P_m}$ are, respectively, the optimal values of intercooling pressure ratio corresponding to the maximum points of the optimized power, optimized efficiency, optimal power, and optimal efficiency. Similarly, $\left(R_{\text{ph,opt}}\right)_{\text{P}_{\text{opt}}}$, $\left(R_{\text{ph,opt}}\right)_{\eta_{\text{opt}}}$, $\left(R_{\text{ph,opt}}\right)_{\text{P}_{\text{m}}}$, and $\left(R_{\text{phi,opt}}\right)_{\eta_{\text{m}}}$ are, respectively, the optimal values of reheat pressure ratio corresponding to the maximum points of the optimized power and efficiency as well as the optimal power and efficiency. However, $(R_{\text{pi , opt}})$ and $(R_{\text{ph , opt}})$ are, respectively, the optimal obtainable values of the intercooling and cycle pressure ratios which fall somewhere between those extreme points mentioned above.

Also the optimal values of these pressure ratios corresponding to the optimized power are lesser than those corresponding to the optimal power. Similarly, the optimal values of these pressure ratios at optimized efficiency are lesser than those corresponding to the optimal efficiency, for the same set of operating parameters. Again the optimal values of the intercooling pressure ratio are more than those of the reheat pressure ratio for all the performance parameters. However, the maximum points of the optimized and optimal efficiencies are higher for the reheat pressure ratio, while it is reverse in the case of the maximum points of the optimized and optimal powers, for the same set of operating conditions. In other words:

$$\begin{pmatrix} \left(R_{\text{pi,opt}}\right)_{\text{P}_{\text{opt}}}, & \left(R_{\text{ph,opt}}\right)_{\text{P}_{\text{opt}}} \\ \left(R_{\text{pi,opt}}\right)_{\eta_{\text{opt}}}, & \left(R_{\text{ph,opt}}\right)_{\eta_{\text{opt}}} \end{pmatrix} \geq \begin{pmatrix} \left(R_{\text{pi,opt}}\right)_{\text{P}_{\text{m}}}, & \left(R_{\text{ph,opt}}\right)_{\text{P}_{\text{m}}} \\ \left(R_{\text{pi,opt}}\right)_{\eta_{\text{m}}}, & \left(R_{\text{ph,opt}}\right)_{\eta_{\text{m}}} \end{pmatrix} \tag{5.45}$$

$$\begin{pmatrix} \left(R_{\text{pi,opt}}\right)_{\text{P}_{\text{opt}}}, & \left(R_{\text{pi,opt}}\right)_{\text{P}_{\text{m}}} \\ \left(R_{\text{pi,opt}}\right)_{\eta_{\text{opt}}}, & \left(R_{\text{pi,opt}}\right)_{\eta_{\text{m}}} \end{pmatrix} \geq \begin{pmatrix} \left(R_{\text{ph,opt}}\right)_{\text{P}_{\text{opt}}}, & \left(R_{\text{ph,opt}}\right)_{\text{P}_{\text{m}}} \\ \left(R_{\text{ph,opt}}\right)_{\eta_{\text{opt}}}, & \left(R_{\text{ph,opt}}\right)_{\eta_{\text{m}}} \end{pmatrix} \tag{5.46}$$

$$\text{while} \left.\begin{matrix} \begin{pmatrix} \left(P_{\text{opt}}\right)_{\text{max}} \\ \left(P_{\text{m}}\right)_{\text{max}} \end{pmatrix}_{\text{R}_{\text{pi}}} \geq \begin{pmatrix} \left(P_{\text{opt}}\right)_{\text{max}} \\ \left(P_{\text{m}}\right)_{\text{max}} \end{pmatrix}_{\text{R}_{\text{ph}}} \\ \begin{pmatrix} \left(\eta_{\text{opt}}\right)_{\text{max}} \\ \left(\eta_{\text{m}}\right)_{\text{max}} \end{pmatrix}_{R_{\text{pi}}} \leq \begin{pmatrix} \left(\eta_{\text{opt}}\right)_{\text{max}} \\ \left(\eta_{\text{m}}\right)_{\text{max}} \end{pmatrix}_{\text{R}_{\text{ph}}} \end{matrix}\right\} \tag{5.47}$$

These conclusions mentioned here and presented by Eqs. (5.43), (5.44), (5.45), and (5.46) may provide some important theoretical instructions for the optimal design and operation of an irreversible modified complex Brayton cycle. It is also important to note the fact that the intercooling and reheat pressure ratios are always larger than unity and less than the cycle pressure ratio, i.e.:

$$1 < R_{\text{Pi}} < R_{\text{p}} \text{ and } 1 < R_{\text{Ph}} < R_{\text{p}} \tag{5.48}$$

According to Eq. (5.48) and Figs. 5.8a and 5.8b, one also can give the optimal criteria for the cycle pressure ratio based on the optimized and optimal performance parameters, such as:

$$\begin{matrix} \left(R_{\text{p,opt}}\right)_{\text{P}_{\text{opt}}} \geq \left(R_{\text{p,opt}}\right) \geq \left(R_{\text{p,opt}}\right)_{\eta_{\text{opt}}} \quad \text{and} \\ \left(R_{\text{p,opt}}\right)_{\eta_{\text{m}}} \leq \left(R_{\text{p,opt}}\right) \leq \left(R_{\text{p,opt}}\right)_{\text{P}_{\text{m}}} \end{matrix} \tag{5.49}$$

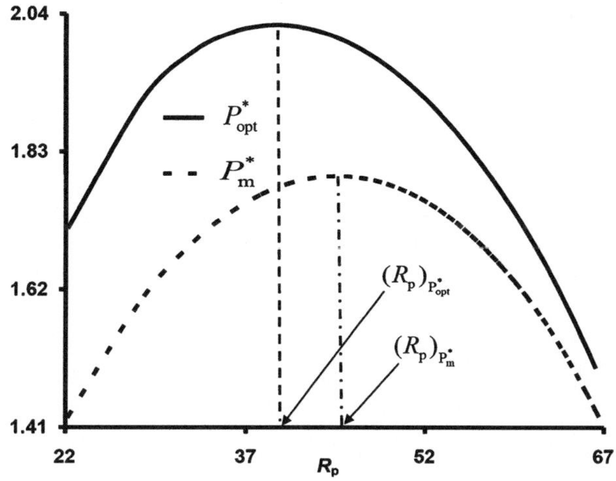

Fig. 5.8a Optimized and optimal power vs cycle pressure ratio

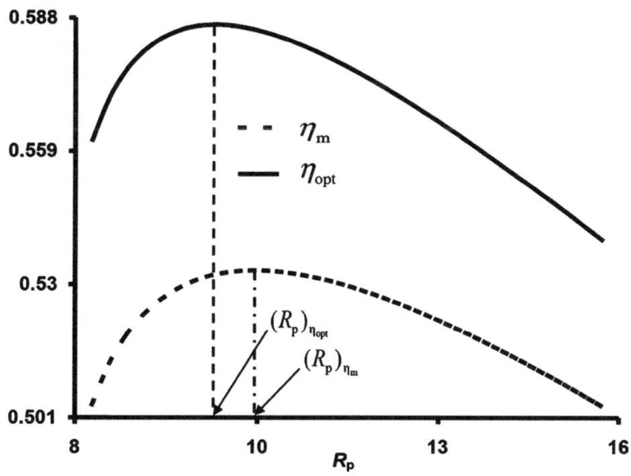

Fig. 5.8b Optimized and optimal efficiency vs cycle pressure ratio

$$\left(R_{p,opt}\right)_{P_m} \geq \left(R_{p,opt}\right) \geq \left(R_{p,opt}\right)_{P_{opt}} \quad \text{and} \quad \left(R_{p,opt}\right)_{\eta_{opt}} \leq \left(R_{p,opt}\right) \leq \left(R_{p,opt}\right)_{\eta_m} \tag{5.50}$$

where $\left(R_{p,opt}\right)_{P_{opt}}$, $\left(R_{p,opt}\right)_{P_m}$, $\left(R_{p,opt}\right)_{\eta_{opt}}$ and $\left(R_{p,opt}\right)_{\eta_m}$ are, respectively, the optimal values of cycle pressure ratio corresponding to the maximum points of the optimized power, optimized efficiency, optimal power, and optimal efficiency, as indicated in Figs. 5.8a and 5.8b. However, $\left(R_{p,opt}\right)_{\eta_m}$ is the optimal obtainable value of the cycle pressure ratio and falls somewhere between these optimal ranges, mentioned above. According to Eqs. (5.28), (5.29) and (5.43), (5.44), (5.45), (5.46), (5.47), (5.48), (5.49), (5.50), one can also give the optimum criteria for other cycle parameters such as the various state points temperature, heat transfer rates to and from the cycle, and so on, for all the cases. The criteria will be helpful for engineers to optimally design and operate an irreversible modified complex Brayton cycle, for a typical set of operating parameters. The physical meaning and explanations about the results obtained so far can be discussed in different ways, based on thermodynamics, energy consumption, irreversibility, etc.

Based on Figs. 5.8a and Eq. (5.46), it is seen that if cycle pressure ratio tends to unity, the heat will flow directly from the source to the sink, so no real cycle is performed. Again, if cycle pressure ratio tends to the reservoirs temperature ratio $(T_{H1}/T_{L1})^{\gamma/(\gamma-1)}$, the heat transfer rates to and from the cycle tends to zero; again, no real cycle is performed. Thus at the above-mentioned limits, the significant cycle cannot be formed, so the power output and, hence, the efficiency tend to zero, which is not the case in real practice. Thus, there are some optimal values of cycle pressure ratio (and hence for the intercooling and reheat ratios) between unity and $(T_{H1}/T_{L1})^{\gamma/(\gamma-1)}$, at which the cycle attains the optimum performance for a typical set of operating parameters.

Similarly, the effects of other cycle parameters such as, the effectiveness of different heat exchangers, the efficiency of compressors and turbines, etc. can also be evaluated for a typical set of other cycle parameters. Based on the previous cases, the general results for these parameters can be predicted. For example, the higher are the effectiveness of these heat exchangers, the lesser will be the external irreversibility associated with the cycle and hence, the higher will be performance of the cycle. Again, as the isentropic efficiencies of different components are increased, the internal irreversibility and hence, the outlet temperatures of these components go down, resulting, the overall better performance of the cycle. Similarly, the effects of heat capacitance rates of the external fluids and the working fluids on the performance parameters can also be predicted, and the readers are advised to make a thorough analysis and compare the predicted results.

It is also important to note that the present cycle model is general, and the results obtained by earlier workers (Vecchiarelli et al. 1997; Göktun and Yavuz 1999; Tyagi 2009; Tyagi et al. 2000, 2007; Erbay et al. 2001; Kaushik et al. 2003; Wang et al. 2003; Tyagi and Kaushik 2005; Negri-di et al. 1995; Bhargava and Peretto 2002) are the special cases of this cycle model. For example, if $R_{pi} = R_{ph} = 1.0$ and $\eta_C = \eta_T < 1.0$ are chosen, the results obtained in this paper are identical to those obtained in the references Vecchiarelli et al. (1997), Göktun and Yavuz (1999), Kumar (2000), Tyagi et al. (2000), Erbay et al. (2001), and Kaushik et al. (2003); if $R_{ph} = R_{pt} = 1.0$ and $\eta_C = \eta_T < 1.0$ are chosen, the results are identical to those obtained in the references Wang et al. (2003), Tyagi and Kaushik (2005), Tyagi (2009), and Tyagi et al. (2007); if $R_{pi} = R_{pt} = 1.0$ and $\eta_C = \eta_T < 1.0$ are chosen, the results are identical to those obtained in reference Bhargava and Peretto (2002); and if $R_{ph} = R_{pi} = R_{pt} = 1.0$ and $\eta_C = \eta_T = \varepsilon_R < 1.0$ are chosen, the results are identical to those obtained in reference Kaushik and Tyagi (2002). Thus the present cycle model is more general, and the results obtained by earlier authors as available in the literature (Vecchiarelli et al. 1997; Göktun and Yavuz 1999; Tyagi 2009; Tyagi et al. 2000, 2007; Erbay et al. 2001; Kaushik et al. 2003; Wang et al. 2003; Tyagi and Kaushik 2005; Negri-di et al. 1995) are the special cases of the present cycle model.

5.4 Conclusion

An irreversible complex and modified complex Brayton cycle models including external and internal irreversibilities for the finite heat capacities of external reservoirs have been studied in detail. The power output and efficiency are maximized with respect to the cycle temperatures, and the optimum operating parameters are calculated for both cases at a typical set of operating parameters. It is found that there exist the optimal values of the intercooling, reheat, and cycle pressure ratios at which the optimized power output and the corresponding thermal efficiency and vice versa attain their maximum values.

The optimal values of these parameters are different for different performance parameters and change as any of the cycle parameters is changed. The optimal operating region to be chosen is situated between these two maximum points, viz. between the maximum point of the optimized power output and the maximum point of the corresponding thermal efficiency. The optimal performance parameters are also found to be increasing function of the effectiveness of different heat exchangers, components efficiency, and the heat capacitance rates of the external fluids, which is found to be reversed in the case of the working fluid heat capacitance rate. The optimum operating parameters, the intercooling, reheat and cycle pressure ratios, etc. corresponding to the different performance parameters are obtained and discussed in detail.

Chapter 6
Finite Time Thermodynamic Analysis of Stirling and Ericsson Power Cycles

6.1 Introduction

In 1816, Robert Stirling and James Stirling conceived and patented an external combustion heat engine, which turned by their name and was continuously improved by them over the next 30 years. It is also a reversible cycle like the Carnot cycle and, hence, also has the highest thermal efficiency working between two (high and low) temperatures of source and sink reservoirs, but it has some advantages over the Carnot cycle. The Stirling heat engine consists of two cylinders with two pistons on each side and a regenerator in the middle, which is used for temporary storage of thermal energy. The Stirling engine is a closed system and also required more innovative hardware for execution. In 1833, the Swedish-born engineer John Ericsson developed a different type of hot air, reciprocating, external combustion engine, which could operate on either an open or closed loop cycle.

Ericsson and Stirling engines have attracted the attention of several generations of engineers and physicists due to their potential to provide high conversion efficiency. However, the use of these engines did not prove to be successful due to relatively poor material technology available at that time. As the world community has become much more environmentally conscious, further attention in these engines has been again received because these engines are inherently clean. Moreover, as a result of advances in material technology, these engines are currently being considered for variety of applications due to their many advantages like low noise, less pollution, and their flexibility as an external combustion engine to utilize a variety of fuels.

The clear distinction between Ericsson and Stirling cycle machines has not been well understood in the literature, and the latter is often being considered the closed cycle version of the former. However, in thermodynamic terms, the Ericsson cycle is also either closed or open cycle. In both cycles, the heat addition/rejection processes can be modelled as occurring at constant temperatures, while the

S.C. Kaushik et al., *Finite Time Thermodynamics of Power and Refrigeration Cycles*,
https://doi.org/10.1007/978-3-319-62812-7_6

regenerative processes are modelled as isobaric (in Ericsson cycle) and isochoric (in Stirling cycle), respectively.

6.2 Ideal Stirling Cycle

The Stirling engine is a gas power cycle that operates between two temperatures, T_H and T_L of source and sink like a Carnot cycle. The Stirling engine has attracted the attention of several generations of engineers and physicists due to its potential to provide high conversion efficiency. For the heat engine to be totally reversible, the temperature difference between the working fluid and the external reservoirs should never exceed a differential amount dT during any heat transfer process. In other words, both the heat rejection and heat addition processes during the cycle must take place isothermally, one at temperature T_H and other at temperature T_L, similar to a Carnot cycle. However, the Stirling cycle differs from the Carnot cycle in a way that the two adiabatic processes of the latter are replaced with two isochoric processes in the Stirling cycle. The cycle also utilizes regeneration, a process during which the heat is transferred to a temporary thermal energy storage device called regenerator by the working fluid during one part of the cycle and the stored heat is transferred back to the working fluid from the regenerator during another part of the cycle, as shown in Fig. 6.1(a). On the other hand, the *P–v* 48 and *T–s* diagrams of Stirling cycle are shown in Figs. 6.1b and 6.1c, respectively.

Process 1–2: The constant temperature (isothermal) heat addition from external source at temperature T_H

Process 2–3: The constant volume regeneration process, the heat addition from the working fluid to the temporary storage/regenerator

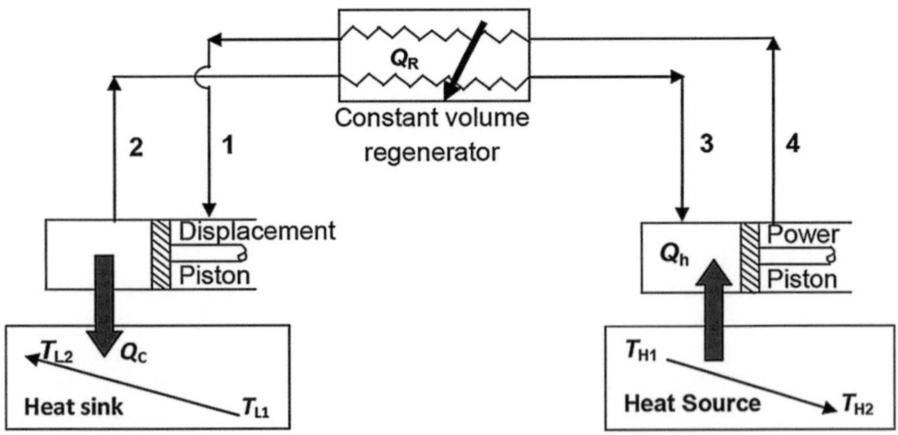

Fig. 6.1a Line diagram of Stirling/Ericsson power cycle

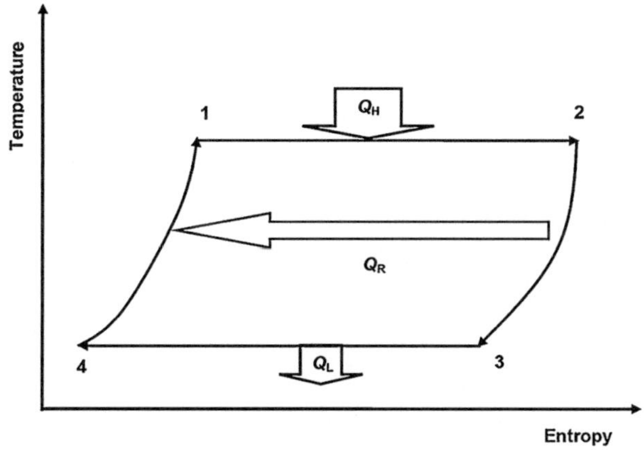

Fig. 6.1b T–s diagram of ideal Stirling power cycle

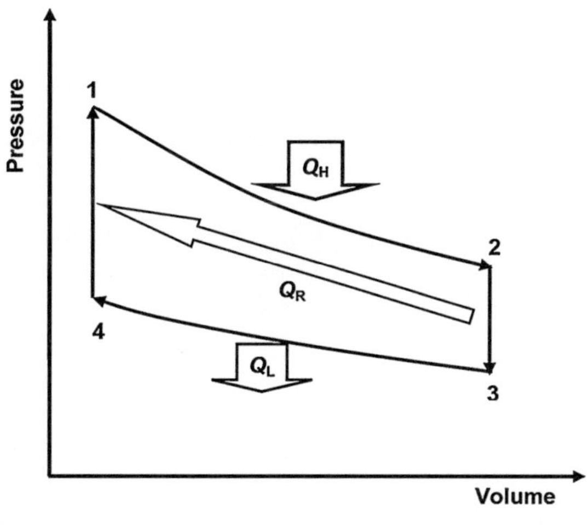

Fig. 6.1c *P–v* diagram of ideal Stirling power cycle

Process 3–4: The constant volume processes, regeneration process, the heat transfer from the temporary storage/regenerator to the working fluid

Process 4–1: The constant temperature heat rejection from the cycle to the external reservoir at temperature T_L

The execution of Stirling requires rather innovative hardware. The actual Stirling engines, including the original one patented by Robert Stirling, are heavy and complicated. The working of Stirling cycle consists of a cylinder with two pistons each side and regeneration in the middle. Initially, the left chamber houses the entire working fluid (a gas) which is at high temperature and pressure. During process 1–2, heat is transferred to the gas from a source at T_H. As the gas expands isothermally, the left piston is moved to the right-hand side at the same rate in such a way that the volume is constant until the entire gas is forced into the right-hand chamber. As the working fluid passes the (porous medium) regenerator, heat is transferred to the regenerator, and the gas temperature drops from T_H to T_L. For further heat transfer process to be reversible, the temperature difference between the gas and the regenerator should not exceed the differential amount dT at any point. Thus the temperature of the regenerator will be T_H on the left end and T_L at the right end of the regenerator when state 3 is reached. During process 3–4, the right-hand side piston moved inwards, compressing the gas, and the heat is transferred from the gas external reservoir in such a way that the temperature of the gas remains constant at T_L, while the pressures rise. Finally, during process 4–1, both pistons are moved to the left at the same rate, keeping the volume constant forcing the entire gas to the left chamber.

The gas temperature rises from T_L to T_H as it passes through the regenerator and picks up thermal energy stored in the regenerator during process 2–3, thereby completing the cycle. Let Q_H be the heat supplied to the cycle at temperature T_H and Q_L be the heat reflected by the cycle at temperature T_L; then:

$$Q_H = Q_{12} = P_1 V_1 \ln \frac{V_2}{V_1} = mRT_H \ln(r_c) \tag{6.1}$$

$$Q_L = Q_{34} = P_1 V_2 \ln \frac{V_2}{V_1} = mRT_L \ln(r_c) \tag{6.2}$$

where $r_c = V_2/V_1$ is the compression ratio of the cycle, R is the universal gas constant, and m is the mass of the working fluid. The net work output of the cycle (W_{net}) and the thermal efficiency are given by:

$$W_{net} = Q_H - Q_L = mR \ln(r_c)(T_H - T_L) \tag{6.3}$$

$$(\eta_{th})_{Stirling} = \frac{W_{net}}{Q_H} = \frac{Q_H - Q_L}{Q_H} = \frac{(T_H - T_L)}{T_H} = 1 - \frac{T_L}{T_H} = \eta_C \tag{6.4}$$

where η_C is the efficiency of a reversible Carnot cycle.

Thus the efficiency of an ideal reversible Stirling cycle is equal to the Carnot cycle when both cycles are working in the same temperature limits. However, it is difficult to obtain 100% efficient regenerator, and hence, it is recommended that a real regenerator is to be considered ($\eta_R < 100\%$). Thus, considering a real regenerator, the efficiency of Stirling cycle can be expressed as:

$$(\eta_{th})_{Stirling} = \frac{R \ln(r_c)(T_H - T_L)}{RT_H \ln(r_c) + (1 - \eta_R)C_v(T_H - T_L)} \tag{6.5}$$

where η_R is the regenerator efficiency and C_v is the specific heat of the working fluid at constant volume. Thus, the efficiency of Stirling cycle with a real regenerator is less than the efficiency of a Carnot cycle, working in the same temperature limits as shown in Eq. (6.5).

6.3 Ideal Ericsson Cycle

The ideal Ericsson cycle is also similar to Stirling cycle except the regeneration is having two constant pressure processes in the former, while there are two constant volume processes in the latter. Therefore, the Ericsson cycle consists of two isothermal processes and two isobaric (constant pressure) processes. The Ericsson cycle is shown on *T–s* and *P–v* diagram of Figs. 6.2a and 6.2b, respectively. The Ericsson cycle is very much like the Stirling cycle except that the two regenerator processes are isobaric (constant pressure) processes unlike the constant volume processes of the Stirling cycle. A steady flow ideal Ericsson cycle is shown in Figs. 6.3a and 6.3b. On the *T–s* and *P–v* diagrams, respectively. It can be seen from Figs. 6.3a and 6.3b that the two isothermal expansion and compression processes are executed in a turbine and a compressor, respectively, and a counter flow heat exchanger serves as a regenerator, where the hot and cold streams enter from opposite direction and exchange heat as shown in Figs. 6.3a and 6.3b. In the ideal Ericsson cycle, the temperature difference between the external reservoirs and the cycle does not exceed a differential amount dT at any point, and the cold fluid leaves the regenerator at the inlet temperature of the hot stream (T_{H}). Let Q_{H} be the amount of heat absorbed by the cycle at temperature T_{H} and Q_{L} be the amount of heat rejected by the cycle at temperature T_{L}; then:

$$Q_{\mathrm{H}} = Q_{12} = P_1 V_1 \ln\frac{P_1}{P_2} = mRT_{\mathrm{H}} \ln\left(r_{\mathrm{p}}\right) \tag{6.6}$$

$$Q_{\mathrm{L}} = Q_{34} = P_1 V_2 \ln\frac{P_1}{P_2} = mRT_{\mathrm{L}} \ln\left(r_{\mathrm{p}}\right) \tag{6.7}$$

Fig. 6.2a *T–s* diagram of ideal Ericsson cycle

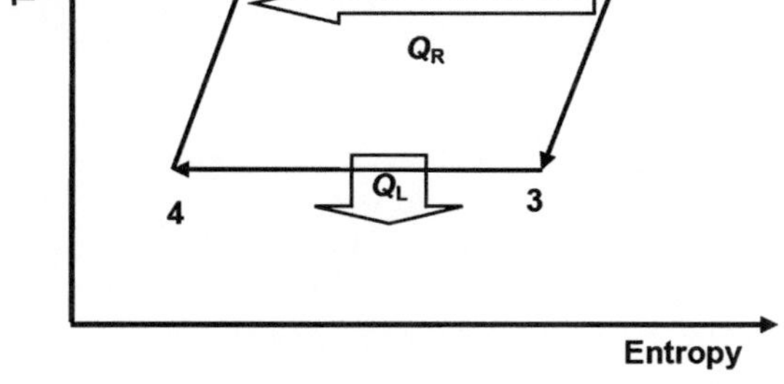

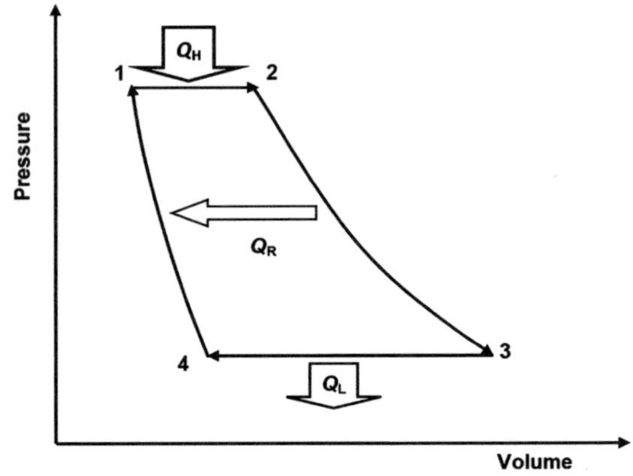

Fig. 6.2b *P–v* diagram of ideal Ericsson cycle

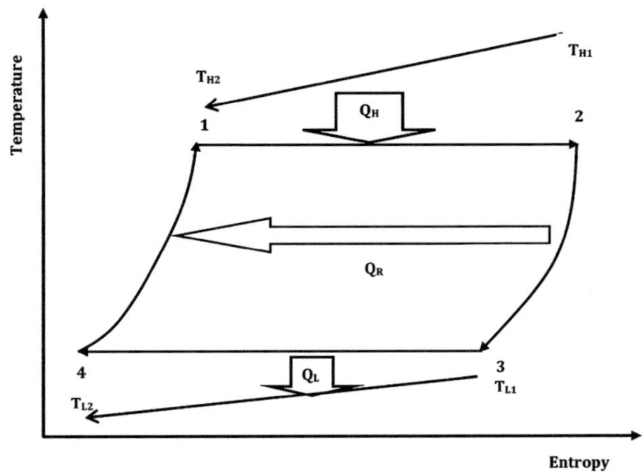

Fig. 6.3a *T–s* diagram of finite time Stirling cycle

where $r_p = P_1/P_2$ is the pressure ratio of the cycle, R is the universal gas constant, and m is the mass of the working fluid. The net work output of the cycle (W_{net}) and the thermal efficiency are given by:

$$W_{net} = Q_H - Q_L = mR \ln(r_p)(T_H - T_L) \tag{6.8}$$

$$(\eta_{th})_{Ericsson} = \frac{W_{net}}{Q_H} = \frac{Q_H - Q_L}{Q_H} = \frac{(T_H - T_L)}{T_H} = 1 - \frac{T_L}{T_H} = \eta_C \tag{6.9}$$

where η_C is the efficiency of a reversible Carnot cycle. Thus, the thermal efficiency of ideal reversible cycle is similar to that of a Carnot cycle working between the same temperature limits. Similar to Stirling cycle, the Ericsson cycle also does not possess the condition of a perfect regeneration.

However, the efficiency of an Ericsson cycle with a real regenerator is less than the Carnot efficiency, as given below:

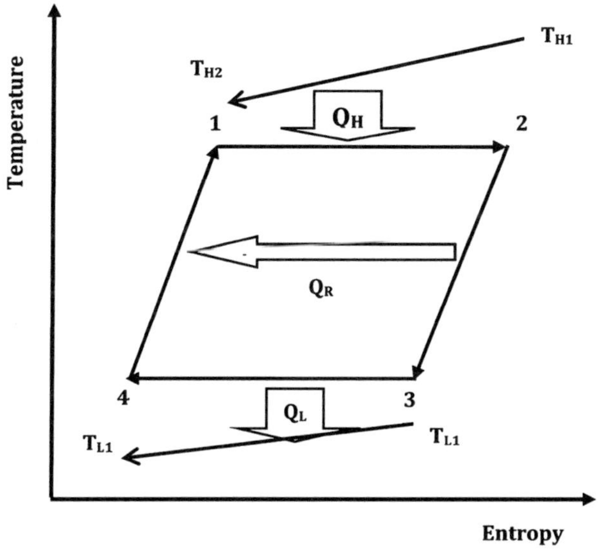

Fig. 6.3b *T–s* diagram of finite time Ericsson cycle

$$(\eta_{\text{th}})_{\text{Ericsson}} = \frac{R\ln(r_{\text{p}})(T_{\text{H}} - T_{\text{L}})}{RT_{\text{H}}\ln(r_{\text{p}}) + (1-\eta_{\text{R}})C_{\text{p}}(T_{\text{H}} - T_{\text{L}})} \tag{6.10}$$

where η_{R} is the regenerator efficiency and C_{p} is the specific heat of the working fluid at constant pressure.

The ideal Stirling and Ericsson cycles are difficult to achieve in real practice because they involve heat transfer through a differential temperature difference in all components including the regenerator. This would require providing infinitely large surface area for heat transfer and/or allowing infinitely large time for the process, and hence, none of these is practical. In this chapter, the concept of finite time thermodynamics has been applied on the Stirling and Ericsson cycles for power generation applications. The effects of major (both internal and the external) irreversibilities on the maximum power output and the corresponding thermal efficiency have been evaluated for these heat engines. Thus, the concept of finite time thermodynamics came into existence in 1975 after the novel work of Curzon and Ahlborn (1975) and has been applied to these engines for different set of operating conditions as given in the forthcoming sections.

6.4 Finite Time Stirling and Ericsson Cycles

The concept of finite time thermodynamics (Curzon and Ahlborn 1975) has been applied for Stirling and Ericsson cycles by a number of authors (Badescu 1992; Blank et al. 1994; Ladas and Ibrahim 1994; Blank and Wu 1995, 1996a, b; Trukhow et al. 1997; Erbay and Yavuz 1997; Chen 1997a; Chen et al. 1998; Senft 1998; Kaushik 1999; Chen and Schouten 1999; Kaushik and Kumar 2000a, 2001; Kumar 2000; Tyagi et al. 2002). A lot of work has been done on these engines using conventional and nonconventional sources of energy using finite time thermodynamic and second law analyses. For example, Badescu (1992) analysed the performance of a solar-powered heat engine operating in the Stirling cycle and studied the influence of design and climatological parameters on both the optimal solar-receiver temperature and overall thermal efficiency. Blank et al. (1994) studied the power optimization of an endoreversible Stirling cycle and provided an estimate of potential performance for a real engine. Ladas and Ibrahim (1994) have presented a finite time analysis of the Stirling engine and discussed the effect of heat transfer contact time and regeneration process on power output and thermal efficiency. Blank and Wu (1995, 1996a, b) examined the power optimization of an extraterrestrial solar radiant Stirling heat engine with an ideal regenerator, endoreversible Ericsson heat engine, and solar radiating Ericsson heat engine for space-conditioning applications. They showed that the optimal power and the corresponding thermal efficiency are based on higher and lower temperature bounds.

Trukhow et al. (1997) investigated the energy balance of an autonomous solar power plant with a Stirling engine and showed that the electric power output is proportional to the direct solar radiation. Erbay and Yavuz (1997) analysed the real Stirling heat engine for maximum power output conditions using polytropic processes as well as regenerative losses. They also determined the efficiency and compression ratio at maximum power density and ascertained the thermal design bounds. Chen (1997a, b) investigated the effect of regenerative losses on maximum power and the corresponding thermal efficiency of an endoreversible Stirling engine with infinite heat capacity of external heat source/sink reservoirs. Chen et al. (1998) studied the efficiency and engine size limits of a solar-driven Stirling heat engine at maximum power output. Senft (1998) studied the theoretical limitations on the performance of a Stirling engine subject to limited heat transfer and external, thermal, and mechanical losses.

In the recent decade, Kaushik (1999) made a comprehensive study and analysed the different aspects of finite time thermodynamics for various power and reverse power cycles for a typical set of operating parameters. Chen and Schouten (1999) presented the comprehensive influences of several major irreversibilities on the performance of Ericsson heat engine. Kaushik and Kumar (2000a, b) presented the effects of regenerative losses on the performance of an endoreversible Stirling heat engine, while the effects of other parameters such as the finite heat capacity of external reservoirs, internal irreversibility parameters, heat leak losses etc. were studied by Kumar (2000) and Kaushik and Kumar (2001) for both Stirling and Ericsson heat engines. Tyagi et al. (2002) presented the ecological optimization of Ericsson and Stirling heat engines considering the environmental conditions as the constraints for all energy conversion systems. However, the combination of Brayton and Stirling cycle (Braysson) was studied by Zhou et al. (2004) for different set of operating conditions.

As mentioned above, real Stirling and Ericsson cycles are not as efficient as an ideal one; however, a real cycle also consists of four processes mentioned above. Yet, there is a finite temperature difference between the cycle and the external reservoirs as shown on the *T–s* diagram of Figs. 6.2a and 6.2b.

6.4.1 Finite Heat Capacity

The external irreversible and internal reversible (endoreversible) Stirling and Ericsson heat engines coupled with a heat source and heat sink reservoirs of finite heat capacity rates and with a real regenerator are depicted on *T–s* diagram of Figs. 6.3a and 6.3b, respectively. This cycle approximates the compression stroke of the real heat engine as an isothermal heat rejection process (3–4) to the low-temperature sink. The heat addition to the working fluid from the regenerator is modelled as the constant volume process (4–1). The expansion stroke producing work is modelled as an isothermal heat addition process (1–2) from a high-temperature heat source. Finally, the heat rejection to the regenerator is modelled as the constant volume process (2–3).

As mentioned earlier, the external source/sink heat transfer processes 1–2 and 3–4 for the real Stirling heat engine must occur in a finite time, which requires that they proceed through a finite temperature difference and are, therefore, defined as being externally irreversible. During the isothermal heat rejection process 3–4, the heat flows from the cycle working fluid (which is maintained at a constant temperature T_c) to the low-temperature heat sink of finite heat capacity (whose temperature increases from T_{L1} to T_{L2}). Similarly, during the isothermal heat addition process 1–2, heat is transferred from the high-temperature heat source of finite heat capacity (whose temperature decreases from T_{H1} to T_{H2}) to the constant cycle temperature (T_h) of the working fluid. If the regenerator is an ideal one, the heat absorbed during the process 2–3 will be equal to the heat rejection during the process 4–1.

However, an ideal regenerator requires an infinite time or infinite area to transfer a finite amount of heat, and hence, a real regenerator is also considered in the present analysis. Similarly, Fig. 6.3b shows the *T–s* diagram of an endoreversible Ericsson heat engine affected by external irreversibility due to heat transfer to and from the external reservoirs, viz. Q_h and Q_c, respectively, to and from the heat engine during heat addition and heat rejection processes. If V_1 and V_2 (for Stirling) and P_1 and P_2 (for Ericsson) are the volumes/pressures of the working substance during the regenerative processes 4–1 and 2–3, respectively, the amount of heat absorbed (Q_h) and heat released (Q_c) by the heat engine working fluid are given as (Kaushik 1999; Tyagi et al. 2002):

$$Q_h = T_h \Delta S = nRT_h \ln(r) \tag{6.11}$$

$$Q_c = T_c \Delta S = nRT_c \ln(r) \tag{6.12}$$

where n is the number of moles of the working substance, ΔS is the change in entropy, R is the universal gas constant, and r is the compression/pressure ratio of the cycle. When the effect of finite heat transfer on the performance of thermodynamic cycles is considered, it is assumed that the heat transfer rate is proportional to the Log Mean Temperature Difference (LMTD) and written as (Kaushik 1999; Tyagi et al. 2002):

$$Q_h = (UA)_H (LMTD)_H t_h = C_H (T_{H1} - T_{H2}) t_h \tag{6.13}$$

$$Q_c = (UA)_L (LMTD)_L t_c = C_L (T_{L2} - T_{L1}) t_c \tag{6.14}$$

where t_h and t_c are heat addition and heat rejection times for the Stirling heat engine, and $(UA)_H$ and $(UA)_L$ are the overall heat transfer coefficient and area products for hot and cold sides, respectively. C_H and C_L are, respectively, the heat capacitance rates of external fluids in heat source/sink reservoirs. $(LMTD)_H$ and $(LMTD)_L$ are defined as:

$$(LMTD)_H = \frac{[(T_{H1} - T_h) - (T_{H2} - T_h)]}{\ln\left[\frac{(T_{H1} - T_h)}{(T_{H2} - T_h)}\right]} \tag{6.15}$$

$$(LMTD)_L = \frac{[(T_C - T_{L1}) - (T_C - T_{L2})]}{\ln\left[\frac{(T_C - T_{L1})}{(T_C - T_{L2})}\right]} \tag{6.16}$$

where T_h and T_c are the temperatures of working fluid on hot and cold side. Thus from Eqs. (6.13), (6.14), (6.15), and (6.16) yields:

$$T_{H2} = T_h + (T_{H1} - T_h) e^{-NTU_H} \tag{6.17}$$

$$T_{L2} = T_c - (T_c - T_{L1}) e^{-NTU_L} \tag{6.18}$$

where NTU_H and NTU_L are the number of heat transfer units; $NTU_H = (UA)_H / C_H$ and $NTU_L = (UA)_L / C_L$. Combining Eqs. (6.13) and (6.14) with Eqs. (6.17) and (6.18) yields:

$$Q_h = C_H \varepsilon_H (T_{H1} - T_h) t_h \tag{6.19a}$$

$$Q_c = C_L \varepsilon_L (T_c - T_{L1}) t_c \tag{6.19b}$$

where ε_H and ε_L are, respectively, the effectiveness of hot- and cold-side heat exchangers and are defined as:

$$\varepsilon_H = 1 - \exp^{-NTU_H} \quad \text{and} \quad \varepsilon_L = 1 - \exp^{-NTU_L} \tag{6.20}$$

It is important to mention that there also exists a finite heat transfer in the regenerative processes besides the two isothermal heat addition and rejection processes. Thus, the regenerative heat transfer (Q_R) is given (Blank et al. 1994; Ladas and Ibrahim 1994; Blank and Wu 1995, 1996a, b; Trukhow et al. 1997; Erbay and Yavuz 1997; Chen 1997a, b; Chen et al. 1998; Senft 1998; Kaushik 1999; Chen and Schouten 1999; Kaushik and Kumar 2000a, b) as:

$$Q_R = nC\varepsilon_R(T_h - T_c) \tag{6.21a}$$

where C is the specific heat of the working substance and ε_R is the effectiveness of the regenerator. Thus, the heat loss in the two regenerative processes (ΔQ per cycle) is given by:

$$\Delta Q = nC(1 - \varepsilon_R)(T_h - T_c) \tag{6.21b}$$

It can be seen from Eq. (6.21b) that ΔQ ;= 0, if $\varepsilon_R = 1.0$. The net heat absorbed (Q_H) from the heat source and released (Q_L) to the heat sink incorporating the regenerative heat loss per cycle are, respectively, given (Blank et al. 1994; Ladas and Ibrahim 1994; Blank and Wu 1995; Blank and Wu 1996a, b; Trukhow et al. 1997; Erbay and Yavuz 1997; Chen 1997a, b; Chen et al. 1998; Senft 1998; Kaushik 1999; Chen and Schouten 1999; Kaushik and Kumar 2000a, b) as:

$$Q_H = Q_h + \Delta Q \tag{6.22}$$

$$Q_L = Q_c + \Delta Q \tag{6.23}$$

To achieve a more realistic case, the time of the regenerative heat transfer processes (t_3 and t_4) should also be considered in the thermodynamic analysis of Stirling engine (Blank et al. 1994). For this purpose, it is assumed that the temperature of the engine working fluid/substance is varying with time in the regenerative processes (Chen 1997a, b; Chen et al. 1998; Senft 1998) as:

$$\frac{dT}{dt} = \pm\alpha \tag{6.24}$$

where α is the proportionality constant, which is independent of the temperature but dependent on the property of the regenerative material. The positive and negative signs, respectively, correspond to heating and cooling processes in the regenerator. Using Eq. (6.24), one can obtain the time (t_R) of the regenerative processes (Blank et al. 1994; Ladas and Ibrahim 1994; Blank and Wu 1995, 1996a, b; Trukhow et al. 1997; Erbay and Yavuz 1997; Chen 1997a, b; Chen et al. 1998; Senft 1998; Kaushik 1999; Chen and Schouten 1999; Kaushik and Kumar 2000a, b) as:

$$t_R = t_2 + t_3 = \frac{2}{\alpha}(T_h - T_c) \tag{6.25}$$

Thus the total cycle time is given by:

$$t_{\text{cycle}} = t_{\text{h}} + t_{\text{c}} + t_{\text{R}} \tag{6.26}$$

For the thermodynamic cycle 1-2-3-4-1, work and the power output per cycle is given by:

$$W = Q_{\text{H}} - Q_{\text{L}} = Q_{\text{h}} - Q_{\text{c}} \tag{6.27a}$$

$$P = \frac{W}{t_{\text{cycle}}} = \frac{(Q_{\text{H}} - Q_{\text{L}})}{t_{\text{cycle}}} = \frac{(Q_{\text{h}} - Q_{\text{c}})}{(t_{\text{h}} + t_{\text{c}} + t_{\text{R}})} \tag{6.27b}$$

Using Eqs. (6.11) and (6.12) and Eqs. (6.19a) and (6.19b), the power output is given by:

$$P = \frac{(T_{\text{h}} - T_{\text{c}})}{\left[\frac{T_{\text{h}}}{k_1(T_{\text{H1}} - T_{\text{h}})} + \frac{T_{\text{c}}}{k_2\left(T_{\text{c}} - T_{\text{L}_1}\right)} + \text{a}_1(T_{\text{h}} - T_{\text{c}})\right]}, \tag{6.28}$$

and the thermal efficiency (η) is given by:

$$\eta = \frac{W}{Q_{\text{H}}} = \frac{(Q_{\text{h}} - Q_{\text{c}})}{(Q_{\text{h}} + \Delta Q)} = \frac{(T_{\text{h}} - T_{\text{c}})}{T_{\text{h}} + a_2(T_{\text{h}} - T_{\text{c}})} \tag{6.29}$$

where $k_1 = C_{\text{H}}\varepsilon_{\text{H}}$, $k_2 = C_{\text{L}}\varepsilon_{\text{L}}$, $a_1 = \frac{2}{n\alpha R \ln(r)}$, and $a_2 = \frac{C(1-\varepsilon_R)}{R \ln(r)}$. For the sake of convenience, two parameters are introduced such that $x = T_{\text{h}}/T_{\text{c}}$ and $y = T_{\text{c}}$ in Eqs. (6.28) and (6.29) yield:

$$P = \frac{(x-1)}{\left[\frac{x}{k_1(T_{\text{H1}} - xy)} + \frac{1}{k_2(y - T_{\text{L1}})} + a_1(x-1)\right]} \tag{6.30}$$

and

$$\eta = \frac{(x-1)}{x + a_2(x-1)} \tag{6.31}$$

It is seen from Eq. (6.30) that the power output (P) is a function of x and y for given parameters like T_{H1}, T_{L1}, ε_{H}, ε_{L}, C_{H}, C_{L}, etc. Thus maximizing power (P) with respect to x and y, we have:
$\frac{\partial P}{\partial x} = 0$ and $\frac{\partial P}{\partial y} = 0$ yields the optimal values of x and y as:

$$x = \sqrt{\frac{T_{\text{H1}}}{T_{\text{L1}}}} \quad \text{and} \quad y = \frac{\left(T_{\text{L1}} + T_{\text{H1}}\sqrt{\frac{k_1}{k_2}}\right)}{\left(1 + \sqrt{\frac{k_1}{k_2}}\right)}$$

Substituting the values of x and y into Eqs. (6.30) and (6.31), the maximum power output is given by:

$$P_{\max} = \frac{K\left[\sqrt{T_{\mathrm{H1}}} - \sqrt{T_{\mathrm{L1}}}\right]^2}{\left[1 + K_{\mathrm{a1}}\left(\sqrt{T_{\mathrm{H1}}} - \sqrt{T_{\mathrm{L1}}}\right)^2\right]} \tag{6.32}$$

and the corresponding thermal efficiency at maximum power will be:

$$\eta_m = \frac{\left(1 - \sqrt{\frac{T_{\mathrm{L1}}}{T_{\mathrm{H1}}}}\right)}{\left[1 + \mathrm{a}_2\left(1 - \sqrt{\frac{T_{\mathrm{L1}}}{T_{\mathrm{H1}}}}\right)\right]} = \frac{\eta_{\mathrm{CA}}}{(1 + \mathrm{a}_2\eta_{\mathrm{CA}})} \tag{6.33}$$

where $K = \frac{k_1k_2}{2}\left(\sqrt{k_1} + \sqrt{k_2}\right)$ and $\eta_{\mathrm{CA}} = 1 - \sqrt{T_{\mathrm{L1}} - T_{\mathrm{H1}}}$ are the Curzon–Ahlborn efficiency of an endoreversible Carnot heat engine for finite heat capacity of external reservoirs. It is seen from Eq. (6.32) that the maximum power output of the Stirling heat engine is not affected by the regenerative heat loss, although it depends upon the time of regenerative processes. Equation 6.33 also clearly shows that when Stirling/Ericsson heat engine with regenerative loss is operating at the maximum power output, its efficiency is different (always smaller) from that of an endoreversible Carnot heat engine efficiency (η_{CA}).

6.4.2 *Infinite Heat Capacity*

$$P_{\max} = \frac{K\left[\sqrt{T_{\mathrm{H}}} - \sqrt{T_{\mathrm{L}}}\right]^2}{\left[1 + K_{a1}\left(\sqrt{T_{\mathrm{H}}} - \sqrt{T_{\mathrm{L}}}\right)^2\right]} \tag{6.34}$$

and the corresponding thermal efficiency at maximum power will be:

$$\eta_{\mathrm{m}} = \frac{\eta_{\mathrm{CA}}}{(1 + \mathrm{a}_2\eta_{\mathrm{CA}})} \quad \text{where} \quad \eta_{\mathrm{CA}} = 1 - \sqrt{\frac{T_{\mathrm{L}}}{T_{\mathrm{H}}}} \tag{6.35}$$

It is seen from Eq. (6.34) that the maximum power output of the Stirling/ Ericsson heat engine is not affected by the regenerative heat loss, although it depends upon the time of regenerative processes. Again, Eq. (6.35) clearly shows that when Stirling heat engine with regenerative loss is operating at the maximum power output, its efficiency is different (always smaller) from that of an endoreversible Carnot heat engine efficiency (η_{CA}).

Some Special Cases

In the investigations of Stirling/Ericsson cycle, there can be different cases based on different assumptions, and the performance of these cycles will change accordingly. Some of the assumptions are given as below:

Case 1 When $\varepsilon_R = 1.0$, i.e. the Stirling/Ericsson engine achieves the condition of perfect/ideal regeneration, although the time of regenerative process is still considered. Then the maximum power output and the corresponding thermal efficiency are given by:

$$P_{\max} = \frac{K\left[\sqrt{T_{H1}} - \sqrt{T_{L1}}\right]^2}{\left[1 + K_{a1}\left[\sqrt{T_{H1}} - \sqrt{T_{L1}}\right]^2\right]} \tag{6.36}$$

and

$$\eta_m = 1 - \sqrt{\frac{T_{L1}}{T_{H1}}} = \eta_{CA} \tag{6.37}$$

Thus for $\varepsilon_R = 1.00$, the performance of the endoreversible Stirling heat engine is with perfect/ideal regenerator in which the time of regeneration is given by Eq. (6.25), and the optimal performance of endoreversible Stirling heat engine is identical to that of endoreversible Carnot heat engine. In such a case, its maximum power output and the corresponding thermal efficiency are, respectively, given by Eqs. (6.36) and (6.37). However, physically for finite regenerative time, ε_R should be less than unity. This shows that in the investigation of the Stirling heat engine, it would be impossible to obtain new conclusions if the regenerator losses were not considered.

Case 2 When $t_R \propto (t_h + t_c)$, i.e. when the time of regenerative processes is directly proportional to the time of two isothermal processes, i.e.:

$$t_R = \gamma(t_h + t_c) \tag{6.38}$$

where γ is the proportionality constant, then:

$$P_{\max} = \frac{K\left[\sqrt{T_{H1}} - \sqrt{T_{L1}}\right]^2}{1 + \gamma} \tag{6.39}$$

Case 3 When $t_R = 0$, i.e. when the regenerative time is negligibly small as compared to the time of the two isothermal processes, i.e. $a_1 = 0$. In such a case, the maximum power output is given by:

$$P_{\max} = K\left[\sqrt{T_{\rm H1}} - \sqrt{T_{\rm L1}}\right]^2 \tag{6.40a}$$

while the corresponding thermal efficiency in both the cases is still given by Eq. (6.33).

Case 4 When $C_{\rm H}$ and $C_{\rm L} = 0$; $\varepsilon_{\rm R}<1.0$ and $t_R = \gamma(t_{\rm h} + t_{\rm c})$, i.e. the heat source and heat sink are of infinite heat capacitance and $t_{\rm R}$ is given by Eq. (6.38). In such a case, the maximum power output becomes:

$$P_{\max} = \frac{K\left[\sqrt{T_{\rm H}} - \sqrt{T_{\rm L}}\right]^2}{1+\gamma} \tag{6.40b}$$

while the corresponding thermal efficiency is still given by Eq. (6.33), whereas the Curzon–Ahlborn efficiency is given by Eq. (6.35). Thus it is seen from Eqs. (6.39) and (6.40a) that the maximum power output of Stirling/Ericsson heat engine is equivalent to that of an endoreversible Carnot heat engine for the same operating conditions in which the time of regeneration process is given by Eqs. (6.25) and (6.36a), but the thermal efficiency will be lower. The effects of both internal and the external irreversibilities on the maximum power output and the corresponding thermal efficiency have also been evaluated for these heat engines. The external irreversibility is due to temperature difference between the reservoirs and the heat engine working fluid and the direct heat leak losses from source to sink reservoir, while the internal irreversibility are due to friction, pressure drop, and the entropy generation within the cycle. The detailed analysis of such cycle is given in the next section as below.

6.5 Irreversible Stirling and Ericsson Cycles

An irreversible Stirling and Ericsson heat engines coupled with a heat source and heat sink reservoirs of finite heat capacity rates and with a real regenerator are depicted on the *T–s* diagram of Figs. 6.4a and 6.4b, respectively. When the working substance of these engines is ideal gas, the Ericsson cycle consists of two isothermal and two isobaric processes, while the Stirling cycle consists of two isothermal and two isochoric processes, respectively. These cycles approximate the compression stroke of real engine as on isothermal process 3–4 with an irreversible heat rejection at constant temperature (T_c) to the heat sink of finite heat capacity whose temperature varies from $T_{\rm L1}$ to $T_{\rm L2}$. The heat addition to the working fluid during the regeneration is modelled as isobaric (in Ericsson cycle) or isochoric (in Stirling cycle) process 4 s–1 s in ideal and 4–1 in real cycles. The work-producing expansion stroke is modelled as isothermal process 1–2 with irreversible heat addition at constant temperature $T_{\rm H}$ from the heat source of finite heat capacity whose temperature varies from $T_{\rm H1}$ to $T_{\rm H2}$. Finally the heat rejection during the

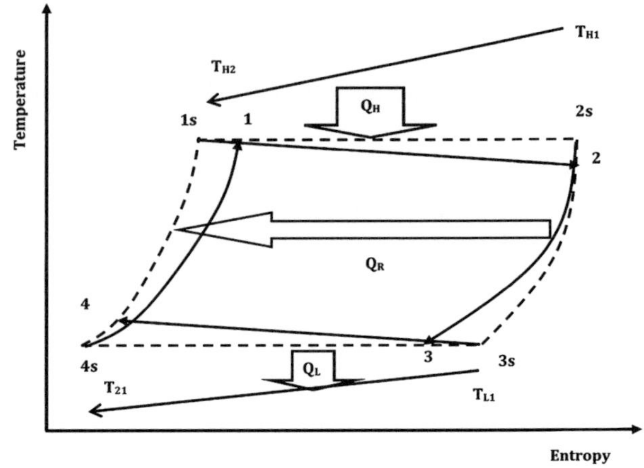

Fig. 6.4a *T*–s diagram of finite time irreversible Stirling cycle

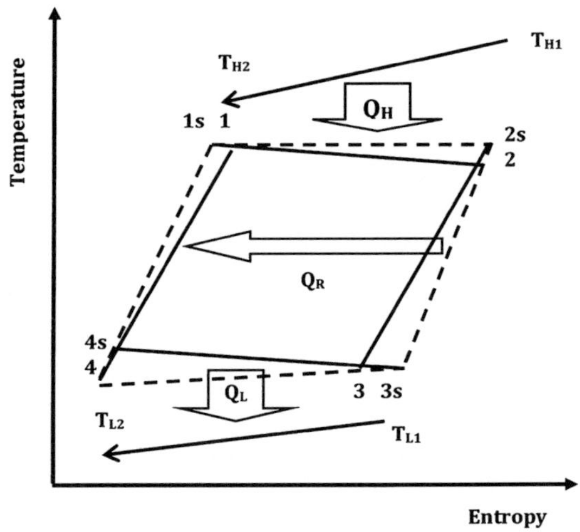

Fig. 6.4b *T*–s Diagram of finite time irreversible Ericsson cycle

regeneration is modelled as isobaric (in Ericsson cycle) or isochoric (in Stirling cycle) process 2 s–3 s in ideal and 2–3 in real cycles.

The work-producing expansion stroke is modelled as isothermal process 3–4 with irreversible heat addition at constant temperature T_h from the heat source of finite heat capacity whose temperature varies from T_{H1} to T_{H2}. Finally, the heat rejection during the regeneration is modelled as isobaric (in Ericsson cycle) or isochoric (in Stirling cycle) process 4–1 s in ideal and 4–1 in real cycles. As mentioned earlier, the external heat transfer processes 1–2 and 3–4 in real cycles must proceed in finite time. This in turn requires that these heat transfer processes occur through finite temperature difference and are, therefore, externally irreversible. Similarly, the entropy change during process 4–1 is more than process 2–3; thus the cycles are also internally irreversible. Since an ideal regeneration requires infinite time/area and hence not practical, it is desirable to consider the real regenerator with finite effectiveness. Let Q_h is the amount of heat absorbed from the heat source at temperature (T_h) and Q_c is the amount of heat released to the heat sink at temperature (T_c). Thus using the heat transfer theory (Kumar 2000; Kaushik and Kumar 2001) yields:

$$\begin{aligned} Q_{\mathrm{h}} = T_{\mathrm{h}}\Delta S = C_{\mathrm{H}}(T_{\mathrm{H1}} - T_{\mathrm{H2}})t_{\mathrm{H}} = (UA)_{\mathrm{H}}(\mathrm{LMTD})_{\mathrm{H}} \\ = C_{\mathrm{H}}\varepsilon_{\mathrm{H}}(T_{\mathrm{H1}} - T_{\mathrm{h}})t_{\mathrm{H}} \end{aligned} \quad (6.41)$$

$$Q_{\mathrm{c}} = T_{\mathrm{c}}\Delta S = C_{\mathrm{L}}(T_{\mathrm{L2}} - T_{\mathrm{L1}})t_{\mathrm{L}} = (UA)_{\mathrm{L}}(\mathrm{LMTD})_{\mathrm{L}} = C_{\mathrm{L}}\varepsilon_{\mathrm{L}}(T_c - T_{\mathrm{L2}})t_{\mathrm{L}} \quad (6.42)$$

where ΔS is the entropy generation; C_H and C_L are the heat capacitance rates of the source and sink reservoirs; t_H and t_L are, respectively, the time for heat addition and heat rejection; $(UA)_H$ and $(UA)_L$ are the heat transfer coefficient-area products; and $(\mathrm{LMTD})_H$ and $(\mathrm{LMTD})_L$ and ε_H and ε_L are, respectively, the Log Mean Temperature Difference and the effectiveness of the heat exchanger on the source and sink side and are defined as (Kumar 2000; Kaushik and Kumar 2001):

$$\begin{aligned} (\mathrm{LMTD})_{\mathrm{H}} &= \frac{(T_{\mathrm{H1}} - T_{\mathrm{h}}) - (T_{\mathrm{H2}} - T_{\mathrm{h}})}{\ln\left(\frac{(T_{\mathrm{H1}} - T_{\mathrm{h}})}{(T_{\mathrm{H2}} - T_{\mathrm{h}})}\right)} \quad \text{and} \\ (\mathrm{LMTD})_{\mathrm{L}} &= \frac{(T_{\mathrm{c}} - T_{\mathrm{L1}}) - (T_{\mathrm{c}} - T_{\mathrm{L2}})}{\ln\left(\frac{(T_{\mathrm{c}} - T_{\mathrm{L1}})}{(T_{\mathrm{c}} - T_{\mathrm{L2}})}\right)} \end{aligned} \quad (6.43)$$

$$\varepsilon_{\mathrm{H}} = 1 - \exp^{-\frac{U_{AH}}{C_{\mathrm{H}}}} \quad \text{and} \quad \varepsilon_{\mathrm{L}} = 1 - \exp^{-\frac{U_{AL}}{C_{\mathrm{L}}}} \quad (6.44)$$

It is also desirable to consider the finite heat transfer through the regenerator. Let ΔQ_R be the regenerative heat loss per cycle during the two regeneration processes, which is proportional to the temperature difference of the working fluid (Kumar 2000; Kaushik and Kumar 2001):

$$\Delta Q_{\mathrm{R}} = nC_{\mathrm{f}}(1 - \varepsilon_{\mathrm{R}})(T_{\mathrm{h}} - T_{\mathrm{c}}) \quad (6.45)$$

where n is the number of moles, C_f is the specific ($C_f = C_p$ for the Ericsson cycle and $C_f = C_v$ of the Stirling cycle) heat of the working fluid, and ε_R the effectiveness of the regenerator.

Further, owing to the influence of irreversibility due to finite heat transfer, the time of two regenerative processes is not negligible as compared to the two isothermal processes. The time of regenerative processes may be considered as proportional to the temperature difference of the working fluid (Kumar 2000; Kaushik and Kumar 2001), as below:

$$t_R = t_{2-3} + t_{4-1} = 2\alpha(T_{\mathrm{h}} - T_{\mathrm{c}}) \quad (6.46)$$

where α is the proportionality constant independent of the temperature difference but depends on the property of the regenerative material and t_3 and t_4 are time taken during processes 2–3 and 4–1, respectively. In addition to the above assumptions, it is also assumed that there is some heat leak loss (Q_0) directly from the heat source to sink and is assumed to be directly proportional to the mean temperature difference between the source sink fluids and the total cycle time (Kumar 2000; Kaushik and Kumar 2001), as given below:

$$Q_0 = k_0(T_{\mathrm{Hm}} - T_{\mathrm{Lm}})t_{\mathrm{cycle}} \tag{6.47}$$

where k_0 is the proportionality constant, t_{cycle} is the total cycle time, and T_{Hm} and T_{Lm} are the mean temperatures of the source and sink reservoirs, respectively (Kumar 2000; Kaushik and Kumar 2001), and are given as below:

$$T_{\mathrm{Hm}} = \frac{1}{2}(T_{\mathrm{H1}} + T_{\mathrm{H2}}) = \frac{1}{2}[(2 - \varepsilon_{\mathrm{H}})T_{\mathrm{H1}} + \varepsilon_{\mathrm{H}} T_{\mathrm{h}}] \tag{6.48}$$

$$T_{\mathrm{Lm}} = \frac{1}{2}(T_{\mathrm{L1}} + T_{\mathrm{L2}}) = \frac{1}{2}[(2 - \varepsilon_{\mathrm{L}})T_{\mathrm{L1}} + \varepsilon_{\mathrm{H}} T_{\mathrm{c}}] \quad \text{and} \quad t_{\mathrm{cycle}} = (t_{\mathrm{H}} + t_{\mathrm{L}} + t_{\mathrm{R}}) \tag{6.49}$$

Thus, rearranging Eqs. (6.7), (6.8), and (6.9), following earlier authors (Kumar 2000; Kaushik and Kumar 2001), yields:

$$Q_0 = [k_{01} + k_{02}(\varepsilon_{\mathrm{H}} T_{\mathrm{h}} - \varepsilon_{\mathrm{L}} T_{\mathrm{c}})]t_{\mathrm{cycle}} \tag{6.50}$$

where $k_{01} = \frac{k_{01}}{2}[(2 - \varepsilon_{\mathrm{H}})T_{\mathrm{H1}} - (2 - \varepsilon_{\mathrm{L}})T_{\mathrm{L1}}]$ and $k_{02} = k_0/2$. Again, using Eqs. (6.41), (6.42), (6.45), and (6.47), the net heat absorbed from the heat source and released to the sink, respectively, is given (Kumar 2000; Kaushik and Kumar 2001) as:

$$Q_{\mathrm{H}} = Q_{\mathrm{h}} + \Delta Q_{\mathrm{R}} + Q_{\mathrm{R}} \quad \text{and} \quad Q_{\mathrm{L}} = Q_{\mathrm{c}} + \Delta Q_{\mathrm{R}} + Q_{\mathrm{R}} \tag{6.51}$$

From the first law of thermodynamics, the net useful work is given (Kaushik and Kumar 2000a, b, 2001; Tyagi et al. 2002) as:

$$W = Q_{\mathrm{H}} - Q_{\mathrm{L}} = Q_{\mathrm{h}} - Q_{\mathrm{c}} \tag{6.52}$$

Again, the power output and the corresponding thermal efficiency for both the cycles are given (Kaushik and Kumar 2000a, b, 2001; Tyagi et al. 2002) as:

$$P = \frac{W}{t_{\mathrm{cycle}}} = \frac{(Q_{\mathrm{h}} - Q_{\mathrm{c}})}{t_{\mathrm{H}} + t_{\mathrm{L}} + t_{\mathrm{R}}} \tag{6.53}$$

$$\eta = \frac{W}{Q_{\mathrm{H}}} = \frac{(Q_{\mathrm{h}} - Q_{\mathrm{c}})}{Q_{\mathrm{h}} + \Delta Q_{\mathrm{R}} + Q_{\mathrm{o}}} \tag{6.54}$$

Also, the second law of thermodynamics for a real cycle (Kaushik and Kumar 2000a, b, 2001; Kumar 2000; Tyagi et al. 2002) yields:

$$\oint \frac{dQ}{T} = \frac{Q_h}{T_h} - \frac{Q_c}{T_c} \leq 0 \tag{6.55}$$

For internal irreversible cycle, introducing the internal irreversibility parameter (R_{K-T}) and using Eq. (6.15) yield (Kumar 2000):

$$\frac{Q_h}{T_h} \Big/ \frac{Q_c}{T_c} = \frac{(S_2 - S_1)}{(S_3 - S_4)} = \frac{1}{R_{K-T}} \Rightarrow R_{K-T}\frac{Q_h}{T_h} - \frac{Q_c}{T_c} = 0 \tag{6.56}$$

where R_{K-T} is greater than unity for real/irreversible cycle. Using Eqs. (6.53) and (6.54), the power output and the corresponding thermal efficiency are given as below (Kumar 2000; Kaushik and Kumar 2001):

$$P = \frac{(x - R_{K-T})}{\left[\frac{x}{k_1(T_{H1}-xy)}\frac{R_{K-T}}{k_2(y-T_{L1})} + bR_{K-T}(x-1)\right]} \tag{6.57}$$

$$\eta = \frac{(x - R_{K-T})}{x + a(x-1) + [k_{01} + k_{02}(\varepsilon_H x - \varepsilon_L)y]\left[\frac{x}{k_1(T_{H1}-xy)}\frac{R_{K-T}}{k_2(y-T_{L1})} + bR_{K-T}(x-1)\right]} \tag{6.58}$$

where $x = T_h/T_c$, $y = T_c$, $a = \frac{2a}{nR\ln(r)}$ $b = \frac{C_f(1-\varepsilon_R)}{R\ln(r)}$, $k_1 = C_H\varepsilon_H$, and $k_2 = C_L\varepsilon_L$. It is further seen from Eq. (6.57) that power output is a function of x and y for given parameters R_{K-T}, ε_H, ε_L, ε_R, T_{H1}, T_{L1}, a, b, etc. Thus Eq. (6.57) and its extremal condition, viz. $\frac{\partial P}{\partial x} = 0$, yield:

$$y_{opt} = \frac{(x_{opt}TL_1 + k_3T_{H1})}{(1 + k_3R_{K-T})x_{opt}} \quad \text{and} \quad x_{opt} = \frac{-B \pm \sqrt{(B^2 - AC)}}{A} \tag{6.59}$$

where $A = 2bT_{L1} - R_{K-T}k_4T_{L1}$; $B = 2T_{H1}T_{L1}b$; $C = 2bT_{H1} + k_4T_{H1}$; $k_3 = \sqrt{\frac{R_{K-T}k_1}{k_2}}$; and $k_4 = (1 + k_3)\left(\frac{1}{k_1}\frac{R_{K-T}}{k_1k_2}\right)$. Substituting the values of x and y into Eqs. (6.57) and (6.58), one can find the expressions for the maximum power output and the corresponding thermal efficiency. It can be seen from these expressions that the maximum power output is not affected by the regenerative loss and the direct heat leak loss, while it is reverse for the thermal efficiency. The cycle model presented in this section is more general, and the results obtained by earlier workers (Badescu 1992; Blank et al. 1994; Ladas and Ibrahim 1994; Blank and Wu 1995, 1996a, b; Trukhow et al. 1997; Erbay and Yavuz 1997; Chen 1997a, b) can also be obtained from this particular model as a special case, including the infinite heat capacity. The readers are also encouraged to take it as an exercise, while some other cases are given as below.

Some Special Cases

The proposed cycle models presented in this chapter are the general cycle models as they include most of the irreversiblities associated with any real cycle. The models presented by other authors (Badescu 1992; Blank et al. 1994; Ladas and Ibrahim 1994; Blank and Wu 1995, 1996a, b; Trukhow et al. 1997; Erbay and Yavuz 1997; Chen 1997a, b; Chen et al. 1998; Senft 1998; Kaushik 1999; Chen and Schouten 1999) are the special cases of the present cycle models. The brief summary of other cycle models for both finite and infinite capacity of external reservoirs, presented by earlier authors, is given as below.

Case 1 When $R_{K-T} = 1.0$, i.e. these cycles are internally reversible. The maximum power output and the corresponding thermal efficiency are given by:

$$P_{\max} = \frac{K\left(\sqrt{T_H} - \sqrt{T_L}\right)^2}{\left[1 + bK\left(\sqrt{T_H} - \sqrt{T_L}\right)^2\right]} \tag{6.60}$$

$$\eta_m = \frac{\eta_{CA}}{\left[(1 + a\eta_{CA}) + \left(\frac{k_{01} + k_{02}k_{04}\left(1 + bKT_H\eta_{CA}^2\right)}{KT_H\eta_{CA}^2}\right)\right]} \tag{6.61}$$

where

$$K = \frac{k_1k_2}{\left(\sqrt{k_1} + \sqrt{k_2}\right)^2}; \quad k_{04} = \frac{\left(\frac{\sqrt{T_HT_L}}{k_2} + \sqrt{T_L}\right)\left(k_1\sqrt{T_H} - k_2\sqrt{T_L}\right)}{\left(1 + \sqrt{k_1/k_2}\right)} \quad \text{and}$$

$$\eta_{CA} = 1 - \sqrt{\frac{T_L}{T_H}}$$

Thus the maximum power output of Ericsson and Stirling heat engines are different from the endoreversible Carnot heat engine but similar to those given by earlier authors (Badescu 1992; Blank et al. 1994; Ladas and Ibrahim 1994; Blank and Wu 1995, 1996a, b).

Case 2 When $t_R = \gamma\,(t_H + t_L)$, i.e. the time of regeneration is directly proportional to the time of two isothermal processes. The maximum power output and the corresponding thermal efficiency will be:

$$P_{\max} = \frac{(x - R_{K-T})}{(1 + \gamma)\left[\frac{x}{k_1(T_H - xy)} + \frac{R_{K-T}}{k_2(y - T_L)} + b(x - 1)\right]} \tag{6.62}$$

$$\eta_{\mathrm{m}} = \frac{(x - R_{\mathrm{K-T}})\lambda(x)}{x + aR_{\mathrm{K-T}}\{(x - y) + (1 + \gamma)[k_{\mathrm{o1}} + k_{\mathrm{o2}}(k_1 x - k_2 y)]\}} \tag{6.63a}$$

where λ(x) is the function of x as given below:

$$\lambda(x) = \frac{1}{\left[\frac{x}{k_1(T_{\mathrm{H}} - xy)R_{\mathrm{K-T}}} + \frac{1}{k_1(y - T_{\mathrm{L}})} + b(x - 1)\right]} \tag{6.63b}$$

The maximum power output of Ericsson and Stirling heat engines are equivalent to the Carnot heat engine in which the time of two adiabatic processes is directly proportional to that of two isothermal processes given in Curzon and Ahlborn (1975), Badescu (1992), Blank et al. (1994), and Ladas and Ibrahim (1994) but the corresponding thermal efficiency much different and less than the equivalent Carnot cycle.

Case 3 When $\varepsilon_{\mathrm{R}} = 1.0$ and $Q_0 = 0$, i.e. these engines possess the condition of perfect regeneration without direct heat leak loss, the maximum power output and the corresponding thermal efficiency are given by:

$$P = \frac{(x - R_{\mathrm{K-T}})}{\left[\frac{x}{k_1(T_{\mathrm{H}} - xy)} + \frac{R_{\mathrm{K-T}}}{k_2(y - T_{\mathrm{L}})} + b(x - 1)\right]} \tag{6.64}$$

$$\eta_{\mathrm{m}} = \frac{(x - R_{\mathrm{K-T}})}{x + R_{\mathrm{K-T}}[k_{01} + k_{02}(\alpha_1 x - \beta_1)y]\left[\frac{x}{k_1(T_{\mathrm{H}} - xy)R_{\mathrm{K-T}}} + \frac{1}{k_2(y - T_{\mathrm{L}})}\right]} \tag{6.65}$$

In this case not only the maximum power output but also the corresponding thermal efficiency of Stirling and Ericsson heat engines are similar to that of equivalent Carnot engine with internal irreversibility (Kaushik 1999; Chen and Schouten 1999; Kaushik and Kumar 2000a, b). However, for finite regeneration time, ε_{R} should be less than unity. This shows that in the investigation of these cycles, it is difficult to obtain new conclusions if the regenerator losses were not considered.

Case 4 When $R_{\mathrm{K-T}} = 1.0$, $Q_0 = 0$, i.e. these cycles are internally reversible, and the heat flow from the source to the sink/environment is negligible. The maximum power output and the corresponding thermal efficiency will be:

$$P_{\max} = \frac{K\left[\sqrt{T_{\mathrm{H}}} - \sqrt{T_{\mathrm{L}}}\right]^2}{\left[1 + Ka_1\left(\sqrt{T_{\mathrm{H}}} - \sqrt{T_{\mathrm{L}}}\right)^2\right]} \tag{6.66}$$

$$\eta_m = \frac{\left(1 - \sqrt{\frac{T_{\mathrm{L}}}{T_{\mathrm{H}}}}\right)}{\left[1 + a_2\left(1 - \sqrt{\frac{T_{\mathrm{L}}}{T_{\mathrm{H}}}}\right)\right]} = \frac{\eta_{\mathrm{CA}}}{(1 + a_2\eta_{\mathrm{CA}})} \tag{6.67}$$

Case 5 When $R_{K-T} = 1.0$, $Q_0 = 0$, and $\varepsilon_R = 1.0$, the Stirling and Ericsson heat engines are internally reversible, which achieves the condition of perfect regeneration without heat leak loss, although the time of regenerative process is finite. Then the maximum power output and the corresponding thermal efficiency are given by:

$$P_{max} = \frac{K\left[\sqrt{T_H} - \sqrt{T_L}\right]^2}{\left[1 + Ka_1\left(\sqrt{T_H} - \sqrt{T_L}\right)^2\right]} \tag{6.68}$$

$$\eta_m = 1 - \sqrt{\frac{T_L}{T_H}} = \eta_{CA} \tag{6.69}$$

Thus for the above assumption, the performance of the endoreversible Stirling and Ericsson heat engines with perfect/ideal regenerator in which the time of regeneration given by Eqs. (6.28) and (6.29) is similar to the optimal performance of endoreversible Carnot cycle given by Curzon and Ahlborn (1975). However, physically for finite regenerative time, ε_R should be less than unity. This shows that in the investigation of the Stirling heat engine, it would be difficult to obtain new conclusions if the regenerator losses were not considered in the analysis.

Case 6 When $R_{K-T} = 1.0$, $Q_0 = 0$, $\varepsilon_R = 1.0$, and $t_R = 0$, the Stirling and Ericsson heat engines are internally reversible, which achieves the condition of perfect regeneration without heat leak loss, and the time of regenerative process is negligible as compared to the two isothermal processes. Then the maximum power output and the corresponding thermal efficiency are given by:

$$P_{max} = K\left[\sqrt{T_H} - \sqrt{T_L}\right]^2 \quad \text{and} \quad \eta_m = 1 - \sqrt{\frac{T_L}{T_H}} = \eta_{CA} \tag{6.70}$$

Thus it is seen from Eq. (6.30) that the maximum power output and the corresponding thermal efficiency of Stirling and Ericsson heat engines are equivalent to that of an endoreversible Carnot heat engine for the same set operating conditions in which the time of regeneration process is negligible small as compared to the two isothermal processes.

6.6 Discussion of Results

In order to have a numerical appreciation of the results, we consider the inlet source and sink temperatures $T_{H1} = 1200$ K, $T_{L1} = 300$ K, the volume ratio $v_1/v_2 = 2.0$ and $p_2/p_1 = 2.64$, the effectiveness of each of the heat exchanger (ε_H, ε_L, and ε_R) in the range 0.40–1.00, the heat capacitance rates (C_H and C_L) of source/sink fluids in the range 0.50–1.50 kW/K, and the internal irreversibility parameter $R_{K-T} = 1.33$. We have studied the effect of each of the parameters on the heat transfers, working fluid temperatures, maximum power output, and the corresponding thermal efficiency of both the heat engines.

6.6.1 *Finite Time Stirling and Ericsson Cycles*

Tables 6.1, 6.2, and 6.3 show the effects of the effectiveness (ε_H, ε_L, and ε_R) of various heat exchangers of the hot side, the cold side, and the regenerator, respectively, on the heat transfers, working fluid temperatures, maximum power output, and the corresponding thermal efficiency for the both heat engines.

Effect of $\boldsymbol{\varepsilon_H}$ Table 6.1 shows that as the effectiveness of the hot-side heat exchanger increases, the heat transfers (Q_h and Q_c) to and from the heat engines, the regenerative heat transfer (Q_R), the working fluid temperatures (T_h and T_c), the maximum power output (P_m), and the thermal efficiency of both heat engines increase. The effect of ε_H is more pronounced for the maximum power output (P_{max}) and less pronounced for the thermal efficiency in both the heat engines.

Effect of $\boldsymbol{\varepsilon_L}$ Table 6.2 shows that as the effectiveness of the cold-side heat exchanger (ε_L) increases, the heat transfers (Q_h and Q_c) to and from the heat engines, the regenerative heat transfer (Q_R), the working fluid temperatures (T_h and T_c), and the thermal efficiency of both heat engines decrease, while the maximum power output (P_{max}) increases. The effect of ε_L is more pronounced for the maximum power output (P_{max}) and less pronounced for the thermal efficiency in both the heat engines.

Effect of $\boldsymbol{\varepsilon_R}$ It is seen from Table 6.3 that as the regenerator effectiveness (ε_R) increases, the regenerative heat transfer (Q_R) and the thermal efficiency (h_m) increase, while the heat transfers (Q_h and Q_c) to and from the heat engine, the working fluid temperatures (T_h and T_c), and the maximum power output (P_{max}) remain constant. The effect of ε_R is more pronounced for the regenerative heat transfer (Q_R) and less pronounced for the thermal efficiency. It is also seen that when there is no heat loss through the regenerator ($\varepsilon_R = 1.00$), the Stirling heat engine attains the efficiency equal to the Curzon–Ahlborn efficiency (η_{CA}) of an endoreversible Carnot heat engine. Thus the effectiveness of the regenerator should be high enough for higher thermal efficiency of both heat engines. However, the Stirling and Ericsson heat engines with ideal regenerator are as efficient as the Carnot heat engine with the same operating condition (with or without internal irreversibility), but it is not practical because an ideal regeneration requires infinite time or infinite regenerative area.

Tables 6.4 and 6.5 show the effects of external heat source/sink reservoirs inlet fluids temperatures (T_{H1} and T_{L1}), on various parameters, i.e. the heat transfers (Q_h and Q_c) to and from the heat engine, the regenerative heat transfer (Q_R), the working fluid temperatures (T_h and T_c), the maximum power output (P_{max}), and the corresponding thermal efficiency (η_m) of Stirling and Ericsson heat engines.

Table 6.1 Effects of ε_H on the heat transfer, power output, working fluid temperatures, and the corresponding thermal efficiency of Stirling and Ericsson heat engine cycles

	(a) Stirling heat engine							(b) Ericsson heat engine						
	Q_h	Q_c	Q_R	P_{max}	T_h	T_c	h_m	Q_h	Q_c	Q_R	P_{max}	T_h	T_c	h_m
ε_H	kJ	kJ	kJ	kW	K	K	%	kJ	kJ	kJ	kW	K	K	%
0.40	138.00	79.65	302.14	25.16	924.95	400.40	26.84	193.27	111.56	422.54	25.18	924.92	400.42	26.99
0.50	140.11	80.87	306.76	28.30	939.06	406.50	26.88	196.22	113.26	428.99	28.33	939.03	406.52	27.02
0.60	141.83	81.86	310.54	31.01	950.62	411.49	26.92	198.63	114.65	434.28	31.04	950.59	411.51	27.05
0.70	143.29	82.70	313.74	33.39	960.38	415.70	26.97	200.67	115.83	438.74	33.43	960.35	415.73	27.08
0.75	143.94	83.07	315.16	34.49	964.74	417.58	26.98	201.59	116.35	440.74	34.53	964.71	417.61	27.09
0.80	144.55	83.42	316.49	35.52	968.81	419.34	27.00	202.43	116.84	442.60	35.56	968.77	419.37	27.11
0.85	145.11	83.75	317.74	36.50	972.62	420.98	27.02	203.23	117.30	444.34	36.55	972.58	421.02	27.12
0.90	145.65	84.06	318.91	37.44	976.20	422.53	27.04	203.98	117.73	445.98	37.49	976.16	422.56	27.13
1.00	146.63	84.62	321.06	39.19	982.77	425.36	27.08	205.35	118.52	448.98	39.24	982.73	425.40	27.16

Table 6.2 Effects of ε_L on the heat transfers, power output, working fluid, and the thermal efficiency of Stirling and Ericsson heat engines

	(a) Stirling heat engine							(b) Ericsson heat engine						
	Q_h	Q_c	Q_R	P_{max}	T_h	T_c	h_m	Q_h	Q_c	Q_R	P_{max}	T_h	T_c	h_m
ε_L	kJ	kJ	kJ	kW	K	K	%	kJ	kJ	kJ	kW	K	K	%
0.40	149.73	86.43	327.82	24.06	1003.58	434.45	27.02	209.70	121.05	458.46	24.08	1003.56	434.47	27.12
0.50	147.71	85.25	323.40	27.49	990.01	428.55	27.01	206.87	119.41	452.26	27.52	989.98	428.58	27.11
0.60	146.03	84.28	319.72	30.52	978.72	423.66	27.00	204.51	118.04	447.12	30.55	978.69	423.68	27.10
0.70	144.59	83.45	316.58	33.23	969.08	419.47	26.99	202.49	116.88	442.72	33.27	969.05	419.50	27.10
0.75	143.94	83.07	315.16	34.49	964.74	417.58	26.98	201.59	116.35	440.74	34.53	964.71	417.61	27.09
0.80	143.33	82.72	313.84	35.68	960.67	415.82	26.98	200.73	115.86	438.88	35.73	960.64	415.85	27.09
0.85	142.76	82.39	312.59	36.83	956.84	414.16	26.97	199.93	115.4	437.13	36.88	956.81	414.19	27.08
0.90	142.22	82.08	311.41	37.93	953.23	412.59	26.97	199.18	114.96	435.48	37.98	953.19	412.62	27.08
1.00	141.23	81.50	309.24	40.00	946.57	409.69	26.96	197.79	114.15	432.44	40.05	946.52	409.72	27.07

Table 6.3 Effects of ε_R on the heat transfers, power output, working fluid, and the thermal efficiency of Stirling and Ericsson heat engines

	(a) Stirling heat engine							(b) Ericsson heat engine						
	Q_h	Q_c	Q_R	P_{max}	T_h	T_c	h_m	Q_h	Q_c	Q_R	P_{max}	T_h	T_c	h_m
ε_R	K	kJ	kJ	kW	K	K	%	kJ	kJ	kJ	kW	K	K	%
0.40	143.94	83.07	157.58	34.49	964.74	417.58	15.89	201.59	116.35	220.37	34.53	964.71	417.61	15.93
0.50	143.94	83.07	196.98	34.49	964.74	417.58	17.71	201.59	116.35	275.46	34.53	964.71	417.61	17.76
0.60	143.94	83.07	236.37	34.49	964.74	417.58	20	201.59	116.35	330.55	34.53	964.71	417.61	20.06
0.70	143.94	83.07	275.77	34.49	964.74	417.58	22.97	201.59	116.35	385.64	34.53	964.71	417.61	23.05
0.75	143.94	83.07	295.46	34.49	964.74	417.58	24.82	201.59	116.35	413.19	34.53	964.71	417.61	24.91
0.80	143.94	83.07	315.16	34.49	964.74	417.58	26.98	201.59	116.35	440.74	34.53	964.71	417.61	27.09
0.85	143.94	83.07	334.86	34.49	964.74	417.58	29.57	201.59	116.35	468.28	34.53	964.71	417.61	29.69
0.90	143.94	83.07	354.56	34.49	964.74	417.58	32.7	201.59	116.35	495.83	34.53	964.71	417.61	32.84
1.00	143.94	83.07	393.95	34.49	964.74	417.58	41.47	201.59	116.35	550.92	34.53	964.71	417.61	41.69

Table 6.4 Effects of T_{H1} on the heat transfers, power output, working fluid, and the thermal efficiency of Stirling and Ericsson heat engines

	(a) Stirling heat engine							(b) Ericsson heat engine						
T_{H1}	Q_h	Q_c	Q_R	P_{max}	T_h	T_c	h_m	Q_h	Q_c	Q_R	P_{max}	T_h	T_c	h_m
K	kJ	kJ	kJ	kW	K	K	%	kJ	kJ	kJ	kW	K	K	%
1000	123.76	78.25	251.21	21.76	829.49	393.37	24.11	173.33	109.60	351.31	21.78	829.47	393.39	24.20
1050	128.84	79.50	267.21	24.78	863.55	399.63	24.92	180.44	111.35	373.69	24.80	863.52	399.66	25.01
1100	133.90	80.72	283.21	27.91	897.43	405.75	25.66	187.52	113.06	396.06	27.93	897.41	405.78	25.76
1150	138.93	81.91	299.19	31.15	931.16	411.73	26.35	194.57	114.72	418.40	31.18	931.13	411.76	26.45
1200	143.94	83.07	315.16	34.49	964.74	417.58	26.98	201.59	116.35	440.74	34.53	964.71	417.61	27.09
1250	148.93	84.21	331.12	37.92	998.18	423.31	27.57	208.57	117.95	463.05	37.97	998.14	423.35	27.69
1300	153.90	85.33	347.08	41.44	1031.49	428.93	28.12	215.53	119.52	485.36	41.50	1031.45	428.97	28.24
1350	158.85	86.42	363.02	45.05	1064.68	434.43	28.64	222.47	121.05	507.65	45.11	1064.63	434.48	28.76
1400	163.78	87.50	378.96	48.73	1097.75	439.84	29.12	229.37	122.56	529.93	48.81	1097.69	439.89	29.24
1450	168.70	88.55	394.88	52.48	1130.71	445.15	29.57	236.26	124.04	552.20	52.57	1130.65	445.20	29.69
1500	173.60	89.59	410.80	56.31	1163.57	450.36	30.00	243.13	125.49	574.46	56.41	1163.50	450.42	30.12

Table 6.5 Effects of T_{L1} on the heat transfers, power output, working fluid, and the thermal efficiency of Stirling and Ericsson heat engines

	(a) Stirling heat engine							(b) Ericsson heat engine						
T_{L1}	Q_h	Q_c	Q_R	P_{max}	T_h	T_c	h_m	Q_h	Q_c	Q_R	P_{max}	T_h	T_c	h_m
K	kJ	kJ	kJ	kW	K	K	%	kJ	kJ	kJ	kW	K	K	%
290	143.13	81.22	317.42	36.08	959.34	408.26	27.48	200.46	113.76	443.90	36.13	959.30	408.29	27.59
295	143.54	82.15	316.29	35.28	962.05	412.93	27.23	201.02	115.06	442.32	35.32	962.02	412.96	27.34
300	143.94	83.07	315.16	34.49	964.74	417.58	26.98	201.59	116.35	440.74	34.53	964.71	417.61	27.09
305	144.34	83.99	314.03	33.71	967.41	422.22	26.74	202.14	117.64	439.15	33.75	967.37	422.25	26.85
310	144.73	84.91	312.90	32.95	970.05	426.83	26.5	202.70	118.93	437.57	32.99	970.02	426.86	26.60
315	145.12	85.83	311.76	32.20	972.68	431.43	26.26	203.24	120.21	435.98	32.24	972.64	431.46	26.36
320	145.51	86.74	310.62	31.47	975.28	436.00	26.01	203.79	121.48	434.39	31.51	975.25	436.03	26.11
325	145.90	87.64	309.48	30.76	977.86	440.56	25.77	204.33	122.76	432.8	30.79	977.83	440.59	25.87
330	146.28	88.55	308.34	30.05	980.42	445.11	25.53	204.86	124.02	431.2	30.08	980.39	445.14	25.63

Effect of T_{H1} It is seen from Table 6.4 that as the external heat source fluid inlet temperature (T_{H1}) increases, the heat transfers (Q_h and Q_c) to and from the heat engines, the regenerative heat transfer (Q_R), the working fluid temperatures (T_h and T_c), the maximum power output (P_{max}), and the thermal efficiency (η_m) of both engines increase. The effect of T_{H1} is more pronounced for the maximum power output (P_{max}) and less pronounced for the working fluid temperature (T_c) on sink side. Thus it is desirable to have a high-temperature heat source from the point of view of higher power output as well as higher thermal efficiency for both heat engines.

Effect of T_{L1} It is seen from Table 6.5 that as the heat sink working fluid inlet temperature (T_{L1}) increases, the heat transfers (Q_h and Q_c) to and from the heat engine and the sink side working fluid temperature (T_c) increase, while the regenerative heat transfer (Q_R), the maximum power output (P_{max}), and the corresponding thermal efficiency (h_m) of both heat engines decrease. The effect of T_{L1} is more pronounced for the maximum power output (P_{max}) and less pronounced for the source side working fluid temperature (T_h) of the both heat engines. Thus it is desirable to have a low-temperature heat sink from the point of view of higher power output (P_{max}) as well as higher thermal efficiency (η_m) for both the heat engines.

Tables 6.6 and 6.7 show the effects of the heat capacitance rates (C_H and C_L) of the source/sink reservoirs on the heat transfers (Q_h and Q_c) to and from the heat engines, the regenerative heat transfer (Q_R), the working fluid temperatures (T_h and T_c), the maximum power output (P_{max}), and the thermal efficiency (η_m) of the Stirling and Ericsson heat engines.

Effect of C_H Table 6.6 (a and b) shows that as the heat capacitance rate (C_H) of the heat source reservoir increases, the heat transfers (Q_h and Q_c) to and from the heat engines, the regenerative heat transfer (Q_R), the working fluid temperatures (T_h and T_c), the maximum power output (P_{max}), and the thermal efficiency of both the heat engines increase. The effect of C_H is more pronounced for the maximum power output (P_{max}) and less pronounced for the thermal efficiency in both engines.

Effect of C_L Table 6.7 shows that as the heat capacitance rate of the heat sink (C_L) reservoir increases, the heat transfers (Q_h and Q_c) to and from the heat engines, the regenerative heat transfer (Q_R), the working fluid temperatures (T_h and T_c), and the thermal efficiency decrease, while the maximum power output (P_{max}) increases. The effect of C_L is more pronounced for the maximum power (P_{max}) output and less pronounced for the thermal efficiency in both engines.

It can be further seen from Tables 6.6 and 6.7 that by increasing C_H, all the performance parameters, viz. Q_H, Q_L, P, η, etc. increase, while it is found to be reversed in the case of C_L for Q_H, Q_L, and η, whereas the power output enhances more in the case of C_L as compared to that of C_H. This shows that by increasing C_L, we have to supply less heat to the engine to get more power than the former case, i.e. when C_L is less than C_H. Although the thermal efficiency slightly decreases by increasing C_L, as power output is the prime object of any heat engine rather than the

Table 6.6 Effects of C_H on the heat transfers, power output, working fluid, and the thermal efficiency of Stirling and Ericsson heat engines

	(a) Stirling heat engine							(b) Ericsson heat engine						
C_H	Q_h	Q_c	Q_R	P_{max}	T_h	T_c	h_m	Q_h	Q_c	Q_R	P_{max}	T_h	T_c	h_m
KW/K	kJ	kJ	kJ	kW	K	K	%	kJ	kJ	kJ	kW	K	K	%
0.40	135.32	78.11	296.26	21.42	906.98	392.64	26.96	189.52	109.4	414.32	21.44	906.95	392.65	27.07
0.50	137.40	79.31	300.81	24.29	920.90	398.65	26.96	192.42	111.07	420.68	24.31	920.87	398.67	27.08
0.60	139.11	80.29	304.57	26.79	932.39	403.61	26.97	194.83	112.46	425.94	26.82	932.36	403.64	27.08
0.70	140.57	81.13	307.77	29.01	942.15	407.83	26.97	196.87	113.63	430.40	29.04	942.12	407.85	27.08
0.80	141.83	81.86	310.54	31.01	950.62	411.49	26.98	198.63	114.65	434.28	31.04	950.59	411.51	27.09
0.90	142.95	82.50	312.98	32.82	958.08	414.71	26.98	200.19	115.55	437.69	32.86	958.05	414.74	27.09
1.00	143.94	83.07	315.16	34.49	964.74	417.58	26.98	201.59	116.35	440.74	34.53	964.71	417.61	27.09
1.10	144.83	83.59	317.13	36.02	970.74	420.17	26.99	202.84	117.08	443.48	36.06	970.71	420.21	27.09
1.20	145.65	84.06	318.91	37.44	976.20	422.53	26.99	203.98	117.73	445.98	37.49	976.16	422.56	27.10
1.30	146.39	84.48	320.55	38.77	981.19	424.68	26.99	205.02	118.33	448.26	38.82	981.16	424.72	27.10
1.40	147.08	84.88	322.05	40.01	985.79	426.67	26.99	205.98	118.89	450.37	40.06	985.76	426.71	27.10
1.50	147.72	85.24	323.45	41.17	990.05	428.51	27.00	206.87	119.40	452.32	41.23	990.01	428.55	27.10
1.60	148.31	85.58	324.75	42.27	994.01	430.22	27.00	207.70	119.88	454.13	42.33	993.97	430.26	27.10
1.70	148.86	85.90	325.96	43.31	997.71	431.81	27.00	208.47	120.32	455.82	43.37	997.67	431.86	27.10

Table 6.7 Effects of C_L on the heat transfers, power output, working fluid, and the thermal efficiency of Stirling and Ericsson heat engines

	(a) Stirling heat engine							(b) Ericsson heat engine						
C_L	Q_h	Q_c	Q_R	P_{max}	T_h	T_c	h_m	Q_h	Q_c	Q_R	P_{max}	T_h	T_c	h_m
KW/K	kJ	kJ	kJ	kW	K	K	%	kJ	kJ	kJ	kW	K	K	%
0.40	152.27	87.89	333.36	20.09	1020.57	441.82	27.00	213.26	123.10	466.21	20.10	1020.55	441.84	27.10
0.50	150.31	86.76	329.08	23.13	1007.44	436.12	26.99	210.51	121.52	460.22	23.14	1007.42	436.15	27.10
0.60	148.67	85.81	325.50	25.84	996.45	431.35	26.99	208.21	120.19	455.21	25.86	996.43	431.38	27.10
0.70	147.26	84.99	322.42	28.28	987.00	427.25	26.99	206.24	119.05	450.89	28.31	986.98	427.28	27.10
0.80	146.03	84.28	319.72	30.52	978.72	423.66	26.99	204.51	118.04	447.12	30.55	978.69	423.68	27.09
0.90	144.93	83.64	317.32	32.58	971.36	420.46	26.99	202.97	117.15	443.76	32.61	971.33	420.49	27.09
1.00	143.94	83.07	315.16	34.49	964.74	417.58	26.98	201.59	116.35	440.74	34.53	964.71	417.61	27.09
1.10	143.04	82.55	313.20	36.26	958.73	414.97	26.98	200.33	115.63	437.99	36.31	958.69	415.01	27.09
1.20	142.22	82.08	311.41	37.93	953.23	412.59	26.98	199.18	114.96	435.48	37.98	953.19	412.62	27.09
1.30	141.47	81.64	309.76	39.50	948.17	410.39	26.98	198.12	114.35	433.17	39.55	948.12	410.42	27.09
1.40	140.77	81.23	308.23	40.98	943.48	408.35	26.98	197.14	113.78	431.03	41.03	943.43	408.38	27.09
1.50	140.12	80.86	306.81	42.38	939.11	406.46	26.98	196.23	113.25	429.04	42.44	939.07	406.49	27.09
1.60	139.51	80.51	305.48	43.71	935.04	404.68	26.98	195.37	112.76	427.18	43.77	934.99	404.72	27.09
1.70	138.94	80.17	304.24	44.97	931.21	403.02	26.98	194.58	112.30	425.44	45.04	931.16	403.06	27.09

thermal efficiency because the thermal efficiency of any ideal heat engine is always more than the actual, the power output of every ideal heat engine is zero. Therefore, it is desirable to have higher C_L rather than C_H from the point of view of lower heat transfers to and from the heat engine as well as higher power output. Alternatively, this can also be explained in terms of irreversibilities produced on source/sink sides of the engine. As C_H increases, total irreversibilities increase, while as C_L increases, total irreversibilities decrease. However, internal irreversibility due to regenerative heat loss is not affected by C_H or C_L. On one hand, external irreversibility on source side decreases while on sink side increases as C_H increases. On the other hand, external irreversibility on source side increases while on sink side decreases as C_L increases. Thus, it is desirable to have higher C_L and lower C_H rather than lower C_L and higher C_H. This particular case is also found to be similar to that of the Brayton cycle discussed in earlier chapter.

6.6.2 Irreversible Stirling and Ericsson Cycles

Most of the parameters for this particular model are similar to those discussed above, while the internal irreversibility parameter being the newly introduced parameter is discussed in this particular section as below, and the readers are encouraged to reproduce the results and compare the two. It can also be noted that the results shall differ quantitatively between the two said cases due to the presence and characteristic of the internal irreversibility parameter in the latter case.

Effect of R_{K-T} Table 6.8 shows that as the internal irreversibility parameter increases (i.e. the internal irreversibility of the cycle decreases), all the parameters of both the engines [viz. the heat transfers (Q_h and Q_c) to and from the heat engines, the regenerative heat transfer (Q_R), the working fluid temperatures (T_h and T_c), the maximum power output (P_{max}), and the thermal efficiency] increase. The effect of R_{K-T} is more pronounced for the maximum power output (P_{max}) and less pronounced for the heat transfer from the heat engines in both cases. It is also found that the effect of internal irreversibility parameter is more pronounced than the other parameters mentioned above. Thus it is desirable to have lower value of R_{K-T}, i.e. lesser entropy generation in these engines, from the point of view of maximum power output as well as for higher thermal efficiency in both the cycles.

Although the thermal efficiency given by Eq. (6.27) is much lesser than the Curzon–Ahlborn, they did not consider any kind of internal irreversibility which can be obtained by substituting values of different parameters. But in the present case, it is observed that the thermal efficiency of a real heat engine is always lesser than the Carnot and the Curzon and Ahlborn (1975) efficiency because there are more irreversibilities in the cycle which were not yet considered. The other causes of irreversibilities in a real cycle are due to various losses such as, pressure loss, friction, heat loss, etc. in the turbine, compressor, regenerator, the connecting pipes, and the direct heat leak loss from the source to the sink. Thus the thermal efficiency

Table 6.8 Effects of R_{K-T} on the heat transfers, power output, working fluid, and the thermal efficiency of Stirling and Ericsson heat engines

	(a) Stirling heat engine							(b) Ericsson heat engine						
	Q_h	Q_c	Q_R	P_{max}	T_h	T_c	η_m	Q_h	Q_c	Q_R	P_{max}	T_h	T_c	η_m
$R_{K\text{-}T}$	kJ	kJ	kJ	kW	K	K	%	kJ	kJ	kJ	kW	K	K	%
2.00	104.89	74.14	392.77	13.20	1054.57	372.68	14.93	146.90	103.84	549.27	13.21	1054.53	372.71	15.00
1.82	113.10	76.22	374.73	17.25	1033.68	383.12	17.59	158.39	106.75	524.03	17.27	1033.64	383.15	17.67
1.67	121.09	78.12	358.13	21.46	1014.48	392.72	20.13	169.58	109.43	500.82	21.49	1014.43	392.75	20.21
1.54	128.88	79.89	342.78	25.76	996.71	401.60	22.53	180.49	111.90	479.36	25.79	996.66	401.63	22.63
1.43	136.49	81.54	328.50	30.11	980.18	409.86	24.82	191.16	114.20	459.39	30.15	980.14	409.90	24.92
1.33	143.94	83.07	315.16	34.49	964.74	417.58	26.98	201.59	116.35	440.74	34.53	964.71	417.61	27.09
1.25	151.23	84.51	302.65	38.86	950.26	424.83	29.04	211.80	118.37	423.24	38.90	950.23	424.86	29.16
1.18	158.38	85.87	290.86	43.21	936.63	431.65	31.00	221.81	120.27	406.77	43.26	936.60	431.67	31.11
1.11	165.39	87.15	279.74	47.53	923.76	438.10	32.85	231.63	122.06	391.22	47.58	923.74	438.11	32.98
1.05	172.27	88.37	269.20	51.81	911.57	444.20	34.61	241.27	123.76	376.50	51.86	911.56	444.21	34.74
1.00	179.04	89.52	259.20	56.04	900.00	450.00	36.29	250.75	125.38	362.52	56.10	900.00	450.00	36.42

(η_m) at maximum power output of any real heat engine can be expressed (theoretically), in terms of source and sink temperatures along with the irreversibility parameter (R_{K-T}) defined by Kaushik and Tyagi (2002). Thus the bound for the efficiency of a real heat engine cycle, having finite heat capacity of external reservoirs, is given by Eq. (3.59) as below:

$$\eta_m = 1 - \sqrt{\frac{R_{K-T}T_{L1}}{T_{H1}}} \quad \text{i.e} \quad \eta_{actual} < \eta_{opt} < \eta_{C-A} < \eta_{Carnot} \tag{6.71}$$

where η_{Carnot} and η_{C-A} are, respectively, the Carnot and Curzon–Ahlborn efficiency and R_{K-T} is the irreversibility parameter, always greater than unity for real heat engines. It is important to note that the present cycle model presented here is more general and the results obtained by earlier workers available in the literature [Curzon and Ahlborn (1975), Badescu (1992), Blank et al. (1994), Ladas and Ibrahim (1994), Blank and Wu (1995, 1996 a, b), Trukhow et al. (1997), Erbay and Yavuz (1997), Chen (1997a, b), Chen et al. (1998) and Senft (1998)] can be drawn directly from this particular model as a special case.

The results obtained in this study are useful to understand the deviation of the actual thermal efficiency from the ideal efficiency due to evolution of the external and internal irreversibilities associated with the system. The external irreversibility is due to the heat transfer rates to and from the cycle, and the internal irreversibility is due to other losses within the cycle, such as pressure loss, friction, heat loss in the regenerator, and so on. The heat capacitance rates (C_H, C_L, and C_w) also play important roles for both the maximum power output and the corresponding thermal efficiency and found to be bounded as $C_L > C_H$ for the better performance of these heat engine cycles.

6.7 Conclusion

In the present chapter, the realistic Stirling and Ericsson cycle models including external and internal irreversibilities have been studied in detail. Also the maximum power output and the corresponding thermal efficiency for different cases are evaluated at different operating conditions and compared with the present cycle models. The effects of different parameters on the maximum power output and the corresponding thermal efficiency and other performance parameters, such as the heat transfer to and from the cycle, the regenerative heat transfer, state point temperature, etc., have been evaluated and presented in the tabular form. Some special cases were also presented in detail and compared with the present cycle model. It is found that the present model is more general, because the cycle models discussed by different authors can be derived as the special case of this particular cycle model. The general trend of this case study exhibits that the bound set by Kaushik and Tyagi (2002) for Carnot and Brayton cycles is also applicable to these cycle models.

Chapter 7
Finite Time Thermodynamics of Vapour Compression Refrigeration, Airconditioning and Heat Pump Cycles

7.1 Introduction

The experimental observations show that heat cannot be transferred from low-temperature reservoir to high-temperature reservoir without external energy input or work. The device that extracts heat from low-temperature reservoir and rejects or transfers it to high-temperature reservoir is called refrigerator (or heat pump), and the cycle followed by the device or system is called refrigeration cycle. These devices are cyclic devices, and the fluid that flows in the device is known as refrigerant. Refrigerator and heat pump are essentially the same device and follow the same thermodynamic cycle, but they differ in their objective function. Removal of heat from the space which needs to be cooled and maintained at low temperature is the main objective of refrigerator, and rejecting this heat to a higher-temperature medium is a necessity of the operation, not the objective. However, producing heating effect is the objective of heat pump, and extracting heat from low-temperature medium is only a necessary part of operation in such system. The refrigeration device may be used for both heating and cooling purposes. In other words, the same device can be used either:

- To absorb heat from a cold body (space to be cooled) at low temperature T_L and to reject it to high-temperature surroundings at temperature $T_H \geq T_a$
- To absorb heat from surroundings at temperature $T_L \leq T_a$ and reject it to a hot reservoir (space to be heated) at temperature T_H

where T_a is the temperature of the surroundings. Figure 7.1 illustrates the comparison between refrigeration and heat pump cycle. It implies that the same machine can be used for both cooling and heating purpose. If the machine is used for cooling, it is called refrigerator, while it is called heat pump if used for heating. The main difference between a refrigerator and a heat pump is the operating temperature. Another essential difference is in their useful objective function, viz. in a refrigerator, the heat exchanger that absorbs heat is connected to the space which needs to

S.C. Kaushik et al., *Finite Time Thermodynamics of Power and Refrigeration Cycles*,
https://doi.org/10.1007/978-3-319-62812-7_7

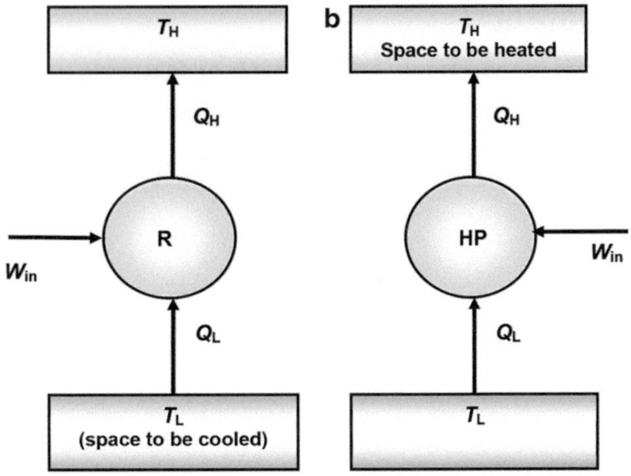

Fig. 7.1 Line diagram of refrigeration (R) and heat pump (HP) cycle

be cooled, while in a heat pump, the heat exchanger that rejects heat is connected to the space which needs to be heated. The other heat exchanger in each case is connected to the surroundings. In other words, a refrigerator that is used for cooling in summer may be used as a heat pump for heating in winter; it may be done either:

- By rotating the machine by 180° to interchange the positions of the two heat exchangers between the space and surroundings
- By exchanging the functions of the two heat exchangers by the operation of valves, for example, a four-way valve in a window airconditioner

In a refrigerator, the objective function is the cooling effect, while in a heat pump, the objective function is the heating effect. The performance of such machine is expressed in terms of coefficient of performance (COP) and defined as:

$$\mathrm{COP_R} = \frac{\text{Desired Output (cooling effect)}}{\text{Required Input (work input)}} = \frac{Q_\mathrm{L}}{W_\mathrm{in}} \tag{7.1}$$

$$\mathrm{COP_{HP}} = \frac{\text{Desired Output (heating effect)}}{\text{Required Input (work input)}} = \frac{Q_\mathrm{H}}{W_\mathrm{in}} \tag{7.2}$$

There are several types of refrigeration cycles; some of the main cycles are as below:

- The reverse Carnot cycle
- Vapour compression refrigeration cycle
- Cascade and multistage refrigeration cycle
- Vapour absorption refrigeration cycle
- Gas refrigeration cycles (Brayton, Stirling, and Eriksson)

In this chapter the main emphasis will be on vapour compression refrigeration cycle along with its detailed finite time thermodynamic analysis.

7.2 The Reverse Carnot Cycle

A French engineer, Sadi Carnot (1824), showed that no heat engine working between the two temperature levels can perform better than the Carnot engine. The efficiency at which the largest possible heat can be converted to work is Carnot efficiency. This Carnot heat engine follows the Carnot cycle which is totally a reversible cycle that consists of two reversible isothermal and two isentropic processes. It has the maximum thermal efficiency for given temperature limits, and it also serves as standard against which actual power cycle can be compared. Since it is a reversible cycle, all four processes that comprise the Carnot cycle can be reversed and called as the reversed Carnot cycle used for cooling and heating applications. This is also called the Carnot refrigeration and/or heat pump cycle.

Also reversing the cycle does reverse the directions of heat and work interactions. The result is a cycle that operates in the counterclockwise direction as shown on the schematic and *T–s* diagram of Figs. 7.2a and 7.2b, respectively. The refrigerator or heat pump that operates on the reversed Carnot cycle is also called the Carnot refrigerator or the heat pump. The reversed Carnot cycle is executed within the saturation dome of a refrigerant as can be seen clearly from Fig. 7.2b. The refrigerant absorbs heat Q_L during process 1–2 at constant temperature T_L from a low-temperature source and is compressed isentropically to state point 3, and as a result its temperature rises to T_H. The refrigerant rejects heat Q_H at constant temperature T_H to a high-temperature sink during process 3–4 and then converts into saturated liquid and expands isentropically to state point 1 and cools down to low temperature T_L, thereby completing the cycle. The coefficient of performance of Carnot refrigerator and heat pump can be expressed in terms of temperature as below (Cengel and Boles 2006):

$$\mathrm{COP_R} = \frac{Q_L}{W_{in}} = \frac{T_L}{T_H - T_L} = \frac{1}{\frac{T_H}{T_L} - 1} \tag{7.3}$$

$$\mathrm{COP_{HP}} = \frac{Q_H}{W_{in}} = \frac{T_H}{T_H - T_L} = \frac{1}{1 - \frac{T_L}{T_H}} \tag{7.4}$$

In reversed Carnot cycle, temperature difference between the refrigerant and surrounding is zero; hence, heat transfer is zero. So to get a non-zero heat transfer, either heat transfer time is infinitely long or heat exchanger area is infinitely large, but both conditions are impractical.

The reversed Carnot cycle is an ideal cycle and all associated processes are reversible. Reversible processes are defined only in the limit of infinitely slow execution. The rate of heat exchange between the system and the environment is

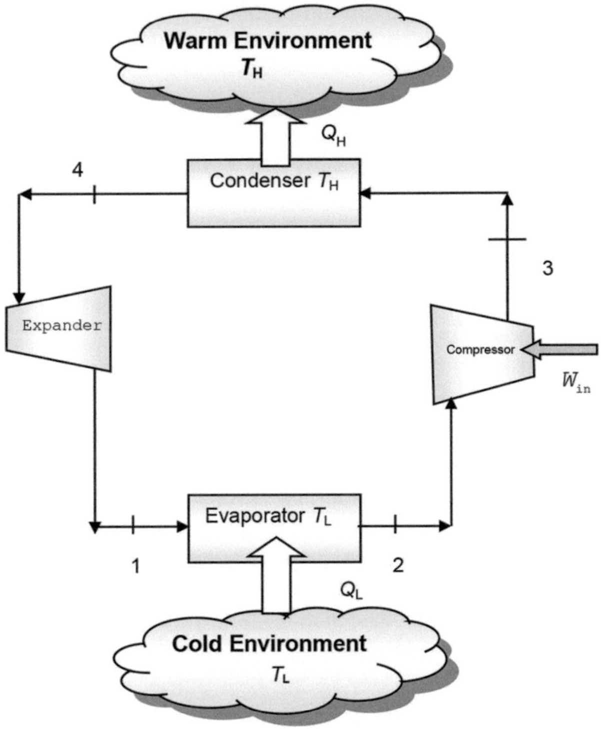

Fig. 7.2a Schematic diagram of the reversed Carnot cycle (Cengel and Boles 2006)

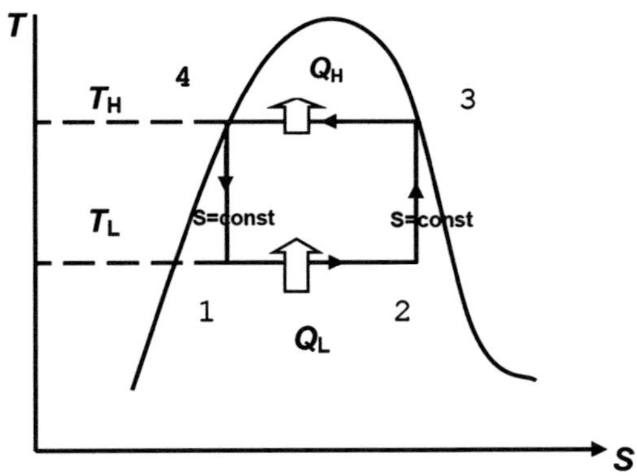

Fig. 7.2b *T–s* diagram of the reversed Carnot cycle

infinitesimally small, and process duration is infinitely long. A consequence of this is zero cooling load of Carnot refrigerator (or zero heating load of Carnot heat pump) for the duration average:

- The main purpose of producing cooling load/heating load is defeated in ideal Carnot system.
- Requirement of infinitely large heat exchange area, making the system uneconomical.

- There are other technical difficulties in this cycle. Reversible isentropic compression/expansion processes 2–3 and 4–1 cannot be approximated closely in practice. This is because process 2–3 requires compression of liquid–vapour mixture to be compressed which requires a compressor that handles two phases. Process 4–1 involves the expansion of high moisture content refrigerant and requires a turbine that can expand the saturated liquid, and either of the two processes is difficult to obtain in the real practice. So there is a need to execute the cycle out of the saturation curve limit, but then the cycle cannot maintain the two isothermal processes 1–2 and 3–4. Therefore, the reversed Carnot cycle is not a practical cycle, while it can serve as the standard against which a real refrigeration cycle may be compared.

7.3 Vapour Compression Cycle

The reversed Carnot cycle is not a practical cycle due to the impracticalities associated with it, like compression of mixture in the compressor and expansion of saturated liquid in the turbine. However, some of them can be eliminated by completely vapourizing the refrigerant before it is being compressed in the compressor and replacing the turbine by a throttling device such as expansion valve and capillary tube. The cycle with these modifications becomes a practical cycle called vapour compression refrigeration cycle (VCR). Since, it is assumed that there is no irreversibility in any process, so it is named ideal vapour compression refrigeration cycle as shown in Fig. 7.3a, b on the schematic and *T–s* diagrams, respectively. The vapour compression cycle is the most widely used refrigeration cycle for both heating and cooling application. The basic processes followed by a VCR cycle are as follows:

Process 1–2: Isobaric heat extraction from the space to be cooled
Process 2–3: Isentropic compression in a compressor
Process 3–4: Isobaric heat rejection to the surroundings
Process 4–1: Isenthalpic expansion in the expansion valve

In the ideal VCR cycle, the saturated vapour enters the compressor at state point 2 as shown in Fig. 7.3 and is compressed to state point 3 isentropically at compressor discharge pressure, and the temperature of the refrigerant also increased. The high-temperature high-pressure (superheated) refrigerant vapour enters the condenser at state point 3. In condenser, heat is rejected isobarically to the surrounding and cools down to saturated liquid to state point 4 where it expands to state point 1 while the enthalpy of the liquid remains constant (isenthalpic process). Again the refrigerant enters the evaporator where it absorbs heat isobarically from the place to be cooled and converts into saturated vapour, thereby completing the cycle. All the four components associated with a VCR cycle are steady-state flow devices; hence, all the four components that make up the cycle can be analysed as steady-state processes. The changes in the kinetic and potential energy are very small as

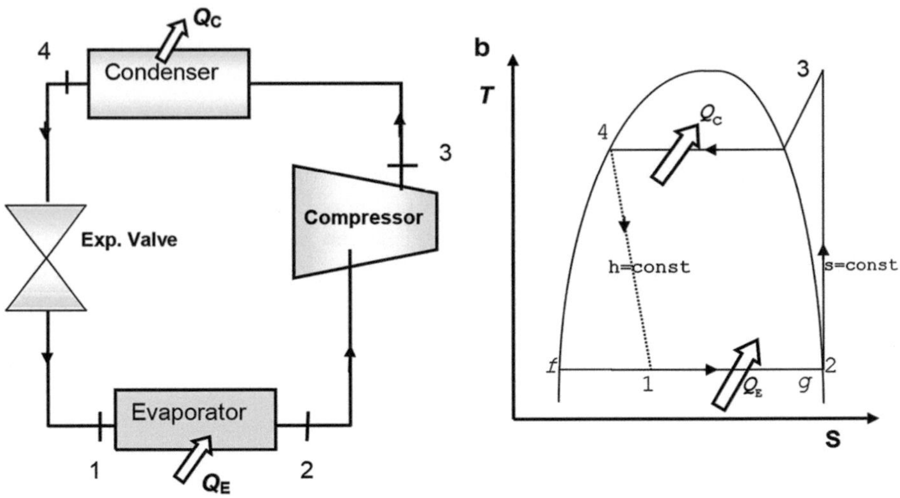

Fig. 7.3 (**a**) Schematic and (**b**) T–s diagrams of a basic VCR cycle

compared to the heat transfer and work terms and hence neglected while evaluating a VCR cycle. Thus, for steady-state flow, energy equation can be written as:

$$\dot{Q}_{\mathrm{E}} = \dot{m}\,(h_2 - h_1) \tag{7.5}$$

$$\dot{Q}_{\mathrm{C}} = \dot{m}\,(h_3 - h_4) \tag{7.6}$$

$$\dot{w}_{\mathrm{in}} = \dot{m}\,(h_3 - h_2) \tag{7.7}$$

$$\mathrm{COP}_{\mathrm{R}} = \frac{\dot{Q}_{\mathrm{E}}}{\dot{w}_{\mathrm{in}}} = \frac{\dot{m}\,(h_2 - h_1)}{\dot{m}\,(h_3 - h_2)} = \frac{(h_2 - h_1)}{(h_3 - h_2)} \tag{7.8}$$

$$\mathrm{COP}_{\mathrm{HP}} = \frac{\dot{Q}_{\mathrm{C}}}{\dot{w}_{\mathrm{in}}} = \frac{\dot{m}\,(h_3 - h_4)}{\dot{m}\,(h_3 - h_2)} = \frac{(h_3 - h_4)}{(h_3 - h_2)} \tag{7.9}$$

where $\dot{m}$ is the mass flow rate of the refrigerant, h_1 and h_2 are the specific enthalpy at state points 1 and 2 which are located at low pressure line and h_3 and h_4 are the specific enthalpy values at state points 3 and 4, located at high pressure lines.

7.4 Finite Time Vapour Compression Cycle

Similar to Carnot heat engine, Carnot cooling/heating system is defined as reversed Carnot cycle that have the highest COP bound working between two temperature levels. The cooling/heating system, which follows reversed Carnot cycle, is an ideal system, all associated process will be reversible, and system will be in thermodynamic equilibrium. In the equilibrium position, temperature of the system working fluid will be the same temperature as of the thermal reservoirs. Thermodynamic description of these equilibrium states and processes is studied in classical thermodynamics. In ideal cooling/heating system, the isothermal heating and cooling processes must be carried out infinitely slowly so that the working substance is in thermal equilibrium with the heat source and sink.

In reversible system, processes are reversible and infinite time executable processes. Although, the performance parameter COP of these ideal systems is highest but because time taken for transferring the heat Q_C or Q_E will be infinite, our desired effects (refrigeration load and heating load) will be zero. Refrigeration load = Refrigerating effect produced over the cycle duration average = Q_E/Time = 0 and Heating load = Heating effect produced over the cycle duration average = Q_C/Time = 0.

In real system, there is some finite temperature difference between the system and surrounding. Considering this finite temperature difference, thermodynamic performance parameters/limits need to be established with finite time and/or finite size thermodynamic processes which are characterized by finite rate of heat exchange between system and environment. Thermodynamic descriptions of these real processes which occur in finite time are studied in finite time thermodynamics (FTT).

The concept of finite time thermodynamics has been applied for refrigeration, airconditioning, and heat pump systems where cooling/heating effect is obtained at the expense of work input. It is desirable to have the optimal coefficient of performance (COP) at minimum power input to the systems. Although, COP is highest in case of Carnot systems but output cooling/heating load is zero because reversible processes in Carnot system are infinite time executable processes. However, practical processes are always irreversible; consequently, like heat engines, the performance of these systems must then be re-examined by means of finite time thermodynamics (Kaushik et al. 2002d). Leff and Teeters (1978) and Wu (1993a, b, c, d) have noted that the straightforward Curzon and Ahlborn (1975) analysis will not be applicable as such for a heat pump/airconditioning system because there is no 'natural maximum' in these systems (as discussed in Appendix 1).

The *R*/AC/HP system operates in cyclic mode with fixed time t_{cycle} allotted for each cycle. Thus, after time t_{cycle} has elapsed, the working fluid returns to the initial state. The following two cases are considered for theoretical analysis:

1. Alternatively connected cycle to thermal reservoirs.

   For the sake of simplicity of theoretical analysis, in most of the cases, it is assumed that system is alternatively connected to the thermal reservoirs (through external heating and cooling fluids). The system operates in cyclic mode with fixed time t_{cycle} allotted for each cycle. Thus, after time t_{cycle} has elapsed, the working fluid returns to the initial stage. The equivalent Carnot cycle consists of two isothermal heat addition/rejection and two isentropic expansion/compression processes. In alternatively connected case, cycle time can be taken $t_{cycle} = t_E + t_C$, where t_E and t_C are, respectively, the time for isothermal heat extraction and rejection, and time required for isentropic expansion and compression is assumed to be negligibly small relative to the time of two isothermal heat addition/rejection processes (Lee and Kim 1992; Kumar 2002).

2. Continuously connected cycle to thermal reservoirs.

   A practical system operates in continuously connected case. In this case, time taken for heat extraction is the same as the time required for heat rejection and which is equal to the cycle time, i.e. $t_{\text{cycle}} = t_{\text{E}} = t_{\text{C}}$.

Finite time thermodynamic analysis for each of both cases is given in the following sections.

7.4.1 Alternatively Connected Cycle to Thermal Reservoirs

There must be an optimal temperature difference between the working fluid and the reservoirs with which it is in contact for a finite amount of heat to be transferred in a finite time. For these systems, we have the reversed Carnot cycle in which T_{C} and T_{E} are the absolute temperatures of the given heat reservoirs, and we wish to find optimal internal working fluid temperature $(T_{\text{c}})_{\text{o}}$ and $(T_{\text{e}})_{\text{o}}$ for maximum COP. There is no natural maximum as we adjust T_{c} and T_{e}, but we can, of course, always ask for such a maximum, for example, about the minimum input power for given heating load or cooling load (and maximum output heating load or cooling load for a given power input). Therefore, for optimizing these systems, Lagrangian undetermined multiplier method (as discussed in Appendix 2) is used.

Infinite Heat Capacity

Blanchard (1980) has studied the case when heat pump cycle is connected to the thermal reservoirs (heat source/sink) of infinite heat capacity as shown schematically in Fig. 7.4a and its *T–s* diagram in Fig. 7.4b. He applied the concept of finite time thermodynamics and gave a formula for optimal coefficient of performance at minimum input power assuming that system is externally irreversible but internally reversible, i.e. endoreversible.

The mathematical derivation is given below with the following assumptions:

1. The cycle is internally reversible.
2. The time taken during two adiabatic processes is directly proportional to the time taken for two isothermal processes.
3. There is finite temperature difference between the cycle and the external reservoirs.

The heat input from low-temperature heat source to the heat pump is given by:

$$Q_{\text{E}} = \alpha_{\text{E}}(T_{\text{E}} - T_{\text{e}})t_{\text{E}} > 0 \tag{7.10}$$

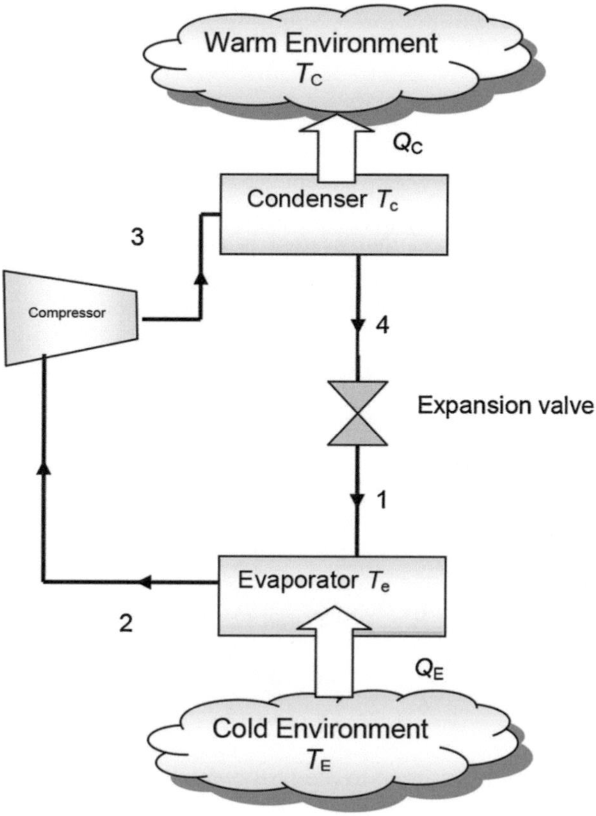

Fig. 7.4a Schematic diagram of heat pump system (Blanchard 1980)

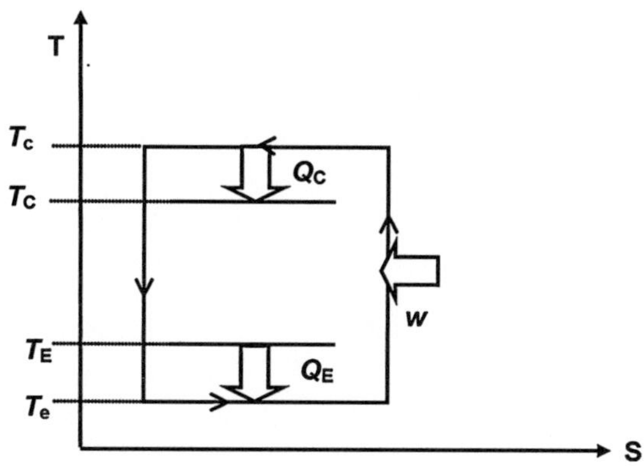

Fig. 7.4b *T–s* diagram of heat pump system

Similarly, the heat output from the heat pump to the heat sink is given by:

$$Q_C = \alpha_C(T_c - T_C)t_C > 0 \tag{7.11}$$

where α_C and α_E are, respectively, effective thermal conductance on condenser and evaporator side. T_c and T_e are the working fluid temperatures on condenser and evaporator side, respectively.

Using the first and second laws of thermodynamics for an internal reversible cycle yields:

$$Q_C = Q_E + W \tag{7.12}$$

$$\frac{Q_C}{T_c} - \frac{Q_E}{T_e} = 0 \tag{7.13}$$

It is assumed that system is alternatively connected to the source/sink heat reservoirs (through external heating and cooling fluids).

Now, the total cycle time ($t = t_E + t_C$) becomes:

$$t = \frac{Q_E}{\alpha_E (T_E - T_e)} + \frac{Q_C}{\alpha_C (T_c - T_C)} \tag{7.14}$$

where work input $W = Q_C - Q_E$. According to the second law of thermodynamics:

$$\frac{Q_E}{T_e} = \frac{Q_C}{T_c} = \frac{W}{(T_c - T_e)} \tag{7.15}$$

Now, using Eqs. (7.14) and (7.15), we have:

$$t = \frac{WT_e}{\alpha_E (T_E - T_e)(T_c - T_e)} + \frac{WT_c}{\alpha_C (T_c - T_C)(T_c - T_e)} \tag{7.16}$$

and the power input to the heat pump system is given by:

$$P = \frac{W}{t} = \frac{\alpha_E \alpha_C xy (T_C + x - T_E + y)}{(T_C + x) y \alpha_E + (T_E - y) \alpha_C x} \tag{7.17}$$

where $x = T_c - T_C$, $y = T_E - T_e$ and $t = t_E + t_C$.

Using Eqs. (7.12), (7.13), and (7.15), we have:

$$\frac{t_E}{t_C} = \frac{\alpha_C (T_c - T_C) T_e}{\alpha_E (T_E - T_e) T_c} \tag{7.18}$$

Using Eqs. (7.11) and (7.18), the output heating load is given by:

$$P_H = \frac{Q_C}{t} = \frac{\alpha_C \alpha_E xy (T_C + x)}{\alpha_E y (T_C + x) + \alpha_C x (T_E - y)} \tag{7.19}$$

We, therefore, define a Lagrangian operator (L) to minimize the power input (P) for a given heating load (P_H) and thus seek the unrestricted extremum of the Lagrangian (Appendix 2):

$$L = P + \lambda P_{\rm H} = \frac{\alpha_{\rm C}\alpha_{\rm E}xy[(1+\lambda)(T_{\rm C}+x) - (T_{\rm E}-y)]}{\alpha_{\rm E}y(T_{\rm C}+x) + \alpha_{\rm C}x(T_{\rm E}-y)} \tag{7.20}$$

with respect to x and y. The Lagrangian multiplier λ is introduced to maintain a given value of output heating load $P_{\rm H}$; therefore $\frac{dL}{dx} = 0$ and $\frac{dL}{dy} = 0$ give:

$$x = \left(1 + \sqrt{\frac{\alpha_{\rm C}}{\alpha_{\rm E}}}\right)\frac{P_{\rm H}}{\alpha_{\rm C}} \tag{7.21}$$

$$y = \frac{T_{\rm E}\sqrt{\frac{\alpha_{\rm C}}{\alpha_{\rm E}}}\left(1+\sqrt{\frac{\alpha_{\rm C}}{\alpha_{\rm E}}}\right)\frac{P_{\rm H}}{\alpha_{\rm C}}}{T_{\rm C} + \left(1+\sqrt{\frac{\alpha_{\rm C}}{\alpha_{\rm E}}}\right)^2\frac{P_{\rm H}}{\alpha_{\rm C}}} \tag{7.22}$$

Using Eqs. (7.21) and (7.22), the resulting optimal values of internal working fluid temperatures, coefficient of performance, and input power to endoreversible heat pump system are given by:

$$(T_{\rm c})_{\rm o} = T_{\rm C} + \left(1 + \sqrt{\frac{\alpha_{\rm C}}{\alpha_{\rm E}}}\right)\frac{P_{\rm H}}{\alpha_{\rm C}} \tag{7.23}$$

$$(T_{\rm e})_{\rm o} = T_{\rm E} - \frac{T_{\rm E}\sqrt{\frac{\alpha_{\rm C}}{\alpha_{\rm E}}}\left(1+\sqrt{\frac{\alpha_{\rm C}}{\alpha_{\rm E}}}\right)\frac{P_{\rm H}}{\alpha_{\rm C}}}{T_{\rm C} + \left(1+\sqrt{\frac{\alpha_{\rm C}}{\alpha_{\rm E}}}\right)^2\frac{P_{\rm H}}{\alpha_{\rm C}}} \tag{7.24}$$

$$({\rm COP})_{\rm o} = \left[1 - \frac{T_{\rm E}}{T_{\rm C} + \left[1+\sqrt{\frac{\alpha_{\rm C}}{\alpha_{\rm E}}}\right]^2\frac{P_{\rm H}}{\alpha_{\rm C}}}\right]^{-1} \tag{7.25}$$

$$P_{\rm o} = \frac{P_{\rm H}\left[T_{\rm C} - T_{\rm E} + \left(1+\sqrt{\frac{\alpha_{\rm C}}{\alpha_{\rm E}}}\right)^2\frac{P_{\rm H}}{\alpha_{\rm C}}\right]}{T_{\rm C} + \left(1+\sqrt{\frac{\alpha_{\rm C}}{\alpha_{\rm E}}}\right)^2\frac{P_{\rm H}}{\alpha_{\rm C}}} \tag{7.26}$$

Similar results can also be obtained for refrigeration and airconditioning cycles. The power input to the airconditioning system is given by:

$$P = \frac{W}{t} = \frac{\alpha_{\rm E}\alpha_{\rm C}xy(T_{\rm C}+x-T_{\rm E}+y)}{({\rm T_C}+x)y\alpha_{\rm E} + (T_{\rm E}-y)\alpha_{\rm C}x} \tag{7.27}$$

where $x = T_{\rm c} - T_{\rm C}$ and $y = T_{\rm E} - T_{\rm e}$, and cooling load is given by:

$$P_{\rm L} = \frac{Q_{\rm E}}{t} = \frac{\alpha_{\rm E}\alpha_{\rm C}xy(T_{\rm E}-y)}{\alpha_{\rm E}y({\rm T_C}+x) + \alpha_{\rm C}x(T_{\rm E}-y)} \tag{7.28}$$

We, therefore, seek the unrestricted extremum of L with respect to x and y:

$$L = P + \lambda P_{\mathrm{L}} = \frac{\alpha_{\mathrm{E}}\alpha_{\mathrm{C}}xy[T_{\mathrm{C}} + x - (1-\lambda)(T_{\mathrm{E}} - y)]}{(T_{\mathrm{C}} + x)y\alpha_{\mathrm{E}} + (T_{\mathrm{E}} - y)\alpha_{\mathrm{C}}x} \tag{7.29}$$

The Lagrangian multiplier λ is introduced to maintain a given value of cooling load P_{L}; therefore,

$$\frac{dL}{dx} = 0 \quad \text{and} \quad \frac{dL}{dy} = 0 \quad \text{give :}$$

$$x = \frac{T_{\mathrm{C}}\sqrt{\frac{\alpha_{\mathrm{E}}}{\alpha_{\mathrm{C}}}}\left(1 + \sqrt{\frac{\alpha_{\mathrm{E}}}{\alpha_{\mathrm{C}}}}\right)\frac{P_{\mathrm{L}}}{\alpha_{\mathrm{E}}}}{T_{\mathrm{E}} - \left(1 + \sqrt{\frac{\alpha_{\mathrm{E}}}{\alpha_{\mathrm{C}}}}\right)^2\frac{P_{\mathrm{L}}}{\alpha_{\mathrm{E}}}} \tag{7.30}$$

$$y = \left(1 + \sqrt{\frac{\alpha_{\mathrm{E}}}{\alpha_{\mathrm{C}}}}\right)\frac{P_{\mathrm{L}}}{\alpha_{\mathrm{E}}} \tag{7.31}$$

By using Eqs. (7.30) and (7.31), the resulting optimal values of internal working fluid temperatures, coefficient of performance, and power input to airconditioning system are given by:

$$(T_{\mathrm{e}})_{\mathrm{o}} = T_{\mathrm{E}} - \left(1 + \sqrt{\frac{\alpha_{\mathrm{E}}}{\alpha_{\mathrm{C}}}}\right)\frac{P_{\mathrm{L}}}{\alpha_{\mathrm{E}}} \tag{7.32}$$

$$(T_{\mathrm{C}})_{\mathrm{O}} = T_{\mathrm{C}} + \frac{T_{\mathrm{C}}\sqrt{\frac{\alpha_{\mathrm{E}}}{\alpha_{\mathrm{C}}}}\left(1 + \sqrt{\frac{\alpha_{\mathrm{E}}}{\alpha_{\mathrm{C}}}}\right)\frac{P_{\mathrm{L}}}{\alpha_{\mathrm{E}}}}{T_{\mathrm{E}} - \left(1 + \sqrt{\frac{\alpha_{\mathrm{E}}}{\alpha_{\mathrm{C}}}}\right)^2\frac{P_{\mathrm{L}}}{\alpha_{\mathrm{E}}}} \tag{7.33}$$

$$(\mathrm{COP})_{\mathrm{o}} = \left[\frac{T_{\mathrm{C}}}{T_{\mathrm{E}} - \left[1 + \sqrt{\frac{\alpha_{\mathrm{E}}}{\alpha_{\mathrm{C}}}}\right]^2\frac{P_{\mathrm{L}}}{\alpha_{\mathrm{E}}}} - 1\right]^{-1} \tag{7.34}$$

$$P_{\mathrm{o}} = \frac{P_{\mathrm{L}}\left[T_{\mathrm{C}} - T_{\mathrm{E}} + \left(1 + \sqrt{\frac{\alpha_{\mathrm{E}}}{\alpha_{\mathrm{C}}}}\right)^2\frac{P_{\mathrm{L}}}{\alpha_{\mathrm{E}}}\right]}{\left[T_{\mathrm{E}} - \left(1 + \sqrt{\frac{\alpha_{\mathrm{E}}}{\alpha_{\mathrm{C}}}}\right)^2\frac{P_{\mathrm{L}}}{\alpha_{\mathrm{E}}}\right]} \tag{7.35}$$

Finite Heat Capacity

Since the expansion in expander and compression in the compressor are not perfectly isentropic processes, internal irreversibilities are present within the system. Thus, in real practice, one has to deal with both external and internal irreversibilities. Kaushik (1999) and Kumar (2002) extended Blanchard's analysis to include internal irreversibility and finite heat capacitance of the external fluids in the source/sink heat reservoirs. Thus, irreversible systems with finite heat capacitance rates of external fluid and fixed effectiveness of the heat exchangers for source/sink heat reservoirs are incorporated in the analysis.

In an airconditioning/heat pump cycle, during isobaric heat rejection process, temperature of the working fluid will vary in superheat region. But the heat rejected in the superheat region is small, and the heat transfer coefficients are smaller, with larger temperature differences compared to the condensing region, so condenser design can be carried out on the basis of condenser temperature with that heat rejection. On *T–s* diagram, the amount of heat rejected by the system which is equal to the area under an isobaric heat rejection process (condensation process) can be made as equal to the area under an isothermal process (a horizontal line) with an entropic average temperature $(T_{ea})_c$ of heat rejection. Calculations of entropic average temperature for some specific refrigerants (viz. *R*-12 and *R*-134a) with varying condensation temperature (considering both cases of isentropic and non-isentropic compression) have been carried, and results are given in Appendix 4.

It is found that entropic average temperature in case of non-isentropic compression is slightly higher than the case of isentropic compression, but in both cases, it is only about 0.5% higher than the corresponding condensation temperature. So, entropic average temperature can be considered as equivalent to condensation temperature for the sake of simplicity of theoretical analysis with little loss of accuracy.

An airconditioning/heat pump system having finite heat capacity thermal reservoirs is shown schematically in Fig. 7.5a. A line diagram and *T–s* diagram of endoreversible heat pump cycle having finite heat capacity thermal reservoirs (which is an equivalent cycle of VC cycle) are shown in Figs. 7.5b and 7.5c, respectively, to achieve simplified theoretical analysis (Blanchard 1980; Wu 1993a, b, c, d; Davis and Wu 1997; Wu et al. 1998a, b, c; Kaushik et al. 2001).

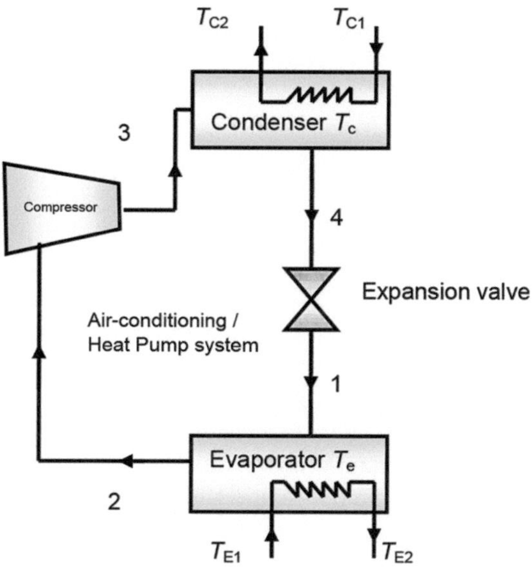

Fig. 7.5a Schematic diagram airconditioning/ heat pump cycle

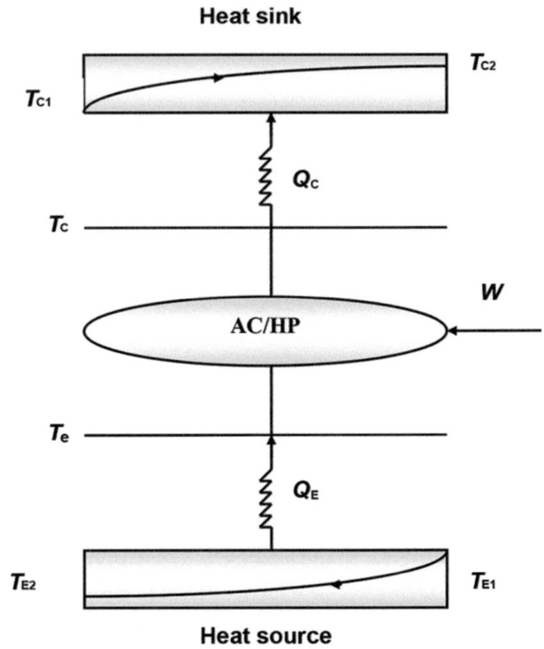

Fig. 7.5b Equivalent line diagram of airconditioning/heat pump cycle

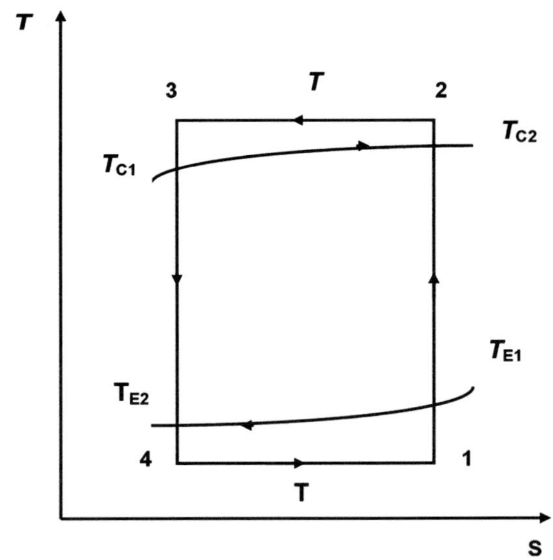

Fig. 7.5c *T–s* diagram of endoreversible airconditioning/heat pump cycle

Heat pump cycle extracts heat from low-temperature source and transfers it to higher-temperature sink with expenditure of some input work. In heat transfer processes, let initial temperature of low-temperature heat source T_{E1} go down to T_{E2} as a result of the heat transfer Q_E to heat pump system. The temperature of the heat sink fluid is initially T_{C1}, and it warms up to T_{C2} as a result of receiving heat Q_C from heat pump system.

The rate of heat input from low-temperature heat source to the heat pump is given by:

$$\dot{Q}_E = \frac{Q_E}{t_E} = U_E A_E (\text{LMTD})_E = \dot{m}_E C_{PE} (T_{E1} - T_{E2}) \tag{7.36}$$

Similarly, the rate of heat output from the heat pump to the heat sink is given by:

$$\dot{Q}_C = \frac{\dot{Q}_C}{t_C} = U_C A_C (\text{LMTD})_C = \dot{m}_C C_{PC} (T_{C2} - T_{C1}) \tag{7.37}$$

where U_C and U_E are heat transfer coefficients and A_C and A_E are the heat transfer areas of the high- and low-temperature heat exchangers, respectively. $\dot{m}$ and C_p are, respectively, the mass flow rate and specific heat of external fluid of thermal reservoirs. Log Mean Temperature Difference (LMTD) for evaporator and condenser side is defined as:

$$(\text{LMTD})_E = \frac{(T_{E1} - T_e) - (T_{E2} - T_e)}{\ln \frac{(T_{E1} - T_e)}{(T_{E2} - T_e)}} \quad \text{and}$$

$$(\text{LMTD})_C = \frac{(T_c - T_{C1}) - (T_c - T_{C2})}{\ln \frac{(T_c - T_{C1})}{(T_c - T_{C2})}}$$

Using LMTD expressions in Eqs. (7.36) and (7.37), we have:

$$T_{E2} = T_e + (T_{E1} - T_e) e^{-\frac{U_E A_E}{\dot{m}_E C_{PE}}} \tag{7.38}$$

$$T_{C2} = T_c - (T_c - T_{C1}) e^{-\frac{U_C A_C}{\dot{m}_C C_{PC}}} \tag{7.39}$$

Using Eqs. (7.36), (7.37), (7.38), and (7.39), we have:

$$Q_E = EE(T_{E1} - T_e) t_E \tag{7.40}$$

$$Q_C = CE(T_c - T_{C1}) t_C \tag{7.41}$$

where $CE = C_C \varepsilon_C$, $EE = C_E \varepsilon_E$, $C_E = \dot{m}_E C_{PE}$, $C_C = \dot{m}_C C_{PC}$, $\varepsilon_E = 1 - e^{-\frac{U_E A_E}{\dot{m}_E C_{PE}}}$, and $\varepsilon_C = 1 - e^{-\frac{U_C A_C}{\dot{m}_C C_{PC}}}$.

CE and *EE* are, respectively, effective thermal conductance (which are products of heat capacitance rates (C_C and C_E) and effectiveness of heat exchanger (ε_C and ε_E)) on condenser and evaporator side.

In a boiling or condensation process, the fluid temperature stays essentially constant, or the fluid acts as if it had infinite specific heat. In these cases, heat exchanger effectiveness relation for evaporator and condenser approach simple equation as given above (Holman 1992; Kays and London 1964).

For alternatively connected case and considering finite heat capacity of external fluid, the total cycle time ($t = t_{\mathrm{E}} + t_{\mathrm{C}}$) becomes:

$$t = \frac{Q_{\mathrm{E}}}{EE(T_{\mathrm{E1}} - T_{\mathrm{e}})} + \frac{Q_{\mathrm{C}}}{CE(T_{\mathrm{c}} - T_{\mathrm{C1}})} \tag{7.42}$$

where $W = Q_{\mathrm{C}} - Q_{\mathrm{E}}$. According to second law of thermodynamics:

$$\frac{Q_{\mathrm{E}}}{T_{\mathrm{e}}} = \frac{Q_{\mathrm{C}}}{T_{\mathrm{c}}} \tag{7.43}$$

Now, using Eqs. (7.42) and (7.43), we have:

$$t = \frac{WT_{\mathrm{e}}}{EE(T_{\mathrm{E1}} - T_{\mathrm{e}})(T_{\mathrm{c}} - T_{\mathrm{e}})} + \frac{WT_{\mathrm{c}}}{CE(T_{\mathrm{c}} - T_{\mathrm{C1}})(T_{\mathrm{c}} - T_{\mathrm{e}})} \tag{7.44}$$

and the power input to the heat pump system is given by:

$$P = \frac{W}{t} = \frac{CE\ EE\ x\ y\ (T_{\mathrm{C1}} + x - T_{\mathrm{E1}} + y)}{(T_{\mathrm{C1}} + x)\ y\ EE + (T_{\mathrm{E1}} - y)\ CE\ x} \tag{7.45}$$

where $x = T_{\mathrm{c}} - T_{\mathrm{C1}}$, $y = T_{\mathrm{E1}} - T_{\mathrm{e}}$, and $t = t_{\mathrm{E}} + t_{\mathrm{C}}$. Using Eqs. (7.40), (7.41), and (7.43), we have:

$$\frac{t_{\mathrm{E}}}{t_{\mathrm{C}}} = \frac{CE\ (T_{\mathrm{c}} - T_{\mathrm{C1}})\ T_{\mathrm{e}}}{EE\ (T_{\mathrm{E1}} - T_{\mathrm{e}})\ T_c} \tag{7.46}$$

Using Eqs. (7.41) and (7.46), the output heating load is given by:

$$P_{\mathrm{H}} = \frac{Q_{\mathrm{C}}}{t} \frac{CE\ EE\ x\ y\ (T_{\mathrm{C1}} + x)}{EE\ y\ (T_{\mathrm{C1}} + x) + CE\ x\ (T_{\mathrm{E1}} - y)} \tag{7.47}$$

We seek the minimum input power (P) for a given heating load ($P_{\mathrm{H}} = Q_{\mathrm{C}}/t$). It can be seen that there is no natural maximum in case of R/AC and HP system (Appendix 1). We are unable to find out the minimum input power to these systems with the simple calculation method as done by Curzon and Ahlborn (1975). The specific problem is to minimize input power for a given heating load. Therefore, for optimizing these systems, Lagrangian undetermined multiplier method is used, in case of heat pump system, to minimize our objective function power input (P) subject to given heating load (P_{H}). Conditional function P_{H} is multiplied by a multiplier λ, and the product λP_{H} is added to the objective function (P) to define Lagrangian operator.

We, therefore, seek the unrestricted extremum of the Lagrangian:

$$L = P + \lambda P_H = \frac{CE\ EE\ x\ y[(1+\lambda)\ (T_{C1}+x) - (T_{E1}-y)]}{EE\ y\ (T_{C1}+x) + CE\ x\ (T_{E1}-y)} \tag{7.48}$$

with respect to x and y. The Lagrangian multiplier λ is introduced to maintain a given value of output heating load P_H; therefore $\frac{dL}{dx} = 0$ and $\frac{dL}{dy} = 0$ give:

$$x = \left(1 + \sqrt{\frac{CE}{EE}}\right)\frac{P_H}{CE} \tag{7.49}$$

$$y = T_{E1}\sqrt{\frac{\frac{CE}{EE}\left(1+\sqrt{\frac{CE}{EE}}\right)P_H}{T_{C1} + \left(1+\sqrt{\frac{CE}{EE}}\right)^2 \frac{P_H}{CE}}} \tag{7.50}$$

By using Eqs. (7.49) and (7.50), the resulting optimal values of internal working fluid temperatures, coefficient of performance, and input power to endoreversible heat pump system are given by:

$$(T_c)_o = T_{C1} + \left(1 + \sqrt{\frac{CE}{EE}}\right)\frac{P_H}{CE} \tag{7.51}$$

$$(T_e)_O = T_{E1} - T_{E1}\sqrt{\frac{\frac{CE}{EE}\left(1+\sqrt{\frac{CE}{EE}}\right)\frac{P_H}{CE}}{T_{C1} + \left(1+\sqrt{\frac{CE}{EE}}\right)^2 \frac{P_H}{CE}}} \tag{7.52}$$

$$(\mathrm{COP})_o = \left[1 - \frac{T_{E1}}{T_{C1} + \left[1+\sqrt{\frac{CE}{EE}}\right]^2 \frac{P_H}{CE}}\right]^{-1} \tag{7.53}$$

$$P_o = \frac{P_H\left[T_{C1} - T_{E1} + \left(1+\sqrt{\frac{CE}{EE}}\right)^2 \frac{P_H}{CE}\right]}{T_{C1} + \left(1+\sqrt{\frac{CE}{EE}}\right)^2 \frac{P_H}{CE}} \tag{7.54}$$

If internal irreversibility such as due to friction, non-isentropic compression/expansion is accounted for the heat pump system, then the two isentropic processes become adiabatic processes with entropy generations. On T–s diagram, the four processes of such a heat pump constitute the cycle 1–2′–3–4′ as shown in Fig. 7.5d. Dotted lines correspond to non-isentropic compression and expansion. Heat absorbed by the reversible heat pump is $Q_E = T_e\ (s_1 - s_4)$, and heat absorbed by irreversible heat pump is $Q'_E = T_e\ (s_1 - s'_4)$, since $s_4 < s'_4$, so $Q'_E < Q_E$ (primes are added to quantities associated with the irreversible heat pump). One can define an irreversible heat absorption parameter C_1, such that $Q'_E = C_1\ Q_E$ with

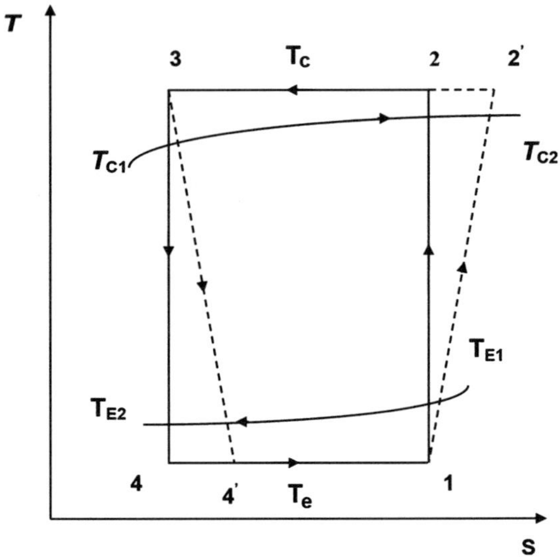

Fig. 7.5d *T–s* diagram of irreversible heat pump/airconditioning cycle

$C_1 = (s_1 - s'_4)/(s_1 - s_4) < 1$. Similarly, an irreversible heat rejection parameter C_2 may be defined as $Q'_c = C_2 Q_c$ with $C_2 = (s'_2 - s_3)/(s_2 - s_3) > 1$. Assuming that the irreversible heat pump operates between the same source and sink temperatures and that the effective thermal conductances remain the same, the second law requires $\int \frac{dQ}{T} = \frac{Q'_E}{T'_e} - \frac{Q'_c}{T'_c} < 0$. This inequality can also be written as:

$$\frac{Q'_E}{T'_e} = R_{\Delta s} \frac{Q'_c}{T'_c} \tag{7.55}$$

where $R_{\Delta s} = \frac{S_1 - S'_4}{S'_2 - S_3} = \frac{c_1}{c_2} < 1$,

As it is seen in the above equation, $R_{\Delta S}$ represents the ratio of entropy differences, and it has a value of less than one for irreversible (real) systems. Thus, it is found that for refrigeration, airconditioning, and heat pump systems, the internal irreversibilities are characterized by a single parameter defined in terms of a ratio of entropy differences. Using computerized refrigerant table (Sonntag et al. 1998), entropy values at various state points can be found out, and hence internal irreversibility parameters are calculated with different compression efficiency and condensation temperature for some specific refrigerants (viz. *R*-12, *R*-134a) (Appendix 4).

Now combining Eqs. (7.42) and (7.55), we have the power input to an irreversible heat pump system:

$$P = \frac{W}{t} = \frac{CE\ EE\ x\ y[T_{C1} + x - (T_{E1} - y)R_{\Delta s}]}{EE\ y\ (T_{C1} + x) + R_{\Delta s}\ CE\ x\ (T_{E1} - y)} \tag{7.56}$$

Equation (7.55) gives:

$$\frac{t_{\mathrm{C}}}{t_{\mathrm{E}}}=\frac{EE\ (T_{\mathrm{E1}}-T_{\mathrm{e}})\mathrm{T_C}}{R_V\ CE\ (T_{\mathrm{c}}-T_{\mathrm{C1}})T_{\mathrm{e}}} \tag{7.57}$$

Now using Eq. (7.57), output heating load is given by:

$$P_{\mathrm{H}}=\frac{Q_{\mathrm{C}}}{t}=\frac{CE\ EE\ x\ y\ (T_{\mathrm{C1}}+x)}{EE\ y\ (T_{\mathrm{C1}}+x)+R_{\Delta\mathrm{s}}\ CE\ x\ (T_{\mathrm{E1}}-y)} \tag{7.58}$$

Let us define the Lagrangian operator:

$$L=P+\lambda P_{\mathrm{H}}=\frac{CE\ EE\ x\ y[(T_{\mathrm{C1}}+x)(1+\lambda)-R_{\Delta\mathrm{s}}(T_{\mathrm{E1}}-y)]}{R_{\Delta s}\ CE\ x\ (T_{\mathrm{E1}}-y)+EE\ y\ (T_{\mathrm{C1}}+x)} \tag{7.59}$$

Differentiating L with respect to x and y and then equating to zero, $dL/dx = 0$ gives:

$$x=\left(1+\sqrt{\frac{R_{\Delta s}\ CE}{EE}}\right)\frac{P_{\mathrm{H}}}{CE} \tag{7.60}$$

and $dL/dy = 0$ gives:

$$y=\frac{T_{\mathrm{E1}}\sqrt{\frac{R_{\Delta s}\ CE}{EE}}\left(1+\sqrt{\frac{R_{\Delta s}\ CE}{EE}}\right)\frac{P_{\mathrm{H}}}{CE}}{T_{\mathrm{C1}}+\left(1+\sqrt{\frac{R_{\Delta s}\ CE}{EE}}\right)^2\frac{P_{\mathrm{H}}}{CE}} \tag{7.61}$$

Now using Eqs. (7.60) and (7.61), optimal working fluid temperature, coefficient of performance, and power input to the irreversible heat pump system are given by:

$$(T_{\mathrm{c}})_{\mathrm{o}}=T_{\mathrm{C1}}+\left(1+\sqrt{\frac{R_{\Delta s}\ CE}{EE}}\right)\frac{P_{\mathrm{H}}}{CE} \tag{7.62}$$

$$(T_{\mathrm{e}})_{\mathrm{o}}=T_{\mathrm{E1}}-\frac{T_{\mathrm{E1}}\sqrt{\frac{R_{\Delta s}\ CE}{EE}}\left(1+\frac{R_{\Delta s}\ CE}{EE}\right)\frac{P_{\mathrm{H}}}{CE}}{T_{\mathrm{C1}}+\left(1+\sqrt{\frac{R_{\Delta s}\ CE}{EE}}\right)^2\frac{P_{\mathrm{H}}}{CE}} \tag{7.63}$$

$$(\mathrm{COP})_{\mathrm{o}}=\left[1-\frac{R_{\Delta s}\ T_{\mathrm{E1}}}{T_{\mathrm{C1}}+\left[1+\sqrt{\frac{R_{\Delta s}\ CE}{EE}}\right]^2\frac{P_{\mathrm{H}}}{CE}}\right]^{-1} \tag{7.64}$$

and

$$P_{\rm o} = \frac{P_{\rm H}\left[T_{\rm C1} - T_{\rm E1}R_{\Delta s} + \left(1 + \sqrt{\frac{CE\ R_{\Delta s}}{EE}}\right)^2 \frac{P_{\rm H}}{CE}\right]}{T_{\rm C1} + \left(1 + \sqrt{\frac{CE\ R_{\Delta s}}{EE}}\right)^2 \frac{P_{\rm H}}{CE}} \tag{7.65}$$

Following the earlier analysis of heat pump system, similar results can also be obtained for airconditioning system. The power input to the airconditioning system is given by (Kaushik et al. 2002b):

$$P = \frac{W}{t} = \frac{CE\ EE\ x\ y\ (T_{\rm C1} + x - T_{\rm E1} + y)}{(T_{\rm C1} + x)y\ EE + (T_{\rm E1} - y)\ CE\ x} \tag{7.66}$$

where

$x = T_{\rm c} - T_{\rm C1}$, $y = T_{\rm E1} - T_{\rm e}$, $CE = C_{\rm C}\,\varepsilon_{\rm C}$, $EE = C_{\rm E}\,\varepsilon_{\rm E}$, $C_{\rm E} = \dot{m}_{\rm E} C_{\rm PE}$ and $C_{\rm C} = \dot{m}_{\rm C}\,C_{\rm PC}$, $\varepsilon_{\rm C} = 1 - \exp^{-\frac{U_{\rm C}\ A_{\rm C}}{\dot{m}_{\rm C}\ C_{\rm PC}}}$, and $\varepsilon_{\rm E} = 1 - \exp^{-\frac{U_{\rm E}\ A_{\rm E}}{\dot{m}_{\rm E}\ C_{\rm PE}}}$; cooling load is given by:

$$P_{\rm L} = \frac{Q_{\rm E}}{t} = \frac{CE\ EE\ x\ y\ (T_{\rm E1} - y)}{EE\ y(T_{\rm C1} + x) + CE\ x(T_{\rm E1} - y)} \tag{7.67}$$

We, therefore, seek the unrestricted extremum of:

$$L = P + \lambda P_{\rm L} = \frac{CE\ EE\ x\ y[(T_{\rm C1} + x) - (1 - \lambda)(T_{\rm E1} - y)]}{EE\ y(T_{\rm C1} + x) + CE\ x(T_{\rm E1} - y)} \tag{7.68}$$

with respect to x and y. The Lagrangian multiplier λ is introduced to maintain a given value of cooling load $P_{\rm L}$. Therefore, $\frac{dL}{dx} = 0$ and $\frac{dL}{dy} = 0$ give:

$$x = T_{\rm C1}\sqrt{\frac{\frac{EE}{CE}\left(1 + \sqrt{\frac{EE}{CE}}\right)\frac{P_{\rm L}}{EE}}{T_{\rm E1} - \left(1 + \sqrt{\frac{EE}{CE}}\right)^2 \frac{P_{\rm L}}{EE}}} \tag{7.69}$$

$$y = \left(1 + \sqrt{\frac{EE}{CE}}\right)\frac{P_{\rm L}}{EE} \tag{7.70}$$

By using Eqs. (7.69) and (7.70), the resulting optimal values of internal working fluid temperatures, coefficient of performance, and power input to an endoreversible airconditioning system are given by:

$$(T_{\rm e})_{\rm o} = T_{\rm E1} - \left(1 + \sqrt{\frac{EE}{CE}}\right)\frac{P_{\rm L}}{EE} \tag{7.71}$$

$$(T_c)_O = T_{C1} + T_{C1}\sqrt{\frac{\frac{EE}{CE}\left(1+\sqrt{\frac{EE}{CE}}\right)\frac{P_L}{EE}}{T_{E1}-\left(1+\sqrt{\frac{EE}{CE}}\right)^2\frac{P_L}{EE}}} \quad (7.72)$$

$$(\mathrm{COP})_o = \left[\frac{T_{C1}}{T_{E1}-\left(1+\sqrt{\frac{EE}{CE}}\right)^2\frac{P_L}{EE}} - 1\right]^{-1} \quad (7.73)$$

$$P_o = \frac{P_L\left[T_{C1} - T_{E1} + \left(1+\sqrt{\frac{EE}{CE}}\right)^2\frac{P_L}{EE}\right]}{T_{E1}-\left(1+\sqrt{\frac{EE}{CE}}\right)^2\frac{P_L}{EE}} \quad (7.74)$$

Following earlier analysis, the power input to an irreversible airconditioning system is given by:

$$P = \frac{W}{t} = \frac{CE\ EE\ x\ y\ [T_{C1} + x - (T_{E1} - y)R_{\Delta s}]}{EE\ y(T_{C1}+x) + R_{\Delta s}\ CE\ x(T_{E1}-y)} \quad (7.75)$$

and cooling load is given by:

$$P_L = \frac{Q_E}{t} = \frac{CE\ EE\ x\ y(T_{E1}-y)}{EE\ y(T_{C1}+x) + R_{\Delta s}\ CE\ x(T_{E1}-y)} \quad (7.76)$$

where $R_{\Delta S}$ is internal irreversibility parameter, which represents non-isentropic compression/expansion in terms of the ratio of entropy differences. For performance optimization, we have to minimize power input to the system. Now let us define a Lagrangian operator:

$$L = P + \lambda P_L = \frac{CE\ EE\ x\ y\ [(T_{C1}+x) - R_{\Delta s}(1-\lambda)(T_{E1}-y)]}{R_{\Delta s}\ CEx(T_{E1}-y) + EE\ y(T_{C1}+x)} \quad (7.77)$$

Here, power input is an objective function, which has to be minimized, and cooling load is a constraint function, which is given, kept constant during optimization. It is thus seen that L is a function of x and y. Therefore, we differentiate L w.r.t x and y and then equate to zero, i.e. $dL/dx = 0$ which gives:

$$x = T_{C1}\sqrt{\frac{\frac{EE}{R_{\Delta s}\ CE}\left(1+\frac{EE}{R_{\Delta s}\ CE}\right)\frac{P_L}{EE}}{T_{E1}-\left(1+\sqrt{\frac{EE}{R_{\Delta s}\ CE}}\right)^2\frac{P_L}{EE}}} \quad (7.78)$$

and $dL/dy = 0$ which gives:

$$y = \left(1+\sqrt{\frac{EE}{R_{\Delta s}\ CE}}\right)^2\frac{P_L}{EE} \quad (7.79)$$

By using Eqs. (7.78) and (7.79), the optimal values of internal working fluid temperature, coefficient of performance, and power input to an irreversible airconditioning system are given by:

$$(T_e)_o = T_{E1} - \left(1 + \sqrt{\frac{EE}{R_{\Delta s}\ CE}}\right)\frac{P_L}{EE} \tag{7.80}$$

$$(T_c)_o = T_{C1} + T_{C1}\sqrt{\frac{\frac{EE}{R_{\Delta s}\ CE}\left(1 + \sqrt{\frac{EE}{R_{\Delta s}\ CE}}\right)\frac{P_L}{EE}}{T_{E1} - \left(1 + \sqrt{\frac{EE}{R_{\Delta s}\ CE}}\right)^2\frac{P_L}{EE}}} \tag{7.81}$$

$$(\mathrm{COP})_o = \left[\frac{\frac{T_{C1}}{R_{\Delta s}}}{T_{C1} - \left(1 + \sqrt{\frac{EE}{R_{\Delta s}\ CE}}\right)^2\frac{P_L}{EE}} - 1\right]^{-1} \tag{7.82}$$

and

$$P_o = \frac{P_L\left[\frac{T_{C1}}{R_{\Delta s}} - T_{E1} + \left(1 + \sqrt{\frac{EE}{R_{\Delta s}\ CE}}\right)^2\frac{P_L}{EE}\right]}{\left[T_{E1} - \left(1 + \sqrt{\frac{EE}{R_{\Delta s}\ CE}}\right)^2\frac{P_L}{EE}\right]} \tag{7.83}$$

Special Cases

1. If internal irreversible parameter $R_{\Delta s} = 1.0$, then heat pump cycle becomes endoreversible, and Eqs. (7.62), (7.63), (7.64), and (7.65) approach the expressions given by Eqs. (7.51), (7.52), (7.53), and (7.54).
2. For reversible heat pump cycle, heating load $P_H = 0$ and internal irreversible parameter $R_{\Delta s} = 1.0$, and under these conditions, the optimal coefficient of performance of an irreversible heat pump system given by Eq. 7.64 reduces to Carnot coefficient of performance as given by:

$$\mathrm{COP} = \left[1 - \frac{T_{E1}}{T_{C1}}\right]^{-1} = \frac{T_{C1}}{T_{C1} - T_{E1}}$$

3. If internal irreversible parameter $R_{\Delta s} = 1.0$, then airconditioning cycle becomes endoreversible, and Eqs. (7.80), (7.81), (7.82), and (7.83) approach the expressions given by Eqs. (7.71), (7.72), (7.73), and (7.74).
4. For reversible airconditioning cycle, cooling load $P_L = 0$ and internal irreversible parameter $R_{\Delta s} = 1.0$, and under these conditions, the optimal coefficient of performance of an irreversible airconditioning cycle given by Eq. 7.82 reduces to Carnot coefficient of performance as given by:

$$\mathrm{COP} = \left[\frac{T_{\mathrm{C1}}}{T_{\mathrm{E1}}} - 1\right]^{-1} = \frac{T_{\mathrm{E1}}}{T_{\mathrm{C1}} - T_{\mathrm{E1}}}$$

7.4.2 *Continuously Connected Cycle to Thermal Reservoirs*

Earlier in Sect. 4.1, finite time thermodynamic analysis of *R*/AC and HP systems alternatively connected to thermal reservoirs is carried out, and analytical expressions for optimal performance are found out. However, a practical system should approach the performance of a continuously connected case, and therefore a detailed analysis has also been carried out for a continuously connected case when the system is assumed to be continuously connected to source/sink heat reservoirs. Thus, in continuously connected case, the time taken for heat extraction is the same as the time required for heat rejection and which is equal to the cycle time, i.e. $t_{\mathrm{cycle}} = t_{\mathrm{E}} = t_{\mathrm{C}}$. Now, the power input to the system is given by:

$$P = C_{\mathrm{C}}\varepsilon_{\mathrm{C}}(T_{\mathrm{c}} - T_{\mathrm{C1}}) - C_{\mathrm{E}}\varepsilon_{\mathrm{E}}(T_{\mathrm{E1}} - T_{\mathrm{e}}) \tag{7.84}$$

Second law requires:

$$\oint \frac{d\dot{Q}}{T} = \frac{\dot{Q}_{\mathrm{E}}}{T_{\mathrm{e}}} - \frac{\dot{Q}_{\mathrm{C}}}{T_{\mathrm{c}}} \leq 0,$$

In this inequality, equality sign holds for reversible system, and less than is for irreversible case. So, for internally irreversible system, an internal irreversibility parameter is introduced as:

$$\Delta S = \frac{\dot{Q}_{\mathrm{E}}}{T_{\mathrm{e}}} - R_{\Delta s}\frac{\dot{Q}_{\mathrm{C}}}{T_{\mathrm{c}}} = 0 \tag{7.85}$$

where $R_{\Delta \mathrm{S}} = 1$ is for endoreversible case and $R_{\Delta \mathrm{S}} < 1$ for internally irreversible system.

Irreversible Heat Pump Cycle

For minimizing power input, with given heating load and entropy constraints, let us define Lagrangian function:

$$\begin{aligned} L &= P + \lambda\dot{Q}_{\mathrm{C}} + \mu\Delta S \\ &= C_{\mathrm{C}}\varepsilon_{\mathrm{C}}(T_{\mathrm{c}} - T_{\mathrm{C1}}) - C_{\mathrm{E}}\varepsilon_{\mathrm{E}}(T_{\mathrm{E1}} - T_{\mathrm{e}}) + \lambda C_{\mathrm{C}}\varepsilon_{\mathrm{C}}(T_{\mathrm{c}} - T_{\mathrm{C1}}) \\ &\quad + \mu\left[\frac{\dot{Q}_{\mathrm{E}}}{T_{\mathrm{e}}} - R_{\Delta s}\frac{\dot{Q}_{\mathrm{C}}}{T_{\mathrm{c}}}\right] \end{aligned} \tag{7.86}$$

where λ and μ are Lagrangian multipliers. $\frac{\partial L}{\partial T_e} = 0$ and $\frac{\partial L}{\partial T_c} = 0$ give the optimal values of T_e and T_c. Now, using these optimal values of working fluid temperatures, we can have optimal COP of heat pump system as given by (Kumar 2002):

$$COP = \left(1 - \frac{R_{\Delta s} T_{E1}}{T_{C1} + \frac{\dot{Q}_C}{K}}\right)^{-1} \tag{7.87}$$

where $K = \frac{CEEE}{R_{\Delta s}CE+EE}$, $EE = C_E\ \varepsilon_E$, and $CE = C_C\ \varepsilon_C$and where EE and CE are, respectively, effective thermal conductance on evaporator and condenser side.

Thus, by employing the Lagrangian multiplier method, for a given specified cooling/heating load, the COP has been maximized.

Irreversible Refrigeration/Airconditioning Cycle

A similar analysis is extended for refrigeration/airconditioning cycle, and the input power is minimized with given cooling load and entropy constraints. For this purpose Lagrangian function is defined as follows:

$$\begin{aligned} L &= P + \lambda \dot{Q}_E + \mu \Delta S \\ &= C_C \varepsilon_C (T_c - T_{C1}) - C_E \varepsilon_E (T_{E1} - T_e) + \lambda C_E \varepsilon_E (T_{E1} - T_e) \\ &\quad + \mu \left[\frac{\dot{Q}_E}{T_e} - R_{\Delta s} \frac{\dot{Q}_C}{T_c}\right] \end{aligned} \tag{7.88}$$

where λ and μ are Lagrangian multipliers. $\frac{\partial L}{\partial T_e} = 0$ and $\frac{\partial L}{\partial T_c} = 0$ give the optimal values of T_e and T_c. Now, using these optimal values of working fluid temperatures, the optimal COP of refrigeration/airconditioning system is obtained as follows:

$$\text{COP} = \left(\frac{\frac{T_{C1}}{R_{\Delta s}}}{T_{E1} - \frac{\dot{Q}_E}{K}}\right)^{-1} \tag{7.89}$$

where $K = \frac{CE\ \ EE\ \ R_{\Delta s}}{R_{\Delta s}CE+EE}$.

Equations 7.87, 7.88, and 7.89 are, respectively, the optimal coefficients of performance of heat pump and refrigeration/airconditioning system continuously connected to the source/sink heat reservoirs. The coefficient of performance for alternatively connected case is having similar form to the COP as that of continuously connected case but slightly differs because of a different form of K which depends on effectiveness, heat capacitance rates, and internal irreversibility.

For the same input parameters, a comparative study of alternatively and continuously connected cases has also been carried out, and computed results are tabulated in Appendix 3. It is found that COP predicted by continuously connected case is slightly higher than the COP of alternatively connected case, and this difference of COP decreases with increasing internal irreversibility (decreasing $R_{\Delta s}$ value).

7.5 Modified Vapour Compression Cycle

If we combine superheating of vapour with liquid subcooling, we have a liquid–vapour (L–V) heat exchanger. A liquid–vapour heat exchanger may be installed as shown in Fig. 7.6a. On *T–s* diagram during the process 1–1′, the refrigerant vapour from the evaporator is superheated in the regenerative heat exchanger with consequent subcooling of liquid from the condenser during process 3–3′. Considering cycle 1′–2′–2–3–3′–4′–1 of Fig. 7.6b, the entropic average temperature of heat rejection becomes (Kumar 2002):

$$(T_{\mathrm{ea}})_{\mathrm{c}} = \frac{h_2' - h_3'}{s_2' - s_3'}$$

In Fig. 7.6b, non-isentropic compression/expansion processes are shown by dotted lines and constitute a cycle 1′–2″–2′–2–3–3′–4′–1. Now, entropic average temperature of heat rejection becomes:

$$(T_{\mathrm{ea}})_{\mathrm{c}} = \frac{h_2'' - h_3''}{s_2'' - s_3''}$$

Similarly, entropic average temperature of evaporation becomes:

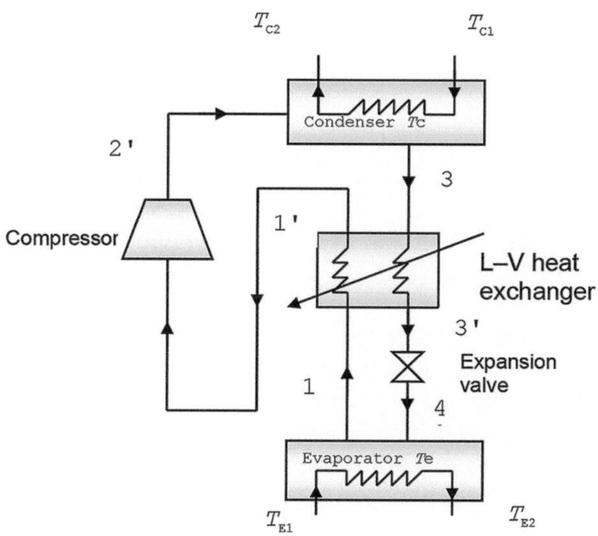

Fig. 7.6a Schematic diagram of vapour compression refrigeration/heat pump cycle with a liquid–vapour heat exchanger

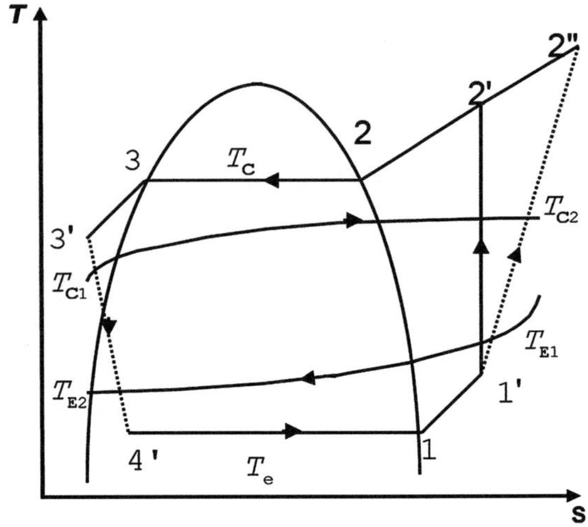

Fig. 7.6b *T–s* diagram of vapour compression refrigeration/heat pump cycle with liquid–vapour heat exchanger

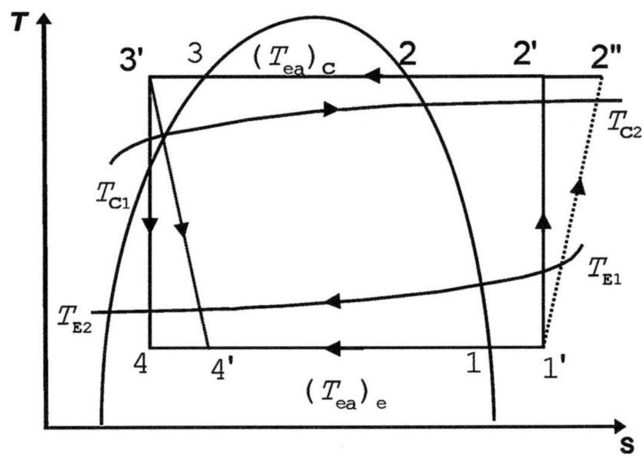

Fig. 7.6c *T–s* diagram of equivalent Carnot refrigeration/heat pump cycle

$$(T_{\mathrm{ea}})_{\mathrm{e}} = \frac{h_1' - h_4'}{s_1' - s_4'}$$

Using an entropic average temperature, the cycle shown in Fig. 7.6b can be modified to an equivalent Carnot cycle as shown in Fig. 7.6c, to achieve the theoretical formula/analysis with little loss of accuracy. The detailed calculations of entropic average temperature (T_{ea}) and internal irreversibility parameter ($R_{\Delta \mathrm{S}}$) values for various refrigerants with liquid–vapour heat exchanger within VC *R*/AC/ HP systems are also carried out and presented in Appendix 5. It is found that internal irreversibility parameter of L–V heat exchanger case is having higher value and reaches towards endoreversible case (i.e. $R_{\Delta \mathrm{S}} = 1$) than the corresponding value of the parameters in case of without L–V heat exchanger.

7.6 Comparison of Theoretical and Experimental Performance

In order to have a comparison of our theoretical results with the available experimental performance data published in Chua et al. (1996), a simple comparative study for the same set of operating conditions is carried out and presented in Appendix 6. Theoretical predicted performance is coming only about 2.4% lower than the experimental value that may be because of our calculated value of internal irreversibility parameter which is lower than the actual value because of the simplifying assumption of taking saturated states at the inlet point of compressor and outlet point of condenser.

7.7 Discussion of Results

In order to have numerical appreciation of the theoretical analysis of heat pump/airconditioning cycle alternatively connected to thermal reservoirs, the effect of various input parameters on the coefficient of performance of heat pump and airconditioning systems is studied, and results are shown in Tables 7.1, 7.2, 7.3, 7.4, 7.5, 7.6, 7.7, and 7.8, respectively.

7.7.1 *Heat Pump Cycle*

During the variation of any one parameter, all other parameters are assumed to be constant as given below:

Table 7.1 Effect of source/sink inlet fluid temperature (T_{E1}/T_{C1}), on working fluid temperatures, power input, and performance of the heat pump system

T_{E1}	$(T_c)_O$	$(T_e)_O$	$(COP)_O$	P_O	T_{C1}	$(T_c)_O$	$(T_e)_O$	$(COP)_O$	P_O
K	K	K	–	kW	K	K	K	–	kW
278	318.00	271.18	6.79	0.44	300	308.00	275.84	9.58	0.31
279	318.00	272.15	6.94	0.43	302	310.00	275.88	9.09	0.33
280	318.00	273.13	7.09	0.42	304	312.00	275.93	8.65	0.35
281	318.00	274.10	7.24	0.41	306	314.00	275.97	8.26	0.36
282	318.00	275.08	7.41	0.40	308	316.00	276.01	7.90	0.38
283	318.00	276.06	7.58	0.40	310	318.00	276.06	7.58	0.40
284	318.00	277.03	7.76	0.39	312	320.00	276.10	7.29	0.41
285	318.00	278.01	7.95	0.38	314	322.00	276.14	7.02	0.43
286	318.00	278.98	8.15	0.37	316	324.00	276.18	6.78	0.44
287	318.00	279.96	8.36	0.36	318	326.00	276.22	6.55	0.46

Table 7.2 Effect of evaporator/condenser heat capacitance rate (C_E/C_C), on working fluid temperatures, power input, and performance of the heat pump system

C_E	$(T_c)_O$	$(T_e)_O$	$(COP)_O$	P_O	C_C	$(T_c)_O$	$(T_e)_O$	$(COP)_O$	P_O
kW/K	K	K	–	kW	kW/K	K	K	–	kW
0.6	319.16	272.80	7.24	0.41	0.6	321.83	275.23	6.59	0.46
0.7	318.78	273.93	7.35	0.41	0.7	320.50	275.49	6.90	0.43
0.8	318.47	274.80	7.44	0.40	0.8	319.47	275.71	7.16	0.42
0.9	318.22	275.49	7.52	0.40	0.9	318.66	275.90	7.39	0.41
1.0	318.00	276.06	7.58	0.40	1.0	318.00	276.06	7.58	0.40
1.1	317.81	276.53	7.64	0.39	1.1	317.45	276.19	7.75	0.39
1.2	317.65	276.92	7.69	0.39	1.2	316.98	276.32	7.91	0.38
1.3	317.51	277.27	7.74	0.39	1.3	316.59	276.43	8.05	0.37
1.4	317.38	277.56	7.78	0.39	1.4	316.24	276.52	8.17	0.37
1.5	317.27	277.83	7.81	0.38	1.5	315.93	276.61	8.28	0.36

Table 7.3 Effect of evaporator/condenser side heat exchanger effectiveness ($\varepsilon_E/\varepsilon_C$), on working fluid temperatures, power input, and performance of the heat pump system

ε_E	$(T_c)_O$	$(T_e)_O$	$(COP)_O$	P_O	ε_C	$(T_c)_O$	$(T_e)_O$	$(COP)_O$	P_O
–	K	K	–	kW	–	K	K	–	kW
0.20	321.75	263.70	6.60	0.45	0.20	332.75	273.65	4.90	0.61
0.30	320.32	269.03	6.94	0.43	0.30	326.32	274.47	5.74	0.52
0.40	319.48	271.84	7.16	0.42	0.40	322.98	275.02	6.34	0.47
0.50	318.90	273.59	7.32	0.41	0.50	320.90	275.41	6.80	0.44
0.60	318.47	274.80	7.44	0.40	0.60	319.47	275.71	7.16	0.42
0.70	318.14	275.69	7.54	0.40	0.70	318.43	275.95	7.45	0.40
0.75	318.00	276.06	7.58	0.40	0.75	318.00	276.06	7.58	0.40
0.80	317.87	276.38	7.62	0.39	0.80	317.62	276.15	7.70	0.39
0.90	317.65	276.92	7.69	0.39	0.90	316.98	276.32	7.91	0.38
1.00	317.46	277.37	7.75	0.39	1.00	316.46	276.46	8.09	0.37

Table 7.4 Effect of internal irreversibility parameter ($R_{\Delta S}$), on working fluid temperatures, power input, and performance of the heat pump system

$R_{\Delta S}$	$(T_c)_O$	$(T_e)_O$	$(COP)_O$	P_O
–	K	K	–	kW
0.50	316.83	278.78	1.78	1.69
0.55	316.97	278.49	1.92	1.56
0.60	317.10	278.20	2.10	1.43
0.65	317.22	277.92	2.31	1.30
0.70	317.35	277.64	2.56	1.17
0.75	317.46	277.37	2.88	1.04
0.80	317.58	277.10	3.29	0.91
0.85	317.69	276.84	3.84	0.78
0.90	317.79	276.57	4.59	0.65
0.95	317.90	276.31	5.72	0.52
1.00	318.00	276.06	7.58	0.40

Table 7.5 Effect of source/sink inlet fluid temperature (T_{E1}/T_{C1}), on working fluid temperatures, power input, and performance of the airconditioning system

T_{E1}	$(T_c)_O$	$(T_e)_O$	$(COP)_O$	P_O	T_{C1}	$(T_c)_O$	$(T_e)_O$	$(COP)_O$	P_O
K	K	K	–	kW	K	K	K	–	kW
265	328.22	257.00	3.61	0.83	310	319.58	267.00	5.08	0.59
267	328.14	259.00	3.75	0.80	312	321.64	267.00	4.89	0.61
269	328.06	261.00	3.89	0.77	314	323.70	267.00	4.71	0.64
271	327.98	263.00	4.05	0.74	316	325.76	267.00	4.54	0.66
273	327.90	265.00	4.21	0.71	318	327.82	267.00	4.39	0.68
275	327.82	267.00	4.39	0.68	320	329.88	267.00	4.25	0.71
277	327.75	269.00	4.58	0.66	322	331.95	267.00	4.11	0.73
279	327.67	271.00	4.78	0.63	324	334.01	267.00	3.98	0.75
281	327.60	273.00	5.00	0.60	326	336.07	267.00	3.87	0.78
283	327.53	275.00	5.24	0.57	328	338.13	267.00	3.75	0.80

Table 7.6 Effect of condenser/evaporator heat capacitance rate (C_C/C_E), on working fluid temperatures, power input, and performance of the airconditioning system

C_C	$(T_c)_O$	$(T_e)_O$	$(COP)_O$	P_O	C_E	$(T_c)_O$	$(T_e)_O$	$(COP)_O$	P_O
kW/K	K	K	–	kW	kW/K	K	K	–	kW
0.6	332.81	265.84	3.97	0.76	0.6	329.47	263.17	3.97	0.76
0.7	331.05	266.22	4.11	0.73	0.7	328.92	264.50	4.11	0.73
0.8	329.72	266.53	4.22	0.71	0.8	328.48	265.53	4.22	0.71
0.9	328.67	266.78	4.31	0.70	0.9	328.12	266.34	4.31	0.70
1.0	327.82	267.00	4.39	0.68	1.0	327.82	267.00	4.39	0.68
1.1	327.12	267.19	4.46	0.67	1.1	327.57	267.55	4.46	0.67
1.2	326.53	267.35	4.52	0.66	1.2	327.35	268.02	4.52	0.66
1.3	326.03	267.49	4.57	0.66	1.3	327.15	268.41	4.57	0.66
1.4	325.59	267.62	4.62	0.65	1.4	326.98	268.76	4.62	0.65
1.5	325.21	267.73	4.66	0.64	1.5	326.83	269.07	4.66	0.64

Table 7.7 Effect of condenser/evaporator side heat exchanger effectiveness ($\varepsilon_C/\varepsilon_E$), on working fluid temperatures, power input, and performance of the airconditioning system

ε_C	$(T_c)_O$	$(T_e)_O$	$(COP)_O$	P_O	ε_E	$(T_c)_O$	$(T_e)_O$	$(COP)_O$	P_O
–	K	K	–	kW	–	K	K	–	kW
0.20	348.07	263.25	3.10	0.97	0.20	333.53	252.25	3.10	0.97
0.30	338.90	264.68	3.57	0.84	0.30	331.22	258.68	3.57	0.84
0.40	334.34	265.52	3.86	0.78	0.40	329.93	262.02	3.86	0.78
0.50	331.58	266.10	4.06	0.74	0.50	329.09	264.10	4.06	0.74
0.60	329.72	266.53	4.22	0.71	0.60	328.48	265.53	4.22	0.71
0.70	328.37	266.86	4.34	0.69	0.70	328.02	266.57	4.34	0.69
0.75	327.82	267.00	4.39	0.68	0.75	327.82	267.00	4.39	0.68
0.80	327.34	267.13	4.44	0.68	0.80	327.65	267.38	4.44	0.68
0.90	326.53	267.35	4.52	0.66	0.90	327.35	268.02	4.52	0.66
1.00	325.87	267.54	4.59	0.65	1.00	327.09	268.54	4.59	0.65

Table 7.8 Effect of internal irreversibility parameter ($R_{\Delta S}$), on working fluid temperatures, power input, and performance of the airconditioning system

$R_{\Delta S}$	$(T_c)_O$	$(T_e)_O$	$(COP)_O$	P_O
–	K	K	–	kW
0.50	335.26	265.34	0.65	4.58
0.55	333.92	265.61	0.78	3.86
0.60	332.81	265.84	0.92	3.26
0.65	331.87	266.04	1.09	2.76
0.70	331.05	266.22	1.29	2.33
0.75	330.34	266.38	1.53	1.96
0.80	329.72	266.53	1.83	1.64
0.85	329.16	266.66	2.21	1.36
0.90	328.67	266.78	2.71	1.11
0.95	328.22	266.90	3.40	0.88
1.00	327.82	267.00	4.39	0.68

$$T_{C1} = 310\ \text{K}, T_{E1} = 283\ \text{K}, C_C = C_E = 1.00\ \text{kW/K},$$
$$\varepsilon_C = \varepsilon_E = 0.75\ R_{\Delta S} = 1.0, \text{and } P_H = 3.0\ \text{kW}.$$

It is seen from Table 7.1 that by increasing the inlet temperature of the external fluid on source side (T_{E1}), optimal power input to the system decreases, and consequently optimal coefficient of performance increases. By increasing inlet temperature of external fluid on sink side (T_{C1}) and temperature of optimal working fluid on high-temperature side, optimal power input to the system increases, and hence optimal coefficient of performance of the system decreases, whereas temperature of optimal working fluid on low-temperature side increases slowly.

Table 7.2 shows similar effects of heat capacitance rates of external fluid in source/sink thermal reservoirs (C_E, C_C). By increasing any one of them, source-side optimal working fluid temperature increases and sink-side temperature decreases, both of which result in lower optimal power input and consequently higher optimal coefficient of performance.

Table 7.3 shows the effect of effectiveness of source-/sink-side heat exchanger (ε_E, ε_C). By increasing any one of them while the other is being kept as constant, temperature of sink-side working fluid decreases while source-side working fluid temperature increases with increasing effectiveness, so optimal power input to the system decreases, and consequently optimal coefficient of performance of the system increases because heating load is given.

Table 7.4 shows the effect of internal irreversibility parameter ($R_{\Delta S}$); by increasing this parameter, sink-side optimal working fluid temperature increases whereas source-side working fluid temperature decreases, so optimal power input to the system decreases, and hence optimal coefficient of performance of the system increases. The case $R_{\Delta S} = 1.0$ corresponds to the endoreversible case.

In a practical heat pump system, there is a need to understand the deviation of actual performance from the ideal reversible performance. The inclusion and cause of the external and internal irreversibilities in the theoretical model help to fill the gap. An estimate of irreversibility parameter ($R_{\Delta S}$) can be obtained from practical units, and the effect of change in $R_{\Delta S}$ can then be studied. Though a fair comparison is rather difficult, certainly the present model can be used to assist designers and practising engineers to optimize the performance of heat pump systems.

7.7.2 Airconditioning Cycle

In order to have a numerical appreciation of the theoretical analysis of vapour compression airconditioning system, the effect of various input parameters on the optimal coefficient of performance of the airconditioning system is studied, and results are shown in Tables 7.5, 7.6, 7.7, and 7.8. During the variation of any one parameter, all other parameters are assumed to be constant as given below:

$$T_{C1} = 318\ \mathrm{K}\ T_{E1} = 275\ \mathrm{K}\ C_C = C_E = 1.00\ \mathrm{kW/K}$$

$$\varepsilon_C = \varepsilon_E = 0.75\ R_{\Delta S} = 1.0, \text{and } \mathrm{P_L} = 3.0\ \mathrm{kW}$$

Table 7.5 shows that by increasing the inlet temperature of the external source-side fluid, the optimal power input to the system decreases, and consequently the optimal coefficient of performance increases. Although by increasing sink inlet temperature, the optimal temperature of internal working fluid on high-temperature side and the power input to the system both increase, and hence the optimal coefficient of performance of the system decreases, whereas the temperature of the optimal internal working fluid on the low-temperature side remains constant.

Table 7.6 shows the effect of heat capacitance rates of external fluid on source/sink side. By increasing any one of them, the sink-side optimal internal working fluid temperature and the power input to the system both decrease; consequently optimal coefficient of performance of the system increases, and the source-side optimal internal working fluid temperature is also increased.

Table 7.7 shows the effect of effectiveness (ε_E, ε_C) of source-/sink-side heat exchangers of the airconditioning system. By increasing either value with the other being kept constant, the optimal power input to the system decreases, and consequently the optimal coefficient of performance of the system increases because cooling load is fixed. The optimal temperature of the sink-side internal working fluid decreases, while the source-side internal working fluid temperature increases with increasing effectiveness.

Table 7.8 shows the effect of the internal irreversibility parameter ($R_{\Delta S}$). By increasing this parameter (decreasing internal irreversibility), the source-side optimal internal working fluid temperature increases, whereas the sink-side optimal internal working fluid temperature and the power input to the system decrease, and hence optimal coefficient of performance of the system increases. The case $R_{\Delta S} = 1.0$ corresponds to the endoreversible case.

7.8 Conclusion

Finite time thermodynamic analysis of an irreversible heat pump/airconditioning system has been carried out using the method of Lagrangian multiplier by minimizing power input for a given heating/cooling load condition. It can be seen that COP is not only the function of thermal reservoir temperatures but also heat capacitance of external fluids, effectiveness of heat exchangers, and internal irreversibility parameters. The effect of inlet temperatures of external fluids in heat source/sink side, finite heat capacitance of external fluids, effectiveness of the heat exchangers, and internal irreversibility parameters, on the coefficient of performance of refrigeration airconditioning and heat pump systems, have been studied. The internal irreversibility in a heat pump/airconditioning system has been characterized by a single parameter representing the ratio of two entropy differences, and it is calculated for some specific refrigerants (viz. R-12 and R-134a). It is found that internal irreversibility parameter increases and approaches unity (endoreversible) with increasing compression efficiency/condensation temperature. This parameter appears in both the equations for minimum input power and coefficient of performance of these systems. The equations clearly show that a system with internal irreversibility needs more input power and has less coefficient of performance than endoreversible system. It is seen that internal irreversibility has more drastic effect on performance than any other parameter. A comparative study of alternatively and continuously connected case shows that coefficient of performance of continuously connected case is slightly higher than the corresponding COP of alternatively connected case. However, the difference in COP decreases as the internal irreversibility increases (value of R_{DS} decreases). It can be seen from the study of VC system with liquid–vapour (L–V) heat exchanger that with increasing compression efficiency and effectiveness of L–V heat exchanger, the internal irreversibility parameter for some specific refrigerants (viz. R-12 and R-134a) increases and reaches towards endoreversible case, i.e. $R_{\mathrm{DS}} = 1$, but it is found decreasing with increasing condensation temperature. Coefficient of performance predicted by finite time thermodynamic model (when system is continuously connected to thermal reservoirs) is found to be quite close to the experimental COP. Optimal power input and heat rejection rate are also found to be comparable with the experimental values. In continuing search for a realistic theoretical upper limit for coefficient of performance of the system, new equations have been developed using finite time thermodynamic model.

Chapter 8
Finite Time Thermodynamics of Cascaded Refrigeration and Heat Pump Cycles

8.1 Introduction

The conventional vapour compression refrigeration systems cannot be effectively used for low-temperature applications because the power consumed by the compressor is largely increased, and hence for low-temperature applicatireservoir and cycle-I working fluid, cycon, a cng the analysis of refrigeration/heat pump cyascaded refrigeration system is used. For finding the optimum performance of cascaded systems, we are applying Lagrangian multiplier method in which power input to the system is minimized at given cooling/heating load condition. This optimization method has been made for single stage irreversible airconditioning/heat pump system in earlier Chap. 7. Now similar analysis is being extended for cascaded (double stage) refrigeration/heat pump system, and a comparison of cascaded and single stage refrigeration/heat pump systems is being made in this chapter. The performance of a cascaded endoreversible/irreversible Carnot refrigeration/heat pump cycle is somewhat different from that of a single stage one, because there are changes in the irreversible losses when endoreversible/irreversible refrigeration/heat pump cycles are combined. For reversible refrigeration/heat pump cycles, irreversible loss does not exist, and so there is no change of such loss when reversible cycles are combined.

Chen and Yan (1988) have carried out finite time thermodynamic analysis for combined refrigeration cycle by taking the thermal reservoirs (heat source and heat sink) of infinite heat capacity and considered the system as internally reversible, but in actual case thermal reservoirs will be of finite heat capacity, and system will be externally as well as internally irreversible. Finite heat capacity reservoirs imply that reservoirs are finite in size, and its temperature changes with heat transfer. However, the fact that the equal time duration and identical heat absorption and rejection by the two cycles at the interface do not imply that the working fluids have to be the same in the two cycles because the temperatures of heat rejection and heat absorption at the interface are different. Different working fluids have to be used

S.C. Kaushik et al., *Finite Time Thermodynamics of Power and Refrigeration Cycles*,
https://doi.org/10.1007/978-3-319-62812-7_8

such that identical heat absorption in cycle-I and heat rejection in cycle-II occurs in the same time. On the mode of coupling between two cycles, it is always desirable to find out a more realistic bound on coefficient of performance (COP) considering the thermal irreversibilities at the interface rather than partial coupling of the two cycles by inserting a heat storage reservoir, thereby ignoring these irreversibilities. So in this chapter, we have considered heat transfer irreversibilities at the interface due to finite temperature difference across the common heat exchanger, and analysis is extended including the effect of finite heat capacitance of thermal reservoirs and internal irreversibility.

8.2 Cascade Refrigeration and Heat Pump Cycles

Let us consider two refrigeration/heat pump cycles in series to form a single cascaded refrigeration/heat pump system operating between two thermal reservoirs (heat source/heat sink) of finite heat capacities as shown in Fig. 8.1a.

T–s diagram of cascaded refrigeration/heat pump cycle is shown in Fig. 8.1b. Cycle-II extracts heat from low-temperature thermal reservoir and transfers this heat to cycle-I, with some external work W_2 supplied to the cycle-II. The heat transferred to cycle-I is then rejected to the high-temperature thermal reservoir

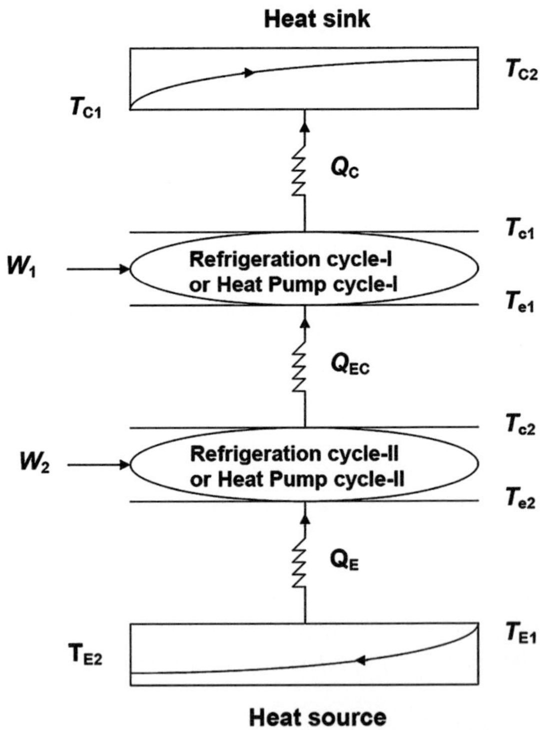

Fig. 8.1a Schematic diagram of cascaded refrigeration/heat pump cycle

Fig. 8.1b *T–s* diagram of cascaded refrigeration heat pump cycle

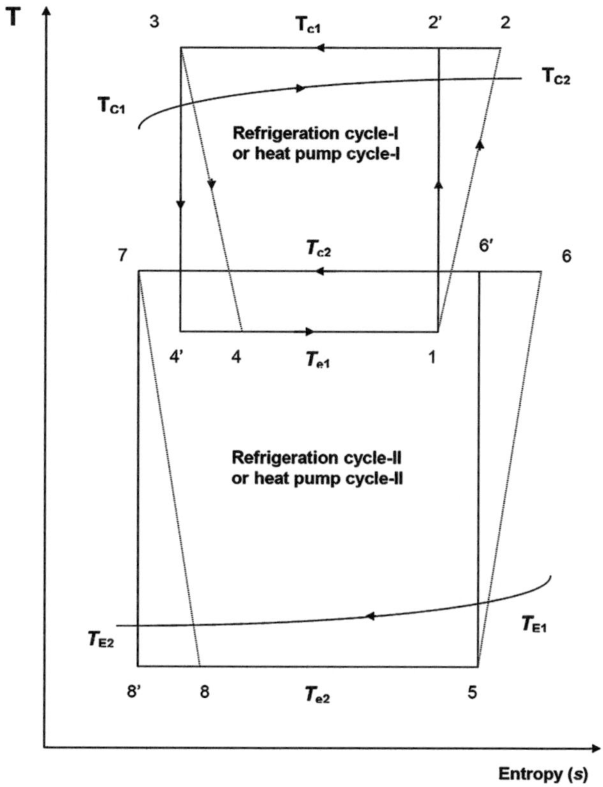

(heat sink) with work input W_1 supplied to cycle-I. We have taken source/sink thermal reservoirs of finite heat capacities, so during heat transfer, the temperatures of low- and high-temperature thermal reservoir will vary from T_{E1} to T_{E2} and T_{C1} to T_{C2}, respectively. Temperatures of the working fluid in the first cycle are taken as T_{c1} and $T_{e1,}$ while the temperatures of the working fluid in the second cycle are T_{c2} and T_{e2} when heat leaves and enters the working fluid, respectively.

In actual case, the temperature of working fluid during heat rejection from the system to the heat sink will not be constant; first it will desuperheat and then condense at constant condensation temperature. But for the sake of simplicity in our analysis, we are taking entropic average temperature which will be little higher than condensation temperature, and thus actual cycle on *T–s* diagram can be considered as modified reversed Carnot refrigeration cycle (Lee and Kim 1992).

The isentropic processes occur in negligible time. This means that they must occur on a time scale that is fast as compared to the slow rates of heat leaks to the environment but slow compared to the rapid internal relaxation of pressure gradients in the working fluid thereby ensuring the need for equilibrium of the working fluid throughout the process. However, isothermal (heat addition/rejection) processes are too slow in comparison to the time of rapid internal relaxation of pressure gradients in the working fluid as well as time of the isentropic processes (Kaushik et al. 2002c).

Therefore, the total cycle time for each stage is nearly the time of isothermal processes, and the time for isentropic processes is negligible. For convenience, we further assume that:

1. The two refrigeration cycles have the same cycle time (τ).
2. The heat exchange between two refrigeration cycles is carried out directly and heat leakage is negligible, so the quantity of heat rejected by the second working fluid is equal to that absorbed by the first working fluid. Thus the time of the heat rejected of the second working fluid is equal to that of the absorbed heat of the first working fluid. We consider this time as t_1.
3. Thus the time of the absorbed heat in the second cycle is also equal to that of the rejected heat in the first one. If we consider this time as t_2 then obviously, it is $t_2 = \tau - t_1$.

8.3 Finite Time Cascade Cycles

The rate of heat extraction from low-temperature heat source by the system (cycle-II) is given by:

$$Q_E = \frac{Q_E}{t_2} = U_E A_E (\text{LMTD})_E = \dot{m}_E C_{PE}(T_{E1} - T_{E2}) \tag{8.1}$$

Similarly, the rate of heat rejection from system (cycle-I) to the heat sink:

$$Q_C = \frac{Q_C}{t_2} = U_C A_C (\text{LMTD})_C = \dot{m}_C C_{PC}(T_{C2} - T_{C1}) \tag{8.2}$$

where

$$(\text{LMTD})_C = \frac{(T_{c1} - T_{C1}) - (T_{c1} - T_{C2})}{\ln \dfrac{(T_{c1} - T_{C1})}{(T_{c1} - T_{C2})}}$$

and

$$(\text{LMTD})_E = \frac{(T_{E1} - T_{e2}) - (T_{E2} - T_{e2})}{\ln \frac{(T_{E1} - T_{e2})}{(T_{E2} - T_{e2})}}$$

Using LMTD expressions in Eqs. (8.1) and (8.2), we have:

$$Q_E = t_2 EE(T_{E1} - T_{e2})$$
$$Q_C = t_2 CE(T_{c1} - T_{C1})$$

where

$$CE = C_C\varepsilon_C,\ EE = C_E\varepsilon_E,\ C_E = \dot{m}_E C_{PE},\ C_C = \dot{m}_C C_{PC},$$
$$\varepsilon_E = 1 - e^{-(N_E)}, \varepsilon_C = 1 - e^{-(N_C)}, N_C = \frac{U_C A_C}{\dot{m}_C C_{PC}} \text{ and } N_E = \frac{U_E A_E}{\dot{m}_E C_{PE}}$$

The effective thermal conductance between high-temperature reservoir and cycle-I working fluid, cycle-I working fluid and cycle-II working fluid, and the cycle-II working fluid and low-temperature reservoir are CE(t), ME(t), and EE(t), respectively. The dependence between these quantities and the time t is taken as follows:

$$\begin{aligned} CE(t) &= CE\, t_1 \le t \le \tau & \mathrm{ME}(t) &= 0\, t_1 \le t \le \tau & EE(t) &= EE\, t_1 \le t \le \tau \\ &= 0 \quad 0 \le t \le t_1 & &= \mathrm{ME}\ 0 \le t \le t_1 & &= 0 \quad 0 \le t \le t_1 \end{aligned}$$

where, CE and EE are defined as given above and ME is effective thermal conductance between the working fluids of cycle-I and cycle-II. Now heat transfer can be written as:

$$Q_C = \int_0^\tau CE(t)[T_{c1}(t) - T_{C1}]\,dt \tag{8.3}$$

$$Q_E = \int_0^\tau EE(t)[T_{E1} - T_{e2}(t)]\,dt \tag{8.4}$$

From the first law of thermodynamics, we have:

$$W = \int_0^\tau [CE(t)(T_{c1}(t) - T_{C1}) - EE(t)(T_{E1} - T_{e2}(t))]\,dt \tag{8.5}$$

where Q_C, Q_E, and W are the quantities of heat rejected, heat absorbed, and total input work in cascaded refrigeration/heat pump system, respectively.

Irreversible Refrigeration/Heat Pump Cycle
If internal irreversibility such as friction, non-isentropic compression, and expansion is accounted for the system, the two isentropic processes become adiabatic processes with entropy generation. This internal irreversibility of the system can be characterized by an internal irreversibility parameter for each cycle of cascaded refrigeration/heat pump system, representing non-isentropic expansion/compression in terms of the ratio of entropy changes/differences. On T–s diagram, the four processes of each cycle of cascaded refrigeration/heat pump system are shown in Fig. 8.1b.

Refrigeration/Heat Pump Cycle-I
Heat rejected by the reversible refrigeration/heat pump cycle-I is $Q'_C = T_{c1}(S'_2 - S_3)$, while heat rejected by irreversible refrigeration/heat pump cycle-I is

$Q_C = T_{c1}(S_2 - S_3)$. Since $S_2 > S'_2$ so $Q'_C < Q_C$ (primes are added to quantities associated with reversible cycle). Thus we can define an irreversible heat rejection parameter C_1 such that:

$$Q'_C = C_1 Q_C \text{ with } C_1 = (S'_2 - S_3)/(S_2 - S_3) < 1$$

Similarly, an irreversible heat absorption parameter C_2 can be defined as:

$$Q_{EC} = C_2\, Q'_{EC} \text{ with } C_2 = (S_1 - S_4)/(S_1 - S'_4) < 1$$

Using second law of thermodynamics:

$$\int \frac{dQ}{T} = \frac{Q_{EC}}{T_{e1}} - \frac{Q_C}{T_{c1}} < 0$$

and for a reversible cycle

$$\frac{Q'_{EC}}{T_{e1}} - \frac{Q'_C}{T_{c1}} = 0$$

so,

$$\frac{Q_{EC}}{T_{e1}} - R_{\Delta S}\frac{Q_C}{T_{c1}} = 0$$

where

$$R_{\Delta S} = \frac{S_1 - S_4}{S_2 - S_3} = C_1 C_2 < 1$$

Refrigeration/Heat Pump Cycle-II
Following the analysis of refrigeration/heat pump cycle-I, we can write for cycle-II:

$$\frac{Q_E}{T_{e2}} - R'_{\Delta S}\frac{Q_{EC}}{T_{c2}} = 0$$

where

$$R'_{\Delta S} = \frac{S_5 - S_8}{S_6 - S_7} = C_3 C_4 < 1$$

where C_3 and C_4 are the irreversible heat absorption and irreversible heat rejection parameters for cycle-II, respectively, and given by:

$$C_3 = (S_5 - S_8)/(S_5 - S'_8) < 1$$

and

$$C_4 = (S_6' - S_7)/(S_6 - S_7) < 1$$

$R_{\Delta S}$ and $R'_{\Delta S}$ are the internal irreversibility parameters for refrigeration/heat pump cycle-I and cycle-II, respectively. $R_{\Delta S} = R'_{\Delta S} = 1.0$ for endoreversible cycle, and $R_{\Delta S} = R'_{\Delta S} < 1.0$ for irreversible cycle. Now taking these parameters into account and using the second law of thermodynamics, we have:

$$\Delta S_1 = \int_0^{\tau} \left[R_{\Delta S} CE(t) \left(\frac{T_{C_1}}{T_{c1}(t)} - 1 \right) + \text{ME}(t) \left(\frac{T_{c_2}(t)}{T_{e1}(t)} - 1 \right) \right] \mathrm{d}t = 0 \tag{8.6}$$

$$\Delta S_2 = \int_0^{\tau} \left[R'_{\Delta S} \text{ME}(t) \left(\frac{T_{e_1}(t)}{T_{c2}(t)} - 1 \right) + EE(t) \left(\frac{T_{E_1}(t)}{T_{e2}(t)} - 1 \right) \right] \mathrm{d}t = 0 \tag{8.7}$$

where ΔS_1 and ΔS_2 are entropy productions in cycle-I and cycle-II of irreversible refrigeration/heat pump system, respectively. Thermodynamic cycle for refrigeration and heat pump system is the same, but the objective function is different, i.e. desired effect is cooling load for refrigerator and heating load for heat pump system.

8.3.1 *Irreversible Refrigeration Cycle*

For seeking minimum input power, for a given cooling load, we introduce modified Lagrangian operator defined as:

$$\begin{aligned} L = & \left[CE(t)(T_{c1}(t) - T_{C1}) - EE(t)(T_{E1} - T_{e2}(t))\right] + \mu\left[EE(t)(T_{E1} - T_{e2}(t))\right] \\ & + \lambda_1 \left[R_{\Delta S} CE(t) \left(\frac{T_{C1}}{T_{c1}(t)} - 1 \right) + \text{ME}(t) \left(\frac{T_{c2}(t)}{T_{e1}(t)} - 1 \right) \right] \\ & + \lambda_2 \left[R'_{\Delta S} \text{ME}(t) \left(\frac{T_{e1}(t)}{T_{c2}(t)} - 1 \right) + EE(t) \left(\frac{T_{E1}}{T_{e2}(t)} - 1 \right) \right] \end{aligned} \tag{8.8}$$

where μ, λ_1, and λ_2 are Lagrangian undetermined multipliers:

$$\frac{\partial L}{\partial T_{c1}(t)} = 0 \quad \text{and} \quad \frac{\partial L}{\partial T_{c2}(t)} = 0$$

give optimal values as given below:

$$T_{c1}(t) = \sqrt{\lambda_1 T_{e1} R_{\Delta S}} = T_1(\text{say}) \tag{8.9}$$

$$T_{e2}(t) = \sqrt{\frac{\lambda_2 T_{E1}}{1 - \mu}} = T_2(\text{say}) \tag{8.10}$$

Similarly,

$$\frac{\partial L}{\partial T_{e1}(t)} = 0 \quad \text{and} \quad \frac{\partial L}{\partial T_{c2}(t)} = 0 \quad \text{both give}$$

$$\frac{T_{e1}(t)}{T_{c2}(t)} = \sqrt{\frac{\lambda_1}{\lambda_2 R'_{\Delta S}}} = C(\text{say}) \tag{8.11}$$

Substituting the optimal values of working fluid temperatures from Eqs. (8.9) and (8.11) into Eqs. (8.3), (8.4), (8.6), and (8.7), we have:

$$Q_C = CE\,t_2(T_1 - T_{c1}) \tag{8.12}$$

$$Q_E = EE\,t_2(T_{E1} - T_2) \tag{8.13}$$

$$\Delta S_1 = R_{\Delta S} CE\,t_2\left(\frac{T_{C1}}{T_1} - 1\right) + \text{ME}\,t_1\left(\frac{1}{C} - 1\right) = 0 \tag{8.14}$$

$$\Delta S_2 = EE\,t_2\left(\frac{T_{E1}}{T_2} - 1\right) + R'_{\Delta S}\,\text{ME}\,t_1(\text{C} - 1) = 0 \tag{8.15}$$

Now using Eqs. (8.12), (8.13), (8.14), and (8.15), we have:

$$T_1 = \frac{T_{C1}\left(T_{E1} - \frac{Q_E}{EE\,t_2} - \frac{Q_E}{R'_{\Delta S}\,\text{ME}\,t_1}\right)}{T_{E1} - Q_E\left(\frac{1}{R_{\Delta S} R'_{\Delta S}\,CE\,t_2} + \frac{1}{EE\,t_2} + \frac{1}{R'_{\Delta S}\,\text{ME}\,t_1}\right)} \tag{8.16}$$

$$T_2 = T_{E1} - \frac{Q_E}{EE\,t_2} \tag{8.17}$$

$$C = 1 - \frac{\frac{Q_E}{\text{ME}\,t_1}}{\left(T_{E1} - \frac{Q_E}{\text{EE}\,t_2}\right)} \tag{8.18}$$

$$Q_C = \frac{\frac{Q_E\,T_{C1}}{R_{\Delta S} R'_{\Delta S}}}{T_{E1} - Q_E\left(\frac{1}{R_{\Delta S} R'_{\Delta S}\,CE\,t_2} + \frac{1}{EE\,t_2} + \frac{1}{R'_{\Delta S}\,\text{ME}\,t_1}\right)} \tag{8.19}$$

and so COP of the cascaded refrigeration system is given by:

$$\text{COP} = \frac{Q_E}{Q_C - Q_E} \tag{8.20}$$

$$\text{COP} = \left[\frac{\frac{T_{C1}}{R_{\Delta S} R'_{\Delta S}}}{T_{E1} - \text{Q}_E\left(\frac{1}{R_{\Delta S} R'_{\Delta S}\,CE\,t_2} + \frac{1}{EE\,t_2} + \frac{1}{R'_{\Delta S}\,\text{ME}\,t_1}\right)} - 1\right]^{-1} \tag{8.21}$$

$$\frac{\partial \mathrm{COP}}{\partial t_1} = 0 \quad \text{gives,}$$
$$\frac{\tau}{t_1} = 1 + \sqrt{\frac{\mathrm{ME}\left(R_{\Delta \mathrm{S}} R'_{\Delta \mathrm{S}} CE + EE\right)}{R_{\Delta \mathrm{S}} CE\, EE}} \tag{8.22}$$

$$\frac{\partial \mathrm{COP}}{\partial t_2} = 0 \quad \text{gives,}$$
$$\frac{\tau}{t_2} = 1 + \sqrt{\frac{R_{\Delta \mathrm{S}}\, \mathrm{ME}\, EE)}{\mathrm{ME}\left(R_{\Delta \mathrm{S}} R'_{\Delta \mathrm{S}} CE + EE\right)}} \tag{8.23}$$

On substituting the optimal values of t_1 and t_2 from Eqs. (8.22) and (8.23) into Eqs. (8.16), (8.17), (8.18), and (8.21), we have optimal working fluid temperatures and optimal COP of an irreversible refrigeration system as given by:

$$(T_{\mathrm{c1}})_{\mathrm{o}} = \frac{T_{\mathrm{C1}}\left[T_{\mathrm{E1}} - \frac{P_{\mathrm{L}}}{EE}\left(1 + \sqrt{\frac{CE\, EE\, R_{\Delta \mathrm{S}}}{\mathrm{ME}\left(CE\, R_{\Delta \mathrm{S}}\, R'_{\Delta \mathrm{S}} + EE\right)}}\right)\left(1 + \frac{1}{R'_{\Delta \mathrm{S}}}\sqrt{\frac{(EE\left(CE\, R_{\Delta \mathrm{S}}\, R'_{\Delta \mathrm{S}} + EE\right)}{\mathrm{ME}\, EE\, R_{\Delta \mathrm{S}}}}\right)\right]}{T_{\mathrm{E1}} - P_{\mathrm{L}}\frac{\left[\sqrt{CE\, EE\, R_{\Delta \mathrm{S}}} + \sqrt{\mathrm{ME}\left(CE\, R_{\Delta \mathrm{S}}\, R'_{\Delta \mathrm{S}} + EE\right)}\right]^2}{CE\, EE\, \mathrm{ME}\, R_{\Delta \mathrm{S}}\, R'_{\Delta \mathrm{S}}}} \tag{8.24}$$

$$(T_{\mathrm{e2}})_{\mathrm{o}} = T_{\mathrm{E1}} - \frac{P_{\mathrm{L}}}{EE}\left(1 + \sqrt{\frac{CE\, EE\, R_{\Delta \mathrm{S}}}{\mathrm{ME}\left(CE\, R_{\Delta \mathrm{S}}\, R'_{\Delta \mathrm{S}} + EE\right.}}\right) \tag{8.25}$$

$$\left(\frac{T_{\mathrm{e1}}}{T_{\mathrm{c2}}}\right)_{\mathrm{o}} = 1 - \frac{\frac{P_{\mathrm{L}}}{\mathrm{ME}\, R'_{\Delta \mathrm{S}}}\left(1 + \sqrt{\frac{\mathrm{ME}\left(CE\, R_{\Delta \mathrm{S}}\, R'_{\Delta \mathrm{S}} + EE\right.}{CE\, EE\, R_{\Delta \mathrm{S}}}}\right)}{T_{\mathrm{E1}} - \frac{P_{\mathrm{L}}}{EE}\left(1 + \sqrt{\frac{CE\, EE\, R_{\Delta \mathrm{S}}}{\mathrm{ME}\left(CE\, R_{\Delta \mathrm{S}}\, R'_{\Delta \mathrm{S}} + EE\right.}}\right)} \tag{8.26}$$

$$\mathrm{COP}_{\mathrm{o}} = \left[\frac{T_{\mathrm{C1}}/R_{\Delta \mathrm{S}} R'_{\Delta \mathrm{S}}}{T_{\mathrm{E1}} - \frac{P_{\mathrm{L}}}{K}} - 1\right]^{-1} \tag{8.27}$$

where

$$P_{\mathrm{L}} = \frac{Q_{\mathrm{E}}}{\tau}, K = \frac{R_{\Delta \mathrm{S}}\, R'_{\Delta \mathrm{S}}\, CE\, EE\, \mathrm{ME}}{\left[\sqrt{\mathrm{ME}\left(R_{\Delta \mathrm{S}}\, R'_{\Delta \mathrm{S}}\, CE + EE\right)} + \sqrt{CE\, EE\, R_{\Delta \mathrm{S}}}\right]^2}$$

Equation (8.27) is the relation between optimal COP and given rate of refrigeration of a cascaded refrigeration system. If we assume $R_{\Delta \mathrm{S}} = R'_{\Delta \mathrm{S}} = 1.0$ and introduce the equivalent temperature given by:

$$T^*_{\mathrm{E1}} = T_{\mathrm{E1}} - \frac{Q_{\mathrm{E}}}{\tau K}, \tag{8.28}$$

optimal COP of the combined refrigeration system becomes $\mathrm{COP_o} = T^*_{\mathrm{E1}}/(T_{\mathrm{C1}} - T^*_{\mathrm{E1}})$. This equation shows that combined refrigeration cycle operating between the reservoir temperatures T_{C1} and T_{E1} at a given rate of refrigeration is equivalent to a single reversible Carnot refrigeration cycle operating between the reservoirs at temperatures T_{C1} and T^*_{E1}, where T^*_{E1} is the effective source temperature for the cascaded refrigeration cycle caused by the existence of effective thermal resistance of the cascaded system.

Further if we assume that $T^*_{\mathrm{E1}} = T_{\mathrm{E1}}$, then COP expression becomes Carnot COP as given by:

$$(\mathrm{COP})_{r1} = \frac{T_{\mathrm{E1}}}{T_{\mathrm{C1}} - T_{\mathrm{E1}}} \tag{8.29}$$

Similarly, COP for single stage modified Carnot refrigeration cycle with the same operating temperatures T_{C1} and T_{E1} can be given as:

$$\mathrm{COP} = \left[\frac{T_{\mathrm{C1}}/R_{\Delta\mathrm{S}}}{T_{\mathrm{E1}} - \frac{P'_{\mathrm{L}}}{K}} - 1\right]^{-1} \tag{8.30}$$

where

$$P'_{\mathrm{L}} = \frac{Q'_{\mathrm{E}}}{\tau'}, K' = \frac{CE'\,EE'\,\mathrm{R}_{\Delta\mathrm{S}}}{\left[\sqrt{CE'\,R_{\Delta\mathrm{S}}} + \sqrt{EE'}\right]^2}$$

P'_{L} is a rate of refrigeration, and CE' and EE' are effective thermal conductances on heat sink and source side of single stage refrigeration system, respectively. Using Eqs. (8.27) and (8.30), if we equate the performance of single stage and cascaded refrigeration system and assume that $R_{\Delta\mathrm{S}} = R'_{\Delta\mathrm{S}} = 1.0$, then we have:

$$\frac{P_{\mathrm{L}}}{P'_{\mathrm{L}}} = \frac{\mathrm{K}}{\mathrm{K}'} = \frac{\left[\left(CE^{-1} + EE^{-1}\right)^{1/2} + \mathrm{ME}^{-1/2}\right]^{-2}}{\left[CE'^{-1/2} + EE^{-1/2}\right]^{-2}} \tag{8.31}$$

8.3.2 *Irreversible Heat Pump Cycle*

The schematic diagram given in Fig. 8.1a for an irreversible refrigeration cycle holds equally valid for an irreversible heat pump also. However, here we are seeking for minimum input power, with a given heating load rather than cooling load. For this we have introduced a modified Lagrangian operator defined as:

$$\begin{aligned} L = {} & [CE(t)\,(T_{c1}(t) - T_{C1}) - EE(t)(T_{E1} - T_{e1}(t))] \\ & +\mu[CE(t)\,(T_{c1}(t) - T_{C1})] \\ & +\lambda_1\left[R_{\Delta S}\,CE(t)\left(\frac{T_{C1}}{T_{c1}(t)} - 1\right) + \mathrm{ME}(t)\left(\frac{T_{c2}(t)}{T_{e1}(t)} - 1\right)\right] \\ & +\lambda_2\left[R'_{\Delta S}\,\mathrm{ME}(t)\left(\frac{T_{e1}(t)}{T_{c2}(t)} - 1\right) + EE(t)\left(\frac{T_{E1}}{T_{e2}(t)} - 1\right)\right] \end{aligned} \tag{8.32}$$

Now, following the earlier section analysis of cascaded irreversible refrigeration cycle, we have found out optimum working fluid temperature and COP of an irreversible cascaded heat pump system as given below:

$$(T_{e2})_o = \frac{T_{E1}\left[T_{C1} - \frac{P_H}{CE}\left(1 + \sqrt{\frac{CE\,EE\,R_{\Delta S}}{\mathrm{ME}\,(CE\,R_{\Delta S}\,R'_{\Delta S} + EE)}}\right)\left(1 + \sqrt{\frac{CE\,R_{\Delta S}(CE\,R_{\Delta S}\,R'_{\Delta S} + EE)}{\mathrm{ME}\,EE}}\right)\right]}{T_{C1} + P_H \frac{\left[\sqrt{CE\,EE\,R_{\Delta S}} + \sqrt{\mathrm{ME}\,(CE\,R_{\Delta S}\,R'_{\Delta S} + EE)}\right]}{CE\,EE\,\mathrm{ME}}} \tag{8.33}$$

$$(T_{c1})_o = T_{C1} + \frac{P_H}{CE}\left(1 + \sqrt{\frac{CE\,EE\,R_{\Delta S}}{\mathrm{ME}\,(CE\,R_{\Delta S}\,R'_{\Delta S} + EE)}}\right) \tag{8.34}$$

$$\left(\frac{T_{e1}}{T_{c2}}\right)_o = \frac{T_{C1}\,\frac{P_H}{\mathrm{CE}}\left(1 + \sqrt{\frac{CE\,\mathrm{EE}\,R_{\Delta S}}{\mathrm{ME}\,(CE\,R_{\Delta S}\,R'_{\Delta S} + EE)}}\right)}{T_{C1} + \frac{P_H}{CE}\left(1 + \sqrt{\frac{CE\,\mathrm{EE}\,R_{\Delta S}}{\mathrm{ME}\,(CE\,R_{\Delta S}\,R'_{\Delta S} + EE)}}\right)} \tag{8.35}$$

$$\left(1 + \sqrt{\frac{CE\,R_{\Delta S}(CE\,R_{\Delta S}\,R'_{\Delta S} + \mathrm{EE}}{\mathrm{ME\,EE}}}\right)$$

$$\mathrm{COP}_o = \left[1 - \frac{T_{E1}\,R_{\Delta S}R'_{\Delta S}}{T_{C1} + \frac{P_H}{K}}\right] \tag{8.36}$$

where

$$P_H = \frac{Q_C}{\tau}, K = \frac{CE\,EE\,\mathrm{ME}}{\left[\sqrt{\mathrm{ME}\,(R_{\Delta S}\,R'_{\Delta S}\,CE + EE)} + \sqrt{CE\,EE\,R_{\Delta S}}\right]^2}$$

Equation (8.36) is a relation between optimal COP and given heating load of an irreversible cascaded heat pump system. If we assume that $R_{\Delta S} = \mathrm{R'}_{\Delta S} = 1.0$ and introduce an equivalent temperature given by:

$$T_{\mathrm{C1}}^{*} = T_{\mathrm{C1}} + \frac{P_{\mathrm{H}}}{K}, \tag{8.37}$$

now, optimal COP of the cascaded heat pump system becomes $\mathrm{COP_o} = T_{\mathrm{C1}}^{*}/\left(T_{\mathrm{C1}}^{*} - T_{\mathrm{E1}}\right)$. This equation shows that cascaded heat pump cycle operating between the reservoir temperatures T_{C1} and T_{E1} at a given heating load is equivalent to a single reversible Carnot heat pump cycle operating between reservoirs at temperatures T_{C1}^{*} and T_{E1}, where T_{C1}^{*} is the effective heat sink temperature of the cascaded heat pump system caused by the existence of effective thermal resistance of the cascaded system. Further if we assume that $T_{\mathrm{C1}}^{*} = T_{\mathrm{C1}}$, then COP expression becomes Carnot COP as given by:

$$(\mathrm{COP})_{r2} = \frac{T_{\mathrm{C1}}}{T_{\mathrm{C1}} - T_{\mathrm{E1}}}$$

Similarly, COP for single stage modified Carnot heat pump cycle with the same operating temperatures T_{C1} and T_{E1} can be given as:

$$\mathrm{COP} = \left[1 - \frac{R_{\Delta \mathrm{S}}\, T_{E1}}{T_{\mathrm{C1}} + \frac{P_{\mathrm{H}}'}{K'}}\right]^{-1} \tag{8.38}$$

where,

$$P_{\mathrm{H}}' = \frac{Q_{\mathrm{C}}'}{\tau'}, K' = \frac{CE'\, EE'}{\left[\sqrt{\mathrm{R}_{\Delta \mathrm{S}}\, CE'} + \sqrt{EE'}\right]^{2}}$$

P_{H}' is heating load, and CE' and EE' are effective thermal conductances on heat sink and heat source side of single stage heat pump system, respectively. Using Eqs. (8.36) and (8.38), if we equate the performance of single stage and cascaded heat pump system and assume that $R_{\Delta \mathrm{S}} = R_{\Delta \mathrm{S}}' = 1$, we have:

$$\frac{P_{\mathrm{H}}}{P_{\mathrm{H}}'} = \frac{K}{K'} = \frac{\left[\left(CE^{-1} + EE^{-1}\right)^{1/2} + \mathrm{ME}^{-1/2}\right]^{-2}}{\left[CE'^{-1/2} + EE^{-1/2}\right]^{-2}} \tag{8.39}$$

If we further assume that for both single and cascade refrigeration/heat pump system, heat conductances on heat source and heat sink side are the same $CE = EE = CE' = EE' = CE_0$ (say), then Eqs. (8.31) and (8.39) become:

$$\frac{P_{\mathrm{L}}}{P_{\mathrm{L}}'} = \left[\frac{2}{\sqrt{2} + \sqrt{\frac{CE_{\mathrm{o}}}{ME}}}\right]^{2} \tag{8.40}$$

and

$$\frac{P_{\mathrm{H}}}{P'_{\mathrm{H}}} = \left[\frac{2}{\sqrt{2} + \sqrt{\frac{CE_o}{\mathrm{ME}}}}\right]^2 \tag{8.41}$$

and if $CE_0 = \mathrm{ME}$, then:

$$\frac{P_{\mathrm{L}}}{P'_{\mathrm{L}}} = \frac{4}{3 + 2\sqrt{2}} \text{ or } P'_{\mathrm{L}} \approx 1.46 P_{\mathrm{L}}$$

and

$$\frac{P_{\mathrm{H}}}{P'_{\mathrm{H}}} = \frac{4}{3 + 2\sqrt{2}} \text{ or } P'_{\mathrm{H}} \approx 1.46 P_{\mathrm{H}}$$

These above equations show that if all the three effective thermal conductances on heat source/sink side and between the two cycle working fluids are the same, then single stage refrigeration/heat pump system is better than cascaded refrigeration/heat pump system. Because at the same COP, it can produce almost 1.46 times more refrigeration/heating load than cascaded refrigeration/heat pump system. Equations (8.40) and (8.41) also show that if effective thermal conductance between the two cycles or stages of cascaded refrigeration/heat pump system is much smaller than the other source-/sink-side effective thermal conductance, then cascaded refrigeration/heat pump system is never used. Because at the same COP, refrigeration/heating load produced by the cascaded refrigeration/heat pump system will be much less than the single stage refrigeration/heat pump system, respectively.

However, if $CE_0 \ll \mathrm{ME}$, then:

$$\frac{P_{\mathrm{L}}}{P'_{\mathrm{L}}} \approx 2 \text{ and } \frac{P_{\mathrm{H}}}{P'_{\mathrm{H}}} \approx 2$$

These equations show that if heat conductance between two cycles of cascade refrigeration/heat pump system is very large as compared to the other side heat conductance, then the cascaded refrigeration/heat pump system will be better than single stage refrigeration/heat pump system because at the same COP it can produce almost double refrigeration effect/heating load.

8.4 Discussion of Results

8.4.1 Cascaded Refrigeration Cycle

In order to have numerical appreciation of the theoretical analysis of the cascaded refrigeration system, we have studied the effect of various input parameters on the performance of the system, and results are shown in Tables 8.1a, 8.1b, 8.1c, 8.1d, 8.1e, 8.1f, 8.1g, and 8.1h. During the variation of any one parameter, all other parameters are assumed to be constant as given below:

$$T_{\mathrm{C1}} = 318\,\mathrm{K},\quad T_{\mathrm{E1}} = 253\,\mathrm{K},\quad C_{\mathrm{C}} = C_{\mathrm{E}} = 1.00\,\mathrm{kW/K},\quad \varepsilon_{\mathrm{C}} = \varepsilon_{\mathrm{E}} = 0.75,$$
$$\mathrm{ME} = 10.0\,\mathrm{kW/K}\quad R_{\Delta\mathrm{S}} = R^{/}_{\Delta\mathrm{S}} = 0.75\quad \text{and cooling load}\,(P_{\mathrm{L}}) = 3.0\,\mathrm{kW}$$

It can be seen from all Tables 8.1a, 8.1b, 8.1c, 8.1d, 8.1e, 8.1f, 8.1g, and 8.1h that for above typical operating conditions, the COP of a completely reversible system is found to be $(\mathrm{COP})_{\mathrm{r1}} = 3.89$, whereas COP of completely irreversible cascaded refrigeration system is $\mathrm{COP_o} = 0.72$, and for an endoreversible system, it is calculated separately and found to be 3.16.

Table 8.1a shows that by increasing the inlet temperature of the external fluid on source side, optimal power input to the system decreases, and consequently optimal COP increases. Temperature of cycle-I working fluid on hot side decreases and of cycle-II working fluid on cold side increases with increasing source inlet temperature. Table 8.1b shows that by increasing inlet temperature of external fluid on sink

Table 8.1a Effect of T_{E1} on power input, working fluid temperatures, and cooling performance of cascaded refrigeration system

T_{E1}	$(T_{\mathrm{c1}})_{\mathrm{o}}$	$(T_{\mathrm{e2}})_{\mathrm{o}}$	P_{o}	P_{H}	$(\mathrm{COP})_{\mathrm{r1}}$	$\mathrm{COP_o}$
K	K	K	kW	kW	–	–
245	335.11	238.23	4.58	7.58	3.36	0.68
247	334.96	240.23	4.51	7.51	3.48	0.69
249	334.81	242.23	4.45	7.45	3.61	0.70
251	334.67	244.23	4.38	7.38	3.75	0.71
253	334.52	246.23	4.32	7.32	3.89	0.72
255	334.38	248.23	4.26	7.26	4.05	0.73
257	334.24	250.23	4.20	7.20	4.21	0.74
259	334.11	252.23	4.14	7.14	4.39	0.76
261	333.97	254.23	4.08	7.08	4.58	0.77
263	333.84	256.23	4.02	7.02	4.78	0.78

Table 8.1b Effect of T_{C1} on power input, working fluid temperatures, and cooling performance of cascaded refrigeration system

T_{C1}	$(T_{\mathrm{c1}})_{\mathrm{o}}$	$(T_{\mathrm{e2}})_{\mathrm{o}}$	P_{o}	P_{H}	$(\mathrm{COP})_{\mathrm{r1}}$	$\mathrm{COP_o}$
K	K	K	kW	kW	–	–
310	326.11	246.23	4.14	7.14	4.44	0.76
312	328.21	246.23	4.18	7.18	4.29	0.75
314	330.31	246.23	4.23	7.23	4.15	0.74
316	332.42	246.23	4.27	7.27	4.02	0.73
318	334.52	246.23	4.32	7.32	3.89	0.72
320	336.63	246.23	4.37	7.37	3.78	0.72
322	338.73	246.23	4.41	7.41	3.67	0.71
324	340.83	246.23	4.46	7.46	3.56	0.70
326	342.94	246.23	4.50	7.50	3.47	0.69
328	345.04	246.23	4.55	7.55	3.37	0.69

Table 8.1c Effect of C_C on power input, working fluid temperatures, and cooling performance of cascaded refrigeration system

C_C	$(T_{c1})_o$	$(T_{e2})_o$	P_o	P_H	$(COP)_{r1}$	COP_o
K	K	K	kW	kW	–	–
0.6	347.64	246.00	4.63	7.63	3.89	0.69
0.7	342.82	246.07	4.52	7.52	3.89	0.70
0.8	339.30	246.12	4.43	7.43	3.89	0.71
0.9	336.62	246.18	4.37	7.37	3.89	0.72
1.0	334.52	246.23	4.32	7.32	3.89	0.72
1.1	332.83	246.28	4.28	7.28	3.89	0.73
1.2	331.44	246.32	4.25	7.25	3.89	0.73
1.3	330.28	246.37	4.22	7.22	3.89	0.73
1.4	329.29	246.41	4.19	7.19	3.89	0.74
1.5	328.44	246.45	4.17	7.17	3.89	0.74

Table 8.1d Effect of C_E on power input, working fluid temperatures, and cooling performance of cascaded refrigeration system

C_E	$(T_{c1})_o$	$(T_{e2})_o$	P_o	P_H	$(COP)_{r1}$	COP_o
K	K	K	kW	kW	–	–
0.6	334.10	242.19	4.44	7.44	3.89	0.71
0.7	334.23	243.60	4.40	7.40	3.89	0.71
0.8	334.34	244.68	4.37	7.37	3.89	0.72
0.9	334.44	245.54	4.34	7.34	3.89	0.72
1.0	334.52	246.23	4.32	7.32	3.89	0.72
1.1	334.60	246.80	4.30	7.30	3.89	0.73
1.2	334.67	247.28	4.29	7.29	3.89	0.73
1.3	334.73	247.70	4.28	7.28	3.89	0.73
1.4	334.78	248.05	4.27	7.27	3.89	0.73
1.5	334.83	248.36	4.26	7.26	3.89	0.73

Table 8.1e Effect of ε_C on power input, working fluid temperatures, and cooling performance of cascaded refrigeration system

ε_C	$(T_{c1})_o$	$(T_{e2})_o$	P_o	P_H	$(COP)_{r1}$	COP_o
K	K	K	kW	kW	–	–
0.20	397.15	245.77	5.76	8.76	3.89	0.61
0.30	365.80	245.87	5.04	8.04	3.89	0.66
0.40	351.99	245.96	4.73	7.73	3.89	0.69
0.50	344.25	246.05	4.55	7.55	3.89	0.70
0.60	339.30	246.12	4.43	7.43	3.89	0.71
0.70	335.87	246.19	4.35	7.35	3.89	0.72
0.75	334.52	246.23	4.32	7.32	3.89	0.72
0.80	333.36	246.26	4.29	7.29	3.89	0.73
0.90	331.44	246.32	4.25	7.25	3.89	0.73
1.00	329.93	246.38	4.21	7.21	3.89	0.73

side, temperature of optimal working fluid on high-temperature side and power input to the system increase, and hence optimal COP of the system decreases, whereas temperature of optimal working fluid on low-temperature side remains constant.

Table 8.1f Effect of ε_E on power input, working fluid temperatures, and cooling performance of cascaded refrigeration system

ε_E	$(T_{c1})_o$	$(T_{e2})_o$	P_o	P_H	$(COP)_{r1}$	COP_o
K	K	K	kW	kW	–	–
0.20	333.56	230.63	4.83	7.83	3.89	0.66
0.30	333.78	237.42	4.59	7.59	3.89	0.69
0.40	334.00	240.97	4.48	7.48	3.89	0.70
0.50	334.18	243.17	4.41	7.41	3.89	0.71
0.60	334.34	244.68	4.37	7.37	3.89	0.72
0.70	334.47	245.78	4.33	7.33	3.89	0.72
0.75	334.52	246.23	4.32	7.32	3.89	0.72
0.80	334.57	246.62	4.31	7.31	3.89	0.72
0.90	334.67	247.28	4.29	7.29	3.89	0.73
1.00	334.74	247.82	4.27	7.27	3.89	0.73

Table 8.1g Effect of $R_{\Delta S}$ on power input, working fluid temperatures, and cooling performance of cascaded refrigeration system

$R_{\Delta S}$	$(T_{c1})_o$	$(T_{e2})_o$	P_o	P_H	$(COP)_{r1}$	COP_o
K	K	K	kW	kW	–	–
0.50	341.99	246.59	8.22	11.22	3.89	0.38
0.55	339.96	246.50	7.14	10.14	3.89	0.44
0.60	338.27	246.43	6.25	9.25	3.89	0.50
0.70	335.60	246.29	4.87	7.87	3.89	0.64
0.75	334.52	246.23	4.32	7.32	3.89	0.72
0.80	333.58	246.17	3.84	6.84	3.89	0.81
0.85	332.74	246.12	3.43	6.43	3.89	0.91
0.90	331.99	246.07	3.06	6.06	3.89	1.03
0.95	331.32	246.02	2.73	5.73	3.89	1.15
1.00	330.71	245.98	2.43	5.43	3.89	1.30

Table 8.1h Effect of $R'_{\Delta S}$ on power input, working fluid temperatures, and cooling performance of cascaded refrigeration system

$R'_{\Delta S}$	$(T_{c1})_o$	$(T_{e2})_o$	P_o	P_H	$(COP)_{r1}$	COP_o
K	K	K	kW	kW	–	–
0.50	344.33	246.05	8.36	11.36	3.89	0.38
0.55	341.58	246.09	7.23	10.23	3.89	0.43
0.60	339.33	246.12	6.31	9.31	3.89	0.50
0.70	335.88	246.19	4.88	7.88	3.89	0.64
0.75	334.52	246.23	4.32	7.32	3.89	0.72
0.80	333.35	246.26	3.83	6.83	3.89	0.81
0.85	332.33	246.29	3.41	6.41	3.89	0.92
0.90	331.43	246.32	3.03	6.03	3.89	1.03
0.95	330.63	246.35	2.70	5.70	3.89	1.16
1.00	329.91	246.38	2.40	5.40	3.89	1.30

Tables 8.1c and 8.1d show that as heat capacitance rates of external source-/sink-side fluid increase, the optimal power input to the system decreases; consequently optimal COP of the system increases in both cases since cooling load is fixed. The

variation in heat capacitance rates shows different changes in working fluid temperatures. As sink-side heat capacitance rate of external fluid increases, temperature of cycle-I and cycle-II working fluid on hot and cold side decreases, although by increasing the source-side heat capacitance rate of external fluid, temperature of cycle-I and cycle-II working fluids on hot and cold side increases.

Tables 8.1e and 8.1f show the effect of effectiveness (ε_E, ε_C) of source-/sink-side heat exchanger of the cascaded refrigeration system. By increasing any one of them while other being kept as constant, optimal power input to the system decreases, and consequently optimal COP of the system increases because cooling load is fixed. As sink-side effectiveness of heat exchanger increases, temperature of cycle-I and cycle-II working fluid on source and sink side decreases, although by increasing the source-side effectiveness of heat exchanger, temperature of cycle-I working fluid on sink side decreases slowly, while cycle-II working fluid on source side increases.

Tables 8.1g and 8.1h show the effect of internal irreversibility parameters $R_{\Delta S}$ and $R^{/}_{\Delta S}$, respectively. By increasing any one of them (decreasing internal irreversibility), optimal power input to the system decreases, and hence the optimal COP of the system increases. Temperature of cycle-I working fluid on sink side decreases with increasing any one of them, but cycle-II working fluid on heat source side slowly decreases with increasing $R_{\Delta S}$ and slowly increases with increasing $R^{/}_{\Delta S}$.

8.4.2 Cascaded Heat Pump Cycle

In order to have numerical appreciation of the theoretical analysis of the cascaded heat pump system, we have studied the effect of various input parameters on the COP of the cascaded heat pump system and results are shown in Tables 8.2a, 8.2b, 8.2c, 8.2d, 8.2e, 8.2f, 8.2g and 8.2h. During the variation of any one parameter, all other parameters are assumed to be constant as given below:

Table 8.2a Effect of T_{E1} on power input, working fluid temperatures, and heating performance of cascaded heat pump system

T_{E1}	$(T_{c1})_o$	$(T_{e2})_o$	P_L	P_o	$(COP)_{r2}$	COP_o
K	K	K	kW	kW	–	–
280	332.76	277.77	1.40	1.60	6.83	1.88
282	332.76	279.76	1.41	1.59	7.13	1.89
284	332.76	281.74	1.42	1.58	7.45	1.90
286	332.76	283.73	1.43	1.57	7.81	1.91
288	332.76	285.71	1.44	1.56	8.20	1.93
290	332.76	287.70	1.45	1.55	8.63	1.94
292	332.76	289.68	1.46	1.54	9.11	1.95
294	332.76	291.66	1.47	1.53	9.65	1.96
296	332.76	293.65	1.48	1.52	10.25	1.98
298	332.76	295.63	1.49	1.51	10.93	1.99

Table 8.2b Effect of T_{C1} on power input, working fluid temperatures, and heating performance of cascaded heat pump system

T_{C1}	$(T_{c1})_o$	$(T_{e2})_o$	P_L	P_o	$(COP)_{r2}$	COP_o
K	K	K	kW	kW	–	–
320	324.76	285.66	1.48	1.52	10.00	1.97
322	326.76	285.67	1.47	1.53	9.47	1.96
324	328.76	285.68	1.46	1.54	9.00	1.95
326	330.76	285.70	1.45	1.55	8.58	1.94
328	332.76	285.71	1.44	1.56	8.20	1.93
330	334.76	285.72	1.43	1.57	7.86	1.92
332	336.76	285.74	1.43	1.57	7.55	1.91
334	338.76	285.75	1.42	1.58	7.26	1.90
336	340.76	285.76	1.41	1.59	7.00	1.89
338	342.76	285.78	1.40	1.60	6.76	1.88

Table 8.2c Effect of C_E on power input, working fluid temperatures, and heating performance of cascaded heat pump system

C_E	$(T_{c1})_o$	$(T_{e2})_o$	P_L	P_o	$(COP)_{r2}$	COP_o
kW/K	K	K	kW	kW	–	–
0.6	332.68	284.27	1.44	1.56	8.20	1.92
0.7	332.71	284.78	1.44	1.56	8.20	1.92
0.8	332.73	285.16	1.44	1.56	8.20	1.92
0.9	332.74	285.47	1.44	1.56	8.20	1.92
1.0	332.76	285.71	1.44	1.56	8.20	1.93
1.1	332.77	285.91	1.44	1.56	8.20	1.93
1.2	332.78	286.08	1.44	1.56	8.20	1.93
1.3	332.79	286.22	1.45	1.55	8.20	1.93
1.4	332.80	286.35	1.45	1.55	8.20	1.93
1.5	332.81	286.45	1.45	1.55	8.20	1.93

Table 8.2d Effect of C_C on power input, working fluid temperatures, and heating performance of cascaded heat pump system

C_C	$(T_{c1})_o$	$(T_{e2})_o$	P_L	P_o	$(COP)_{r2}$	COP_o
kW/K	K	K	kW	kW	–	–
0.6	335.73	285.79	1.43	1.57	8.20	1.91
0.7	334.67	285.77	1.43	1.57	8.20	1.92
0.8	333.88	285.75	1.44	1.56	8.20	1.92
0.9	333.26	285.73	1.44	1.56	8.20	1.92
1.0	332.76	285.71	1.44	1.56	8.20	1.93
1.1	332.35	285.70	1.44	1.56	8.20	1.93
1.2	332.00	285.68	1.45	1.55	8.20	1.93
1.3	331.71	285.67	1.45	1.55	8.20	1.93
1.4	331.46	285.66	1.45	1.55	8.20	1.93
1.5	331.24	285.65	1.45	1.55	8.20	1.94

Table 8.2e Effect of ε_E on power input, working fluid temperatures, and heating performance of cascaded heat pump system

ε_E	$(T_{c1})_o$	$(T_{e2})_o$	P_L	P_o	$(COP)_{r2}$	COP_o
–	K	K	kW	kW	–	–
0.20	332.54	279.99	1.41	1.59	8.20	1.89
0.30	332.61	282.52	1.43	1.57	8.20	1.91
0.40	332.66	283.82	1.43	1.57	8.20	1.91
0.50	332.70	284.62	1.44	1.56	8.20	1.92
0.60	332.73	285.16	1.44	1.56	8.20	1.92
0.70	332.75	285.55	1.44	1.56	8.20	1.93
0.75	332.76	285.71	1.44	1.56	8.20	1.93
0.80	332.77	285.85	1.44	1.56	8.20	1.93
0.90	332.78	286.08	1.44	1.56	8.20	1.93
1.00	332.80	286.27	1.45	1.55	8.20	1.93

Table 8.2f Effect of ε_C on power input, working fluid temperatures, and heating performance of cascaded heat pump system

ε_C	$(T_{c1})_o$	$(T_{e2})_o$	P_L	P_o	$(COP)_{r2}$	COP_o
–	K	K	kW	kW	–	–
0.20	344.71	285.93	1.39	1.61	8.20	1.86
0.30	339.36	285.86	1.41	1.59	8.20	1.89
0.40	336.64	285.81	1.43	1.57	8.20	1.91
0.50	334.99	285.77	1.43	1.57	8.20	1.91
0.60	333.88	285.75	1.44	1.56	8.20	1.92
0.70	333.08	285.72	1.44	1.56	8.20	1.92
0.75	332.76	285.71	1.44	1.56	8.20	1.93
0.80	332.48	285.70	1.44	1.56	8.20	1.93
0.90	332.00	285.68	1.45	1.55	8.20	1.93
1.00	331.62	285.67	1.45	1.55	8.20	1.93

Table 8.2g Effect of $R_{\Delta S}$ on power input, working fluid temperatures, and heating performance of cascaded heat pump system

$R_{\Delta S}$	$(T_{c1})_o$	$(T_{e2})_o$	P_L	P_o	$(COP)_{r2}$	COP_o
–	K	K	kW	kW	–	–
0.50	332.66	286.50	0.97	2.03	8.20	1.47
0.55	332.68	286.34	1.06	1.94	8.20	1.55
0.60	332.70	286.19	1.16	1.84	8.20	1.63
0.70	332.74	285.87	1.35	1.65	8.20	1.82
0.75	332.76	285.71	1.44	1.56	8.20	1.93
0.80	332.77	285.55	1.54	1.46	8.20	2.05
0.85	332.79	285.39	1.63	1.37	8.20	2.19
0.90	332.80	285.23	1.73	1.27	8.20	2.36
0.95	332.82	285.08	1.82	1.18	8.20	2.55
1.00	332.83	284.92	1.92	1.08	8.20	2.77

Table 8.2h Effect of $R'_{\Delta S}$ on power input, working fluid temperatures, and heating performance of cascaded heat pump system

$R'_{\Delta S}$	$(T_{c1})_o$	$(T_{e2})_o$	P_L	P_o	$(COP)_{r2}$	COP_o
–	K	K	kW	kW	–	–
0.50	332.81	286.45	0.96	2.04	8.20	1.47
0.55	332.80	286.30	1.06	1.94	8.20	1.55
0.60	332.79	286.15	1.16	1.84	8.20	1.63
0.70	332.77	285.86	1.35	1.65	8.20	1.82
0.75	332.76	285.71	1.44	1.56	8.20	1.93
0.80	332.75	285.56	1.54	1.46	8.20	2.05
0.85	332.74	285.42	1.63	1.37	8.20	2.20
0.90	332.73	285.27	1.73	1.27	8.20	2.36
0.95	332.72	285.13	1.82	1.18	8.20	2.55
1.00	332.72	284.98	1.92	1.08	8.20	2.77

$$T_{C1} = 328\,\text{K}, T_{E1} = 288\,\text{K}, C_C = C_E = 1.00\,\text{kW/K}, \varepsilon_C = \varepsilon_E = 0.75,$$
$$\text{ME} = 10.0\,\text{kW/K}, R_{\Delta S} = R'_{\Delta S} = 0.75\,\text{and heating load}\,(P_H) = 3.0\,\text{kW}$$

It can be seen from all the Tables 8.2a, 8.2b, 8.2c, 8.2d, 8.2e, 8.2f, 8.2g, and 8.2h that for above typical operating conditions, the COP of a completely reversible system is found to be $(COP)_{r2} = 8.20$, whereas COP of completely irreversible cascaded heat pump system is $COP_o = 1.93$, and for an endoreversible system, it is calculated separately and found to be 6.60.

Table 8.2a shows that by increasing the inlet temperature of external fluid on source side, optimal power input to the system decreases, and consequently optimal COP increases. Temperature of cycle-I working fluid on heat sink side remains constant, whereas temperature of cycle-II working fluid on the heat source side increases with increasing inlet temperature of external fluid on source side.

Table 8.2b shows that by increasing inlet temperature of external fluid on sink side, optimal temperature of cycle-I working fluid on heat sink side and power input to the system increase, and hence optimal COP of the system decreases. Temperature of cycle-II working fluid on heat source side slowly increases with increasing inlet temperature of external fluid on sink side.

Tables 8.2c and 8.2d show that as heat capacitance rates of external source-/sink-side fluid increase, the optimal power input to system decreases; consequently optimal COP of the system increases in both cases since heating load is fixed. This variation in heat capacitance rate shows different changes in working fluid temperatures. As sink-side heat capacitance rate of external fluid increases, temperature of cycle-I and cycle-II working fluid on heat source and heat sink side decreases, whereas by increasing the source-side heat capacitance rate of external fluid, temperature of cycle-I and cycle-II working fluid on heat source and heat sink side increases.

Tables 8.2e and 8.2f show the effect of effectiveness of source-/sink-side heat exchanger of the cascaded heat pump system. By increasing any one of them while other being kept as constant, optimal power input to the system decreases, and consequently performance of the system increases because heating load is fixed. The variation in effectiveness shows different changes in working fluid temperatures. As sink-side effectiveness of the heat exchanger increases, temperature of cycle-I and cycle-II working fluid on heat sink and heat source side decreases, whereas by increasing the source-side effectiveness of heat exchanger, temperature of cycle-I and cycle-II working fluid on heat sink and source side increases.

Tables 8.2g and 8.2h show the effect of internal irreversibility parameters $R_{\Delta S}$ and $R'_{\Delta S}$, respectively. By increasing any one of them (decreasing internal irreversibility), optimal power input to the system decreases, and hence the optimal COP of the system increases. Temperature of cycle-I working fluid on heat sink side slowly increases, whereas temperature of cycle-II working fluid on heat source side decreases with increasing internal irreversibility parameter ($R_{\Delta S}$) of the heat pump cycle-I in the cascaded system. Temperature of cycle-I working fluid on heat sink side and of cycle-II working fluid on heat source side decreases with increasing internal irreversibility parameter ($R_{\Delta S}$) of cycle-II of the cascaded heat pump system. The case $R_{\Delta S} = R'_{\Delta S} = 1.0$ corresponds to the endoreversible system (i.e. externally irreversible and internally reversible system).

8.5 Conclusion

Finite time thermodynamic analysis of an irreversible cascaded refrigeration/heat pump system has been carried out using the method of Lagrangian multiplier for minimizing power input to the system with a given cooling load/heating load condition and thermal reservoirs (heat source and sink) of finite heat capacity. Internal irreversibility in the system has been characterized by a single irreversibility parameter, representing the ratio of the entropy changes/differences. It is found that the effect of internal irreversibility is more pronounced as compared to the external irreversibility on the cooling/heating performance of the cascaded refrigeration/heat pump system. Cascaded refrigeration/heat pump system can produce almost double cooling load/heating load as compared to single stage refrigeration/heat pump system, at the same COP for similar operating conditions provided the effective thermal conductance between the working fluids of two refrigeration/heat pump cycles of cascaded system is much higher than the other effective thermal conductance on heat source/sink side.

Chapter 9
Finite Time Thermodynamics of Rankine Cycle Airconditioning and Heat Pump Cycles

9.1 Introduction

In most of the Rankine cycle-based power plants, water/steam is used as the working fluid due to its easy availability, low cost, and safety. The ideal Rankine cycle is a practical attempt to approach the Carnot cycle. The Rankine cycle airconditioning/heat pump system is considered as a heat engine driving an airconditioner/heat pump system.

The main objective of this chapter is to analyse the Rankine cycle airconditioning/heat pump systems coupled with external heat reservoirs of finite heat capacity and having both external and internal irreversibilities (Kaushik et al. 1999b, 2000; Bhardwaj et al. 2001, 2003a).

9.2 Rankine Cycle Airconditioning and Heat Pump Cycles

The basic components of a Rankine heat engine are the pump, constant pressure steam generator, turbine, and constant pressure condenser. Figure 9.1a shows a schematic diagram of a Rankine cycle airconditioning/heat pump system. The output work of heat engine is used as input work to the airconditioning/heat pump cycle. The equivalent cycle model of a Rankine cycle airconditioning/heat pump system is shown in Fig. 9.1b. Thermal reservoirs of finite heat capacity are coupled with the system, so temperatures of these will vary as shown in Fig. 9.1b.

S.C. Kaushik et al., *Finite Time Thermodynamics of Power and Refrigeration Cycles*,
https://doi.org/10.1007/978-3-319-62812-7_9

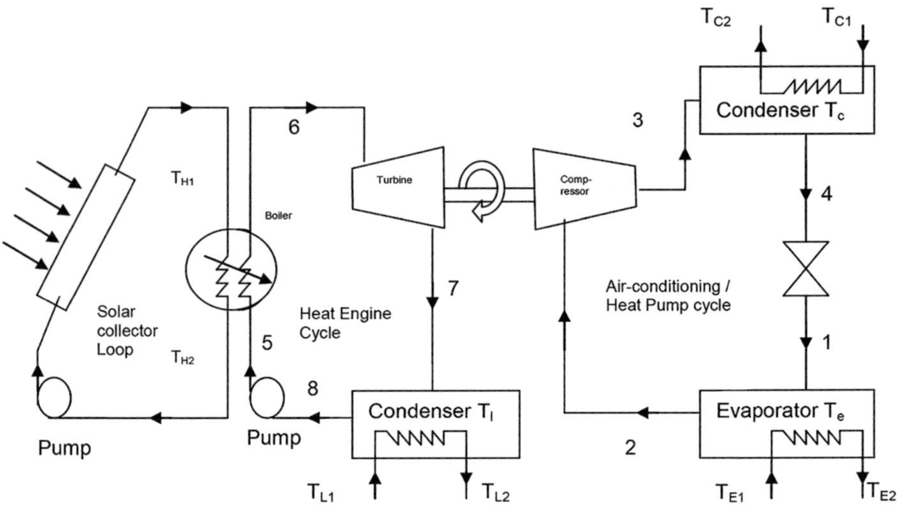

Fig. 9.1a Schematic diagram of Rankine heat engine drive airconditioning/heat pump system

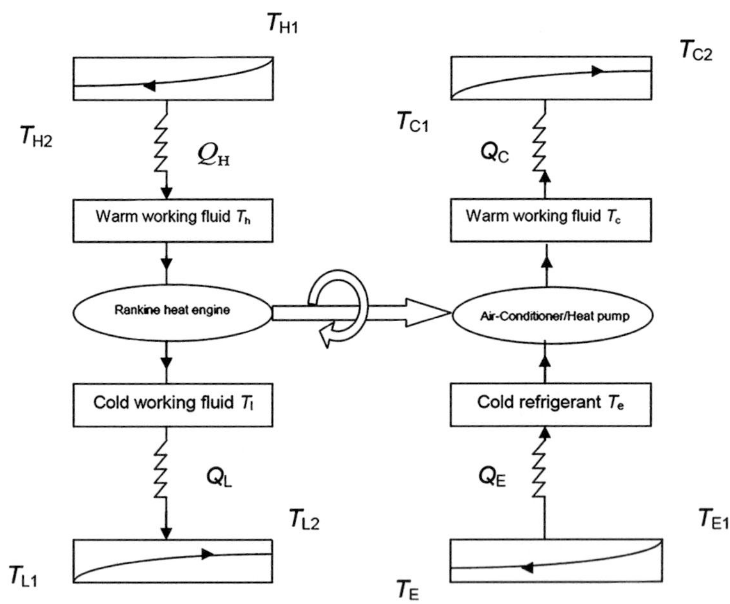

Fig. 9.1b Equivalent cycle diagram of Rankine airconditioning/heat pump system

9.3 Finite Time Thermodynamic Analysis

9.3.1 Rankine Cycle Coupled Airconditioning Cycle

(a) Heat Engine Cycle Analysis

The rate of input heat from heat source to the system is given by:

$$\dot{Q}_{\mathrm{H}} = \frac{Q_{\mathrm{H}}}{t_{\mathrm{H}}} = U_{\mathrm{H}}A_{\mathrm{H}}(\mathrm{LMTD})_{\mathrm{H}} = \dot{m}_{\mathrm{H}}C_{\mathrm{PH}}(T_{\mathrm{H1}} - T_{\mathrm{H2}}) \tag{9.1}$$

Similarly, the rate of heat rejection from system to the heat sink:

$$\dot{Q}_{\mathrm{L}} = \frac{Q_{\mathrm{L}}}{t_{\mathrm{L}}} = U_{\mathrm{L}}A_{\mathrm{L}}(\mathrm{LMTD})_{\mathrm{L}} = \dot{m}_{\mathrm{L}}C_{\mathrm{PL}}(T_{\mathrm{L2}} - T_{\mathrm{L1}}) \tag{9.2}$$

where,

$$(\mathrm{LMTD})_{\mathrm{H}} = \frac{(T_{\mathrm{H1}} - T_{\mathrm{h}}) - (T_{\mathrm{H2}} - T_{\mathrm{h}})}{\ln\dfrac{(T_{\mathrm{H1}} - T_{\mathrm{h}})}{(T_{\mathrm{H2}} - T_{\mathrm{h}})}} \quad \text{and}$$

$$(\mathrm{LMTD})_{\mathrm{L}} = \frac{(T_{\mathrm{l}} - T_{\mathrm{L1}}) - (T_{\mathrm{l}} - T_{\mathrm{L2}})}{\ln\dfrac{(T_{\mathrm{l}} - T_{\mathrm{L1}})}{(T_{\mathrm{l}} - T_{\mathrm{L2}})}}$$

Using LMTD expressions in Eqs. (9.1) and (9.2), we have:

$$Q_{\mathrm{H}} = HE\ (T_{\mathrm{H1}} - T_{\mathrm{h}})t_{\mathrm{H}} \tag{9.3}$$

$$Q_{\mathrm{L}} = LE\ (T_{\mathrm{l}} - T_{\mathrm{L1}})t_{\mathrm{L}} \tag{9.4}$$

where

$$HE = C_{\mathrm{H}}\varepsilon_{\mathrm{H}}, \mathrm{LE} = C_{\mathrm{L}}\varepsilon_{\mathrm{L}}, C_{\mathrm{H}} = m_{\mathrm{H}}C_{\mathrm{PH}}, C_{\mathrm{L}} = m_{\mathrm{L}}C_{\mathrm{PL}},$$

$$\varepsilon_{\mathrm{H}} = 1 - \mathrm{e}^{-\frac{\mathrm{U_H A_H}}{\dot{m}_{\mathrm{H}}\ \mathrm{C_{PH}}}} \text{ and } \varepsilon_{\mathrm{L}} = 1 - \mathrm{e}^{-\frac{\mathrm{U_L A_L}}{\dot{m}_{\mathrm{L}}\ \mathrm{C_{PL}}}}$$

Work output of the heat engine cycle is given by:

$$W = HE(T_{\mathrm{H1}} - T_{\mathrm{h}})t_{\mathrm{H}} - LE(T_{1} - T_{\mathrm{L1}})t_{\mathrm{L}}, \tag{9.5}$$

and using the second law of thermodynamics, an internal irreversibility parameter is introduced as:

$$\Delta S = \frac{Q_{\mathrm{H}}}{T_{\mathrm{h}}} - R_{\Delta \mathrm{s}}\frac{Q_{\mathrm{L}}}{T_{\mathrm{l}}} = 0 \tag{9.6}$$

$R_{\Delta S} = 1$ for endoreversible and $R_{\Delta S} < 1$ for internal irreversible (real) system. For given input heat in the boiler, output work of heat engine can be maximized by defining the modified Lagrangian operator as given by:

$$L = \mathrm{W} - \lambda \mathrm{Q_H} - \mu \Delta \mathrm{S} \tag{9.7}$$

where λ and μ are Lagrangian multipliers.

$\frac{\partial L}{\partial T_h} = 0$ and $\frac{\partial L}{\partial T_l} = 0$ give the optimal values of T_h and T_l. Now, using the optimal value of T_l into Eq. (9.4), we have the thermal efficiency at maximum work output of heat engine:

$$\eta = 1 - \frac{T_{L1}/R_{\Delta S}}{\left(T_{H1} - \frac{Q_H}{HEt_H} - \frac{Q_H}{R_{\Delta S} LE\ t_L}\right)} \tag{9.8}$$

$\frac{\partial \eta}{\partial t_H} = 0$ and $\frac{\partial \eta}{\partial t_L} = 0$ give the optimal values of τ_H and τ_L. Now, after substituting these optimal values of the cycle times into Eq. (9.8), we have the optimal efficiency of the heat engine at maximum work output:

$$\eta = 1 - \frac{T_{L1}/R_{\Delta S}}{T_{H1} - \frac{P_H}{K_1}} \tag{9.9}$$

where,

$$P_H = \frac{Q_H}{\tau_1}, \tau_1 = t_H + t_L, K_1 = \frac{HE\ LE\ R_{\Delta S}}{\left(\sqrt{HE} + \sqrt{R_{\Delta S}\ LE}\right)^2}$$

(b) Airconditioning Cycle Analysis

Following the analysis of Chap. 7, we have the optimal COP of an irreversible airconditioning system operating between reservoirs at temperatures T_{E1} and T_{C1} as given by:

$$\mathrm{COP} = \left[\frac{T_{C1}/R'_{\Delta S}}{\left(T_{E1} - \frac{P_L}{K_2}\right)} - 1\right]^{-1} \tag{9.10}$$

where,

$$P_L = \frac{Q_E}{\tau_2}, \tau_2 = t_E + t_C, K_2 = \frac{EE\ CE\ R'_{\Delta S}}{\left(\sqrt{EE} + \sqrt{R'_{\Delta S} CE}\right)^2}$$

Under the circumstances of the given Q_E, τ_1, and τ_2, the coefficient of performance of an irreversible Rankine cycle airconditioning system is given by:

$$\varepsilon = \left[1 - \frac{T_{L1} / R_{\Delta S}}{\left(T_{H1} - \frac{R_E}{\varepsilon\,\tau_1\,K_1}\right)}\right]\left[\frac{T_{C1} / _R\Delta S'}{\left(T_{E1} - \frac{Q_E}{\tau_2\,K_2}\right)} - 1\right]^{-1} \tag{9.11}$$

Now, let τ_1 and τ_2 vary, while Q_E and $\tau = (\tau_1 + \tau_2)$ remain unvaried; under this constraint conditions, we take ε as an objective function and find its optimum. For this reason, we introduce the Lagrangian operator:

$$L = \varepsilon + \lambda(\tau - \tau_1 - \tau_2) \tag{9.12}$$

From the Euler–Lagrangian equations,

$\frac{\partial L}{\partial \tau_1} = 0$ and $\frac{\partial L}{\partial \tau_2} = 0$ give the optimal value of τ_1 and τ_2. After substituting these optimal values into Eq. (9.11), we have the optimal coefficient of performance of an irreversible Rankine cycle airconditioning system, which satisfies the equation:

$$\varepsilon = \frac{1 - \frac{T_{L1}}{T_{H1}R_{\Delta S}} - \frac{R_L}{\varepsilon T_{H1} T_{E1}}\left[\frac{\varepsilon T_{H1}}{K_2} + \frac{T_{E1}}{K_1}\frac{\varepsilon}{K_2^{1/2}}\left(\sqrt{\frac{T_{L1}}{R_{\Delta S}}}\sqrt{\frac{T_{C1}}{R'_{\Delta S}}}\frac{1}{\sqrt{K_1}} - \frac{T_{L1}}{R_{\Delta S}\sqrt{K_2}}\right)\right]}{\left(\frac{T_{C1}}{T_{E1}R'_{\Delta S}} - 1\right) + \frac{R_L}{\varepsilon T_{H1} T_{E1}}\left[\frac{\varepsilon T_{H1}}{K_2} + \frac{T_{E1}}{K_1} - \frac{1}{K_1^{1/2}}\left(\frac{T_{C1}}{K_1^{1/2} R'_{\Delta S}} - \sqrt{\frac{T_{C1}}{R'_{\Delta S}}}\sqrt{\frac{T_{L1}}{R_{\Delta S}}}\frac{1}{K_2^{1/2}}\right)\right]} \tag{9.13}$$

where cooling load $R_L = Q_E/t$.

Rearranging Eq. (9.13), we have:

$$R_L = \frac{K_1(\varepsilon_R - \varepsilon)T_{H1}\left(\frac{T_{C1}}{R'_{\Delta S}} - T_{E1}\right)}{\left[T_{H1}\frac{K_1}{K_2}(1+\varepsilon) + T_{E1}(1+\varepsilon^{-1}) + 2\sqrt{\frac{T_{C1}}{R'_{\Delta S}}}\sqrt{\frac{T_{L1}}{R_{\Delta S}}}\sqrt{\frac{K_1}{K_2}} - \frac{K_1}{K_2}\frac{T_{L1}}{R_{\Delta S}} - \frac{T_{C1}}{R'_{\Delta S}}\right]} \tag{9.14}$$

where,

$$\varepsilon_R = \frac{T_{E1}\left(T_{H1} - \frac{T_{L1}}{R_{\Delta S}}\right)}{T_{H1}\left(\frac{T_{C1}}{R'_{\Delta S}} - T_{E1}\right)}$$

$\frac{\partial R_L}{\partial \varepsilon} = 0$ gives the value of the coefficient of performance:

$$\varepsilon_{\mathrm{m}} = \frac{-T_{\mathrm{E1}} \pm \sqrt{T_{\mathrm{E1}}^2 + \left\{ T_{\mathrm{H1}} \frac{K_1}{K_2}(1+\varepsilon_{\mathrm{R}}) + 2\sqrt{\frac{T_{\mathrm{C1}}}{R'_{\Delta\mathrm{S}}}}\sqrt{\frac{T_{\mathrm{L1}}}{R_{\Delta\mathrm{S}}}}\sqrt{\frac{K_1}{K_2}} - \frac{K_1}{K_2}\frac{T_{\mathrm{L1}}}{R_{\Delta\mathrm{S}}} - \frac{T_{\mathrm{C1}}}{R'_{\Delta\mathrm{S}}} + T_{\mathrm{E1}} \right\} \varepsilon_{\mathrm{R}} T_{\mathrm{E1}}}}{\left[T_{\mathrm{H1}} \frac{K_1}{K_2}(1+\varepsilon_{\mathrm{R}}) + 2\sqrt{\frac{T_{\mathrm{C1}}}{R'_{\Delta\mathrm{S}}}}\sqrt{\frac{T_{\mathrm{L1}}}{R_{\Delta\mathrm{S}}}}\sqrt{\frac{K_1}{K_2}} - \frac{K_1}{K_2}\frac{T_{\mathrm{L1}}}{R_{\Delta\mathrm{S}}} - \frac{T_{\mathrm{C1}}}{R'_{\Delta\mathrm{S}}} + T_{\mathrm{E1}} \right]} \tag{9.15}$$

at which the rate of refrigeration R_{L} attains the maximum:

$$(R_{\mathrm{L}})_{\mathrm{m}} = \frac{K_1(\varepsilon_{\mathrm{R}} - \varepsilon_{\mathrm{m}}) T_{\mathrm{H1}} \left(\frac{T_{\mathrm{C1}}}{R'_{\Delta\mathrm{S}}} - T_{\mathrm{E1}} \right)}{\left[T_{\mathrm{H1}} \frac{K_1}{K_2}(1+\varepsilon_{\mathrm{m}}) + T_{\mathrm{E1}}\left(1+\varepsilon_{\mathrm{m}}^{-1}\right) + 2\sqrt{\frac{T_{\mathrm{C1}}}{R'_{\Delta\mathrm{S}}}}\sqrt{\frac{T_{\mathrm{L1}}}{R_{\Delta\mathrm{S}}}}\sqrt{\frac{K_1}{K_2}} - \frac{K_1}{K_2}\frac{T_{\mathrm{L1}}}{R_{\Delta S}} - \frac{T_{\mathrm{C1}}}{R'_{\Delta\mathrm{S}}} \right]} \tag{9.16}$$

$(R_{\mathrm{L}})_{\mathrm{m}}$ and ε_{m} are the two important performance parameters of an irreversible Rankine cycle airconditioning system.

(c) Reversible Case

For reversible Rankine cycle airconditioning systems:

$$\text{Cooling load } (R_{\mathrm{L}}) = 0$$

Internal irreversibility parameters $R_{\Delta\mathrm{S}} = R'_{\Delta\mathrm{S}} = 1.0$.

It can be seen from Eq. (9.13) that optimal overall coefficient of performance of irreversible Rankine cycle airconditioning system reduces to Carnot coefficient of performance which is given by:

$$\varepsilon_{\mathrm{CAR}} = \frac{T_{\mathrm{E1}}(T_{\mathrm{H1}} - T_{\mathrm{L1}})}{T_{\mathrm{H1}}(T_{\mathrm{C1}} - T_{\mathrm{E1}})}$$

9.3.2 Rankine Cycle Coupled Heat Pump Cycle

(a) Heat Pump Cycle Analysis

Following the analysis of Chap. 7, we have the optimal COP of an irreversible heat pump system operating between reservoirs at temperatures T_{E1} and T_{C1} as given by:

$$\mathrm{COP} = \left[1 - \frac{T_{\mathrm{E1}} R'_{\Delta\mathrm{S}}}{\left(T_{\mathrm{C1}} + \frac{P_{\mathrm{H}}}{K_2} \right)} \right]^{-1} \tag{9.17}$$

where

$$P_{\mathrm{H}} = \frac{Q_{\mathrm{C}}}{\tau_2}, \tau_2 = t_{\mathrm{E}} + t_{\mathrm{C}}, K_2 = \frac{EE\,CE}{\left(\sqrt{EE} + \sqrt{R'_{\Delta \mathrm{S}} CE}\right)^2}$$

Under the circumstances of the given Q_{C}, τ_1, and τ_2, the optimal overall coefficient of performance of an irreversible Rankine cycle heat pump system is given by:

$$\varepsilon = \left[1 - \frac{T_{\mathrm{L1}}/R_{\Delta \mathrm{S}}}{\left(T_{\mathrm{H1}} + \frac{Q_{\mathrm{C}}}{\varepsilon\,\tau_1 K_1}\right)}\right]\left[1 - \frac{T_{\mathrm{E1}}\,R'_{\Delta \mathrm{S}}}{\left(T_{\mathrm{C1}} + \frac{Q_{\mathrm{C}}}{\tau_2\,K_2}\right)}\right]^{-1} \tag{9.18}$$

Now, let τ_1 and τ_2 vary, while Q_{C} and $\tau = (\tau_1 + \tau_2)$ remain unvaried. Under this constraint conditions, we take ε as objective function and find its optimum. For this reason, we introduce the Lagrangian operator:

$$L = \varepsilon + \lambda(\tau - \tau_1 - \tau_2) \tag{9.19}$$

From the Euler–Lagrangian equations,
$\frac{\partial L}{\partial \tau_1} = 0$ and $\frac{\partial L}{\partial \tau_2} = 0$ give the optimal value of τ_1 and τ_2. After substituting these optimal values into Eq. (9.18), we have the optimal coefficient of performance of an irreversible Rankine cycle heat pump system, which satisfies the equation:

$$\varepsilon = \frac{1 - \frac{T_{\mathrm{L1}}}{T_{\mathrm{H1}}\,R_{\Delta \mathrm{S}}} - \frac{R_{\mathrm{H}}}{\varepsilon\,T_{\mathrm{H1}}\,T_{\mathrm{C1}}}\left[\begin{array}{c}\frac{\varepsilon\,T_{\mathrm{H1}}}{K_2} + \frac{T_{\mathrm{C1}}}{K_1}\frac{\varepsilon}{K_2^{1/2}} \\ \left(\begin{array}{c}\sqrt{\frac{T_{L1}}{R_{\Delta S}}}\sqrt{T_{E1}R'_{\Delta S}} \\ \frac{1}{\sqrt{K_1}} + \frac{T_{\mathrm{L1}}}{R_{\Delta \mathrm{S}}\sqrt{K_2}}\end{array}\right)\end{array}\right]}{\left(1 - \frac{T_{\mathrm{E1}}R'_{\Delta \mathrm{S}}}{T_{\mathrm{C1}}}\right) + \frac{R_{\mathrm{H}}}{\varepsilon\,T_{\mathrm{H1}}\,T_{\mathrm{C1}}}\left[\begin{array}{c}\frac{\varepsilon\,T_{\mathrm{H1}}}{K_2} + \frac{T_{\mathrm{C1}}}{K_1} - \frac{1}{K_1^{1/2}} \\ \left(\begin{array}{c}\frac{T_{\mathrm{E1}}R'_{\Delta \mathrm{S}}}{K_1^{1/2}} + \sqrt{T_{\mathrm{E1}}R'_{\Delta \mathrm{S}}} \\ \sqrt{\frac{T_{\mathrm{L1}}}{R_{\Delta \mathrm{S}}}}\frac{1}{K_2^{1/2}}\end{array}\right)\end{array}\right]} \tag{9.20}$$

where heating load $R_{\mathrm{H}} = Q_{\mathrm{C}}/t$.

Rearranging Eq. (9.20), we have:

$$R_{\mathrm{H}} = \frac{K_1(\varepsilon_{\mathrm{R}} - \varepsilon)T_{\mathrm{H1}}\left(T_{\mathrm{C1}} - T_{\mathrm{E1}}R'_{\Delta \mathrm{S}}\right)}{\left[\begin{array}{l}T_{\mathrm{H1}}\frac{K_1}{K_2}(\varepsilon - 1) + T_{\mathrm{C1}}\left(\varepsilon^{-1} - 1\right) \\ +2\sqrt{T_{\mathrm{E1}}R'_{\Delta \mathrm{S}}}\sqrt{\frac{T_{\mathrm{L1}}}{R_{\Delta \mathrm{S}}}}\sqrt{\frac{K_1}{K_2}} - \frac{K_1\,T_{L1}}{K_2\,R_{\Delta \mathrm{S}}} + T_{\mathrm{E1}}\mathrm{R}'_{\Delta \mathrm{S}}\end{array}\right]} \tag{9.21}$$

where

$$\varepsilon_{\mathrm{R}} = \frac{T_{\mathrm{C1}}\left(T_{\mathrm{H1}} - \frac{T_{\mathrm{L1}}}{R_{\Delta\mathrm{S}}}\right)}{T_{\mathrm{H1}}\left(T_{\mathrm{C1}} - T_{\mathrm{E1}}R'_{\Delta\mathrm{S}}\right)}$$

$\frac{\partial R_{\mathrm{H}}}{\partial \varepsilon} = 0$ gives the value of the coefficient of performance:

$$\varepsilon_{\mathrm{m}} = \frac{-T_{\mathrm{C1}} \pm \sqrt{T_{\mathrm{C1}}^2 + \left\{\begin{array}{l} T_{\mathrm{H1}}\dfrac{K_1}{K_2}(\varepsilon_{\mathrm{R}} - 1) \\ +2\sqrt{T_{\mathrm{E1}}R'_{\Delta\mathrm{S}}}\sqrt{\dfrac{T_{\mathrm{L1}}}{R_{\Delta\mathrm{S}}}}\sqrt{\dfrac{K_1}{K_2}} \\ +\dfrac{K_1}{K_2}\dfrac{T_{\mathrm{L1}}}{R_{\Delta\mathrm{S}}} + T_{\mathrm{E1}}R'_{\Delta\mathrm{S}} - T_{\mathrm{C1}} \end{array}\right\}\varepsilon_{\mathrm{R}}\, T_{\mathrm{C1}}}}{\left[\begin{array}{l} T_{\mathrm{H1}}\dfrac{K_1}{K_2}(\varepsilon_{\mathrm{R}} - 1) \\ +2\sqrt{T_{\mathrm{E1}}R'_{\Delta\mathrm{S}}}\sqrt{\dfrac{T_{\mathrm{L1}}}{R_{\Delta\mathrm{S}}}}\sqrt{\dfrac{K_1}{K_2}} \\ +\dfrac{K_1}{K_2}\dfrac{T_{\mathrm{L1}}}{R_{\Delta\mathrm{S}}} + T_{\mathrm{E1}}R'_{\Delta\mathrm{S}} - T_{\mathrm{C1}} \end{array}\right]} \tag{9.22}$$

at which the rate of heating R_{H} attains the maximum:

$$(R_{\mathrm{H}})_{\mathrm{m}} = \frac{K_1(\varepsilon_{\mathrm{R}} - \varepsilon_{\mathrm{m}})T_{\mathrm{H1}}\left(T_{\mathrm{C1}} - T_{\mathrm{E1}}R'_{\Delta\mathrm{S}}\right)}{\left[\begin{array}{l} T_{\mathrm{H1}}\dfrac{K_1}{K_2}(\varepsilon_{\mathrm{m}} - 1) + T_{\mathrm{C1}}\left(\varepsilon_{\mathrm{m}}^{-1} - 1\right) \\ +2\sqrt{T_{\mathrm{E1}}R'_{\Delta\mathrm{S}}}\sqrt{\dfrac{T_{\mathrm{L1}}}{R_{\Delta\mathrm{S}}}}\sqrt{\dfrac{K_1}{K_2}} + \dfrac{K_1}{K_2}\dfrac{T_{L1}}{R_{\Delta\mathrm{S}}} + T_{\mathrm{E1}}R'_{\Delta\mathrm{S}} \end{array}\right]} \tag{9.23}$$

$(R_{\mathrm{H}})_{\mathrm{m}}$ and ε_{m} are the two important performance parameters of an irreversible Rankine cycle heat pump system.

(b) Reversible Case

For reversible Rankine cycle heat pump systems

Heating load $(R_{\mathrm{H}}) = 0$

Internal irreversibility parameters $R_{\Delta\mathrm{S}} = R'_{\Delta\mathrm{S}} = 1.0$

It can be seen from Eq. (9.20) that optimal overall coefficient of performance of irreversible Rankine cycle heat pump system reduces to Carnot coefficient of performance which is given by:

$$\varepsilon_{\mathrm{CAR}} = \frac{T_{\mathrm{C1}}(T_{\mathrm{H1}} - T_{\mathrm{L1}})}{T_{\mathrm{H1}}(T_{\mathrm{C1}} - T_{\mathrm{E1}})}$$

9.4 Discussion of Results

9.4.1 Rankine Cycle Coupled Airconditioning Cycle

In order to have numerical discussion of theoretical analysis, we have taken input parameters as follows:

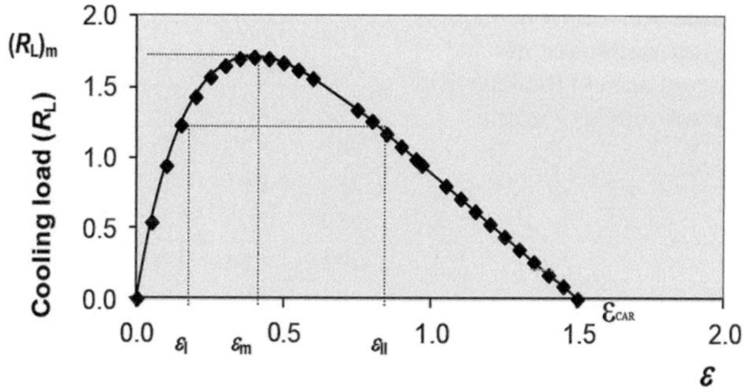

Fig. 9.2a Cooling load versus overall coefficient of performance

Table 9.1 Effect of T_{H1} and T_{L1} on coefficient of performance of Rankine cycle airconditioning system

T_{H1}	ε_{CAR}	ε_I	ε_{II}	T_{L1}	ε_{CAR}	ε_I	ε_{II}
K	–	–	–	K	–	–	–
378	1.37	0.18	0.68	313	1.62	0.13	0.94
380	1.41	0.17	0.73	315	1.57	0.13	0.90
382	1.45	0.16	0.77	317	1.53	0.14	0.85
384	1.48	0.15	0.81	319	1.48	0.15	0.81
386	1.52	0.14	0.85	321	1.44	0.16	0.76
388	1.56	0.13	0.89	323	1.39	0.17	0.71
390	1.59	0.13	0.93	325	1.35	0.18	0.66
392	1.63	0.12	0.96	327	1.30	0.20	0.61
394	1.67	0.12	1.00	329	1.26	0.22	0.55
396	1.70	0.11	1.03	331	1.21	0.25	0.48

$$T_{H1} = 385\ \text{K}, T_{L1} = 318\ \text{K}, T_{E1} = 285\ \text{K}, T_{C1} = 318\ \text{K},$$
$$C_H = C_L = C_E = C_C = 1.0\ \text{kW/K}$$
$$\varepsilon_H = \varepsilon_L = \varepsilon_E = \varepsilon_C = 0.75, R_{\Delta S} = R'_{\Delta S} = 1.0 \text{ and cooling load } R_L = 1.2\ \text{kW}$$

From Eq. (9.14), it is seen that one R_L corresponds to two ε (say ε_I and ε_{II}); where one is smaller than ε_m (say ε_I), the other is larger than ε_m (say ε_{II}), which is sketchily shown in Fig. 9.2a.

Obviously, the optimal value of coefficient of performance is ε_{II} (larger than ε_m) and not ε_I (smaller than ε_m). The important significance of ε_m lies in that it not only shows the optimal value at maximum cooling load but also determines a lower limit to the optimal coefficient of performance. In parametric study, the effect of various input parameters has been studied, and it is found that the internal irreversibility is more prominent for performance reduction than the external irreversibility. The effect of external irreversibility parameters is shown in Tables 9.1, 9.2, 9.3, 9.4, 9.5, and 9.6, whereas the effect of internal irreversibility parameters on the performance of the system is shown in Table 9.7.

For the above typical operating conditions, the COP of a completely reversible system is found to be $\varepsilon_{CAR} = 1.50$. The optimal overall coefficient of

Table 9.2 Effect of T_{E1} and T_{C1} on coefficient of performance of Rankine cycle airconditioning system

T_{E1}	ε_{CAR}	ε_{I}	ε_{II}	T_{C1}	ε_{CAR}	ε_{I}	ε_{II}
K	–	–	–	*K*	–	–	–
283	1.41	0.15	0.77	313	1.77	0.14	0.98
285	1.50	0.14	0.83	315	1.65	0.14	0.91
287	1.61	0.14	0.89	317	1.55	0.14	0.86
289	1.73	0.14	0.96	319	1.46	0.15	0.81
291	1.88	0.14	1.04	321	1.38	0.15	0.76
293	2.04	0.14	1.13	323	1.31	0.15	0.72
295	2.23	0.14	1.23	325	1.24	0.15	0.68
297	2.46	0.13	1.34	327	1.18	0.15	0.64
299	2.74	0.13	1.47	329	1.13	0.15	0.61
300	2.90	0.13	1.54	331	1.08	0.16	0.58

Table 9.3 Effect of C_H and C_L on coefficient of performance of Rankine cycle airconditioning system

C_H	ε_{CAR}	ε_{I}	ε_{II}	C_L	ε_{CAR}	ε_{I}	ε_{II}
kW/*K*	–	–	–	*kW*/*K*	–	–	–
1.0	1.50	0.14	0.83	1.0	1.50	0.14	0.83
1.1	1.50	0.14	0.85	1.1	1.50	0.14	0.85
1.2	1.50	0.13	0.86	1.2	1.50	0.13	0.86
1.3	1.50	0.12	0.87	1.3	1.50	0.12	0.87
1.4	1.50	0.12	0.88	1.4	1.50	0.12	0.88
1.5	1.50	0.11	0.89	1.5	1.50	0.11	0.89
1.6	1.50	0.11	0.89	1.6	1.50	0.11	0.89
1.7	1.50	0.10	0.90	1.7	1.50	0.10	0.90
1.8	1.50	0.10	0.91	1.8	1.50	0.10	0.91
2.0	1.50	0.10	0.92	2.0	1.50	0.10	0.92

Table 9.4 Effect of C_E and C_C on coefficient of performance of Rankine cycle airconditioning system

C_E	ε_{CAR}	ε_{I}	ε_{II}	C_C	ε_{CAR}	ε_{I}	ε_{II}
kW/*K*	–	–	–	*kW*/*K*	–	–	–
1.0	1.50	0.14	0.83	1.0	1.50	0.14	0.83
1.1	1.50	0.14	0.85	1.1	1.50	0.14	0.85
1.2	1.50	0.14	0.86	1.2	1.50	0.14	0.86
1.3	1.50	0.14	0.88	1.3	1.50	0.14	0.88
1.4	1.50	0.14	0.89	1.4	1.50	0.14	0.89
1.5	1.50	0.14	0.90	1.5	1.50	0.14	0.90
1.6	1.50	0.14	0.91	1.6	1.50	0.14	0.91
1.7	1.50	0.14	0.91	1.7	1.50	0.14	0.91
1.8	1.50	0.14	0.92	1.8	1.50	0.14	0.92
2.0	1.50	0.13	0.94	2.0	1.50	0.13	0.94

Table 9.5 Effect of ε_H and ε_L on coefficient of performance of Rankine cycle airconditioning system

ε_H	ε_{CAR}	ε_I	ε_{II}	ε_L	ε_{CAR}	ε_I	ε_{II}
–	–	–	–	–	–	–	–
0.30	1.50	0.38	0.53	0.30	1.50	0.38	0.53
0.40	1.50	0.25	0.69	0.40	1.50	0.25	0.69
0.50	1.50	0.20	0.75	0.50	1.50	0.20	0.75
0.60	1.50	0.17	0.79	0.60	1.50	0.17	0.79
0.70	1.50	0.15	0.82	0.70	1.50	0.15	0.82
0.75	1.50	0.14	0.83	0.75	1.50	0.14	0.83
0.80	1.50	0.14	0.84	0.80	1.50	0.14	0.84
0.90	1.50	0.13	0.86	0.90	1.50	0.13	0.86
0.95	1.50	0.12	0.87	0.95	1.50	0.12	0.87
1.00	1.50	0.12	0.87	1.00	1.50	0.12	0.87

Table 9.6 Effect of ε_E and ε_C on coefficient of performance of Rankine cycle airconditioning system

ε_E	ε_{CAR}	ε_I	ε_{II}	ε_C	ε_{CAR}	ε_I	ε_{II}
–	–	–	–	–	–	–	–
0.30	1.50	0.18	0.62	0.30	1.50	0.18	0.62
0.40	1.50	0.16	0.70	0.40	1.50	0.16	0.70
0.50	1.50	0.15	0.75	0.50	1.50	0.15	0.75
0.60	1.50	0.15	0.79	0.60	1.50	0.15	0.79
0.70	1.50	0.15	0.82	0.70	1.50	0.15	0.82
0.75	1.50	0.14	0.83	0.75	1.50	0.14	0.83
0.80	1.50	0.14	0.84	0.80	1.50	0.14	0.84
0.90	1.50	0.14	0.86	0.90	1.50	0.14	0.86
0.95	1.50	0.14	0.87	0.95	1.50	0.14	0.87
1.00	1.50	0.14	0.88	1.00	1.50	0.14	0.88

Table 9.7 Effect of $R_{\Delta S}$ and $R'_{\Delta S}$ on coefficient of performance of Rankine cycle airconditioning system

$R_{\Delta S}$	ε_{CAR}	ε_I	ε_{II}	$R'_{\Delta S}$	ε_{CAR}	ε_I	ε_{II}
–	–	–	–	–	–	–	–
0.960	1.50	0.27	0.45	0.900	1.50	0.22	0.29
0.965	1.50	0.23	0.52	0.920	1.50	0.18	0.38
0.970	1.50	0.21	0.58	0.930	1.50	0.17	0.43
0.975	1.50	0.19	0.63	0.940	1.50	0.17	0.47
0.980	1.50	0.18	0.67	0.950	1.50	0.16	0.52
0.985	1.50	0.17	0.71	0.960	1.50	0.16	0.57
0.990	1.50	0.16	0.76	0.970	1.50	0.15	0.62
0.995	1.50	0.15	0.79	0.980	1.50	0.15	0.68
0.996	1.50	0.15	0.80	0.990	1.50	0.15	0.75
1.000	1.50	0.14	0.83	1.000	1.50	0.14	0.83

performance of a completely irreversible Rankine cycle airconditioning system is calculated using Eq. (9.13), and for $R_{\Delta S} = R'_{\Delta S} = 0.97$, it is found to be $\varepsilon_{II} = 0.35$, whereas for endoreversible system ($R_{\Delta S} = R'_{\Delta S} = 1.0$), it is found to be $\varepsilon_{II} = 0.83$.

9.4.2 Rankine Cycle Coupled Heat Pump Cycle

In order to have numerical discussion of theoretical analysis, we have taken input parameters as follows:

$$T_{H1} = 385\text{ K}, T_{L1} = 318\text{ K}, T_{E1} = 285\text{ K}, T_{C1} = 318\text{ K},$$
$$C_H = C_L = C_E = C_C = 1.0\text{ kW/K}$$
$$\varepsilon_H = \varepsilon_L = \varepsilon_E = \varepsilon_C = 0.75, R_{\Delta S} = R'_{\Delta S} = 1.0 \text{ and heating load } R_H = 1.2\text{ kW}$$

From Eq. (9.21), it is seen that, similar to airconditioning system, one R_H corresponds to two ε (say ε_I and ε_{II}); where one is smaller than ε_m (say ε_I), the other is larger than ε_m (say ε_{II}), which is sketchily shown in Fig. 9.2b. Obviously, the optimal value of coefficient of performance is ε_{II} (larger than ε_m) and not ε_I (smaller than ε_m). The important significance of ε_m lies in that it not only shows the optimal value at maximum heating load but also determines a lower limit to the optimal coefficient of performance.

In parametric study, the effect of various input parameters has been studied, and it is found that the internal irreversibility is more prominent for performance reduction than the external irreversibility. The effect of external irreversibility parameters is shown in Tables 9.8, 9.9, 9.10, 9.11, 9.12, and 9.13, whereas the effect of internal irreversibility parameters on the performance of the system is shown in Table 9.14. For the above typical operating conditions, the COP of a completely reversible system is found to be $\varepsilon_{CAR} = 1.68$. The optimal overall coefficient of performance of a completely irreversible Rankine cycle heat pump system is calculated using Eq. (9.20), and for $R_{\Delta S} = R'_{\Delta S} = 0.97$, it is found to be $\varepsilon_{II} = 0.64$, whereas for endoreversible system ($R_{\Delta S} = R'_{\Delta S} = 1.0$), it is found to be $\varepsilon_{II} = 1.07$.

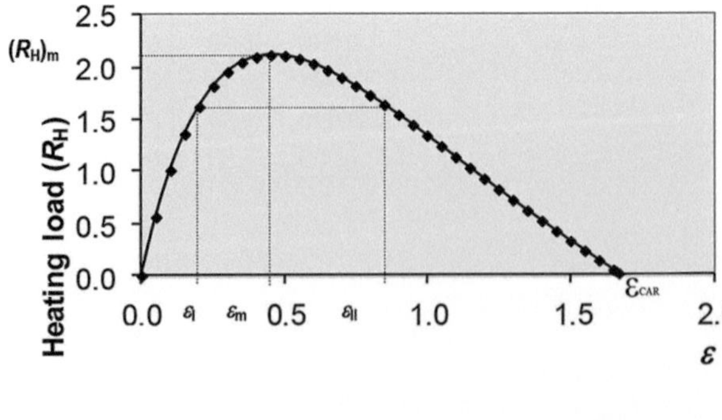

Fig. 9.2b Heating load versus overall coefficient of performance

Table 9.8 Effect of T_{H1} and T_{L1} on coefficient of performance of Rankine cycle heat pump system

T_{H1}	ε_{CAR}	ε_{I}	ε_{II}	T_{L1}	ε_{CAR}	ε_{I}	ε_{II}
K	–	–	–	–	–	–	–
378	1.53	0.15	0.91	313	1.80	0.11	1.19
380	1.57	0.14	0.96	315	1.75	0.12	1.14
382	1.61	0.13	1.00	317	1.70	0.12	1.09
384	1.66	0.13	1.05	319	1.65	0.13	1.04
386	1.70	0.12	1.09	321	1.60	0.14	0.99
388	1.74	0.12	1.13	323	1.55	0.14	0.94
390	1.78	0.11	1.17	325	1.50	0.15	0.89
392	1.82	0.11	1.21	327	1.45	0.16	0.84
394	1.86	0.11	1.25	329	1.40	0.17	0.78
396	1.90	0.10	1.29	331	1.35	0.19	0.72

Table 9.9 Effect of T_{E1} and T_{C1} on coefficient of performance of Rankine cycle heat pump system

T_{E1}	ε_{CAR}	ε_{I}	ε_{II}	T_{C1}	ε_{CAR}	ε_{I}	ε_{II}
K	–	–	–	K	–	–	–
283	1.58	0.13	1.01	313	1.95	0.12	1.21
285	1.68	0.13	1.07	315	1.83	0.13	1.15
287	1.79	0.13	1.13	317	1.72	0.13	1.09
289	1.91	0.12	1.20	319	1.63	0.13	1.04
291	2.05	0.12	1.28	321	1.55	0.13	1.00
293	2.21	0.12	1.36	323	1.48	0.13	0.95
295	2.41	0.12	1.46	325	1.41	0.13	0.91
297	2.64	0.12	1.58	327	1.35	0.13	0.88
299	2.91	0.12	1.71	329	1.30	0.13	0.85
300	3.07	0.12	1.78	331	1.25	0.13	0.82

Table 9.10 Effect of C_H and C_L on coefficient of performance of Rankine cycle heat pump system

C_H	ε_{CAR}	ε_{I}	ε_{II}	C_L	ε_{CAR}	ε_{I}	ε_{II}
kW/K	–	–	–	kW/K	–	–	–
1.0	1.68	0.13	1.07	1.0	1.68	0.13	1.07
1.1	1.68	0.12	1.08	1.1	1.68	0.12	1.08
1.2	1.68	0.11	1.09	1.2	1.68	0.11	1.09
1.3	1.68	0.11	1.10	1.3	1.68	0.11	1.10
1.4	1.68	0.10	1.11	1.4	1.68	0.10	1.11
1.5	1.68	0.10	1.11	1.5	1.68	0.10	1.11
1.6	1.68	0.10	1.12	1.6	1.68	0.10	1.12
1.7	1.68	0.09	1.13	1.7	1.68	0.09	1.13
1.8	1.68	0.09	1.13	1.8	1.68	0.09	1.13
2.0	1.68	0.09	1.14	2.0	1.68	0.09	1.14

Table 9.11 Effect of C_E and C_C on coefficient of performance of Rankine cycle heat pump system

C_E	ε_{CAR}	ε_I	ε_{II}	C_C	ε_{CAR}	ε_I	ε_{II}
kW/K	–	–	–	kW/K	–	–	–
1.0	1.68	0.13	1.07	1.0	1.68	0.13	1.07
1.1	1.68	0.13	1.08	1.1	1.68	0.13	1.08
1.2	1.68	0.12	1.09	1.2	1.68	0.12	1.09
1.3	1.68	0.12	1.10	1.3	1.68	0.12	1.10
1.4	1.68	0.12	1.11	1.4	1.68	0.12	1.11
1.5	1.68	0.12	1.12	1.5	1.68	0.12	1.12
1.6	1.68	0.12	1.13	1.6	1.68	0.12	1.13
1.7	1.68	0.12	1.14	1.7	1.68	0.12	1.14
1.8	1.68	0.12	1.15	1.8	1.68	0.12	1.15
2.0	1.68	0.12	1.16	2.0	1.68	0.12	1.16

Table 9.12 Effect of ε_H and ε_L on coefficient of performance of Rankine cycle heat pump system

ε_H	ε_{CAR}	ε_I	ε_{II}	ε_L	ε_{CAR}	ε_I	ε_{II}
–	–	–	–	–	–	–	–
0.30	1.68	0.26	0.87	0.30	1.68	0.26	0.87
0.40	1.68	0.20	0.95	0.40	1.68	0.20	0.95
0.50	1.68	0.17	1.00	0.50	1.68	0.17	1.00
0.60	1.68	0.15	1.03	0.60	1.68	0.15	1.03
0.70	1.68	0.13	1.06	0.70	1.68	0.13	1.06
0.75	1.68	0.13	1.07	0.75	1.68	0.13	1.07
0.80	1.68	0.12	1.08	0.80	1.68	0.12	1.08
0.90	1.68	0.11	1.09	0.90	1.68	0.11	1.09
0.95	1.68	0.11	1.10	0.95	1.68	0.11	1.10
1.00	1.68	0.11	1.10	1.00	1.68	0.11	1.10

Table 9.13 Effect of ε_E and ε_C on coefficient of performance of Rankine cycle heat pump system

ε_E	ε_{CAR}	ε_I	ε_{II}	ε_C	ε_{CAR}	ε_I	ε_{II}
–	–	–	–	–	–	–	–
0.30	1.68	0.14	0.79	0.30	1.68	0.14	0.79
0.40	1.68	0.14	0.89	0.40	1.68	0.14	0.89
0.50	1.68	0.13	0.95	0.50	1.68	0.13	0.95
0.60	1.68	0.13	1.00	0.60	1.68	0.13	1.00
0.70	1.68	0.13	1.03	0.70	1.68	0.13	1.03
0.75	1.68	0.13	1.06	0.75	1.68	0.13	1.06
0.80	1.68	0.13	1.07	0.80	1.68	0.13	1.07
0.90	1.68	0.13	1.08	0.90	1.68	0.13	1.08
0.95	1.68	0.12	1.09	0.95	1.68	0.12	1.09
1.00	1.68	0.12	1.11	1.00	1.68	0.12	1.11

Table 9.14 Effect of $R_{\Delta S}$ and $R'_{\Delta S}$ on coefficient of performance of Rankine cycle heat pump system

$R_{\Delta S}$	ε_{CAR}	ε_I	ε_{II}	$R'_{\Delta S}$	ε_{CAR}	ε_I	ε_{II}
–	–	–	–	–	–	–	–
0.960	1.68	0.19	0.71	0.90	1.68	0.14	0.58
0.965	1.68	0.18	0.76	0.92	1.68	0.13	0.64
0.970	1.68	0.17	0.81	0.93	1.68	0.13	0.68
0.975	1.68	0.16	0.85	0.94	1.68	0.13	0.72
0.980	1.68	0.15	0.90	0.95	1.68	0.13	0.76
0.985	1.68	0.14	0.94	0.96	1.68	0.13	0.81
0.990	1.68	0.14	0.99	0.97	1.68	0.13	0.86
0.995	1.68	0.13	1.03	0.98	1.68	0.13	0.92
0.996	1.68	0.13	1.03	0.99	1.68	0.13	0.99
1.000	1.68	0.13	1.07	1.00	1.68	0.13	1.07

9.5 Conclusion

In this chapter, the important features of the irreversible Rankine cycle airconditioning/heat pump system have been modelled by coupling it with thermal reservoirs of finite heat capacitance and effectiveness and taking into account the internal irreversibilities existing in the real systems. Obviously, internal irreversibility plays more prominent role in affecting the performance of the system than external irreversibility. But out of two internal irreversibility parameters, internal irreversibility parameter of Rankine heat engine has more drastic effect on performance than the internal irreversibility parameter of vapour compression cooling/heating cycle.

Chapter 10
Finite Time Thermodynamics of Brayton Refrigeration Cycle

10.1 Introduction

Most of the refrigeration cycles discussed so far involve change in phase of the working fluid (refrigerant) while undergoing through different stages. However, the gas refrigeration cycle is one in which the working fluid (refrigerant) remains as gas throughout the cycle, such as Brayton refrigeration cycle and Stirling/Ericsson refrigeration cycle. The gas refrigeration cycle deviates from the reversed Carnot cycle/vapour compression refrigeration (VCR) cycle because the heat transfer to and from the gas cycle is not isothermal. In fact, the gas temperature varies considerably during heat transfer processes. Consequently, the gas refrigeration cycle has much lower coefficient of performance (COP) as compared to the VCR and reversed Carnot cycle. Also, gas refrigeration systems have a number of important applications over the vapour compression, vapour absorption, and hybrid refrigeration systems. Some of them are used to achieve very low temperature for many important applications, such as liquefaction of gases/air, aircraft cabin cooling, superconductivity-related R&D works, and other specialized applications. Among others, the Brayton refrigeration cycle is one of the important gas refriger-ation cycles being used for space airconditioning applications with its unique application for aircraft cooling, liquefaction of gases, and cryogenic applications.

10.2 Ideal Brayton Refrigeration Cycle

As a general illustration, the Brayton refrigeration cycle is a reverse of closed Brayton power cycle as shown on the schematic and *T–s* diagrams of Figs. 10.1a and 10.1b, respectively. The refrigerant gas, which may be air, enters the compressor at state point 1, where the temperature is somewhat below the temperature of the cold region (i.e. heat source) T_L and is compressed in the compressor to state point 2.

S.C. Kaushik et al., *Finite Time Thermodynamics of Power and Refrigeration Cycles*,
https://doi.org/10.1007/978-3-319-62812-7_10

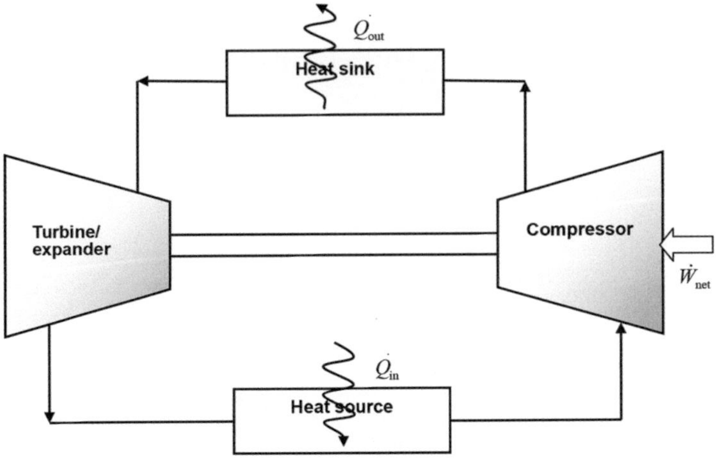

Fig. 10.1a Schematic of an ideal Brayton refrigeration cycle

Fig. 10.1b *T–s* diagram of an ideal Brayton refrigeration cycle

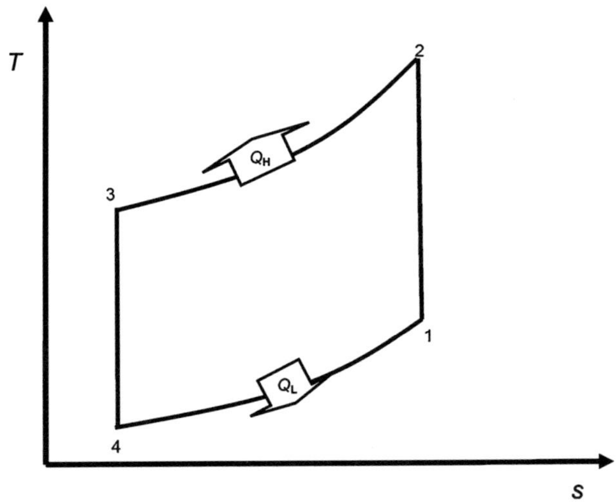

The gas then cooled up to state point 3, where the temperature of the gas refrigerant approaches to the temperature of the sink/warm region, T_H. After releasing the heat to the sink/surroundings, the gas refrigerant is expanded in an expander up to the state point 4, where the exit temperature is well below than that of the cold region/heat source (T_L) as can be seen from Fig. 10.1b. At state point 4, the gas starts absorbing heat from the heat source (cold region) as it passes from state 4 to state 1, thereby, completing the cycle. Thus, the ideal Brayton refrigeration cycle shown on the *T–s* diagram of Fig. 10.1b is denoted by 1–2–3–4–1 in which all the processes in the compressor and expander (turbine) are assumed to be adiabatic. However, the actual compression and expansion processes, in general, are not adiabatic because there is some less useful work during the actual processes (3–4), respectively. But, for simplicity of the analysis, the irreversibilities due to pressure drops have been ignored in this particular model of Brayton refrigeration cycle.

The method of analysis of the Brayton refrigeration cycle is similar to that of Brayton power cycle. Thus, for steady-state operation, the heat transfer rates to and from the cycle, power consumption by compression, and power production by expander are given below:

$$\dot{Q}_{\text{in}} = \dot{m}\,(h_1 - h_4) = \dot{m}\,c_{\text{p}}(T_1 - T_4) \tag{10.1}$$

$$\dot{Q}_{\text{out}} = \dot{m}\,(h_2 - h_3) = \dot{m}\,c_{\text{p}}(T_2 - T_3) \tag{10.2}$$

$$W_{\text{c}} = \dot{m}\,(h_2 - h_1) = \dot{m}\,c_{\text{p}}(T_2 - T_1) \tag{10.3}$$

$$W_{\text{t}} = \dot{m}\,(h_3 - h_4) = \dot{m}\,c_{\text{p}}(T_3 - T_4) \tag{10.4}$$

where $\dot{m}$ is the mass flow rate and $\dot{C}_{\text{p}}$ is the specific heat of the working fluid/air being used in the cycle. It is also important to note the work/power consumed by the cycle like the VCR cycle.

Thus the net work input and COP of the cycle are given as below:

$$W_{\text{net}} = W_{\text{c}} - W_{\text{t}} = \dot{m}\,c_{\text{p}}\{(T_2 - T_1) - (T_3 - T_4)\} \tag{10.5}$$

$$\text{COP} = \frac{Q_{\text{in}}}{W_{\text{net}}} = \frac{(h_1 - h_4)}{(h_2 - h_1) - (h_3 - h_4)} = \frac{T_1 - T_4}{(T_2 - T_1) - (T_3 - T_4)} \tag{10.6}$$

However, the gas refrigeration cycles have few special characteristics such as simple and lighter weight of components and they can incorporate regeneration. These two desirable characteristics make them suitable for two important applications such as aircraft cooling and liquefaction of gases and some applications despite their lower COP as compared to VCR/reversed Carnot cycle. For aircraft cooling, the low-temperature atmosphere air is passed through ram to partially heat up before compressing through the compressor, and it is again cooled down by the low-temperature ambient air and then further expanded through the turbine at the desired temperature and pressure which is then directly fed to the turbine as shown in Figs. 10.1a and 10.1b. The schematic and *T–s* diagrams of regeneration gas refrigeration cycle are shown in Figs. 10.1a and 10.1b, respectively. The regenerative cooling is achieved by incorporating a counterflow heat exchanger into the cycle; the high pressure gas is further cooled down to the lowest temperature than that of the cycle without regeneration. Lowering the turbine inlet automatically lowers the turbine exit temperatures, which is the lowest temperature of the cycle. Also, extremely low temperature can be achieved by repeating this process, which is generally used for gas liquefaction and cryogenic applications.

The cooling/heating system which follows reversed Carnot cycle is ideal system, all associated process will be reversible, and system will be in thermodynamic equilibrium. In the equilibrium position, temperature of the system working fluid will be the same as of the thermal reservoirs. Thermodynamic description of these equilibrium states and processes is studied in classical thermodynamics. In ideal cooling/heating system, the isothermal heating and cooling processes must be carried out infinitely slowly so that the working substance is in thermal equilibrium with the heat source and sink. Amount of heat extracted from heat source to the system is Q_L, amount of heat rejected from the system to the sink is Q_H, and work input to the system is W. In reversible system, processes are reversible and infinite time executable processes. Although, the performance parameter COP of these ideal systems is the highest time taken for transferring the heat, Q_H or Q_L will be infinite, and hence our desired effects (refrigeration load and heating load) will be zero. Refrigeration load = refrigerating effect produced over the cycle duration average = Q_L/time tends to zero, and hence, the desired output (heating/cooling load) which is the effect produced over the cycle duration average also tends to zero (Q_H/Time = 0).

10.3 Finite Time Brayton Refrigeration Cycle

The actual Brayton refrigeration cycle is not as efficient as the ideal one, because in real system, the processes followed by the cycle are not reversible leading to loss of potential to do work. There are two types of irreversibilities. The internal irreversibility is due to entropy generation, friction, pressure drop, etc., in different components, like turbine/expansion, compression, etc., while the external irreversibility is due to finite temperature difference between the cycle and the external reservoirs as mentioned by Curzon and Ahlborn (1975), for heat engines. Later on, it was extended by a number of authors for different RAC system (Blanchard 1980; Leff and Teeters 1978; Tyagi et al. 2006a). It is also desirable to have the corresponding results for the COP of the refrigeration, airconditioning, and heat pump systems. Leff and Teeters (1978) have noted that the straightforward C–A calculations will not work for reversed Carnot cycle because there is no 'natural maximum'. Blanchard (1980) has applied the Lagrangian method of undetermined multiplier to find out the COP of endoreversible Carnot heat pump operated at minimum power input for a given heating load.

However, in real system, there is some finite temperature difference between the system and surrounding. Considering this finite temperature difference, thermodynamic performance parameters/limits need to be established with finite time and/or finite size thermodynamic processes which are characterized by finite rate of heat exchange between system and environment. Thermodynamic descriptions of these real processes which occur in finite time are studied in FTT. There must be an optimal temperature difference between the working fluid and the reservoirs with which it is in contact for a finite amount of heat to be transferred in a finite time. For these systems, we have the reversed Carnot cycle in which T_C and T_E are the absolute temperatures of the given heat reservoirs and we wish to find optimal internal working fluid temperature $(T_c)_o$ and $(T_e)_o$ for maximum COP. There is no natural maximum as we adjust T_c and T_e, but one can, of course, always seek for such a maximum, for example, about the minimum input power for given heating load (or maximum output heating load for a given power input). Blanchard (1980) has studied the case when heat pump system is connected to the thermal reservoirs (heat source/sink) of infinite heat capacity and gave a formula for optimal COP at minimum input power assuming that system is externally irreversible but internally reversible, i.e. endoreversible.

In this chapter, the concept of FTT has been applied on the Brayton refrigeration cycle for useful applications. The effects of major (both internal and the external) irreversibilities on the maximum cooling and the corresponding COP have been evaluated for different set of operating conditions as given in the forthcoming sections.

10.3.1 Infinite Heat Capacity

The finite time Brayton refrigeration cycle is shown on the *T–s* diagram of Fig. 10.2. In this case of Brayton refrigeration system, the refrigerant enters the compressor at state point 1 and is compressed in the compressor to state point 2 in a compressor. The gas then cooled up to state point 3 by rejecting the heat to the heat sink of infinite heat capacity whose temperature remains constant at temperature T_H. After releasing the heat to the sink/surroundings, the gas refrigerant is expanded in an expander up to the state point 4. At state point 4, the gas enters the heat exchanger and starts absorbing heat from the heat source (cold region) of infinite heat capacity whose temperature remains constant at temperature T_L, as it passes from state 4 to state 1, thereby, completing the cycle. Thus, the ideal Brayton refrigeration cycle shown on the *T–s* diagram of Fig. 10.2 is denoted by 1–2–3–4–1. However, for simplicity of the analysis, the internal irreversibility due to various losses such as non-isentropic compression and expansion processes, pressure drop, etc. has been ignored in this particular model of Brayton refrigeration cycle. The heat transfer rates ($\dot{Q}_L$ and $\dot{Q}_H$) to and from the cycle can be written as (Tyagi et al. 2004a, b, c):

$$\dot{Q}_H = C_w(T_3 - T_4) = (UA)_H(LMTD)_H \tag{10.7}$$

$$\dot{Q}_L = C_w(T_2 - T_1) = (UA)_L(LMTD)_L \tag{10.8}$$

where $(UA)_H$ and $(UA)_L$ are, respectively, the heat transfer coefficient-area products for sink- and source-side heat exchangers, C_w is the heat capacitance rate of the working, and LMTDs are the Log Mean Temperature Differences on the sink- and source-side heat exchangers, respectively, and given as (Tyagi et al. 2004a, b, c):

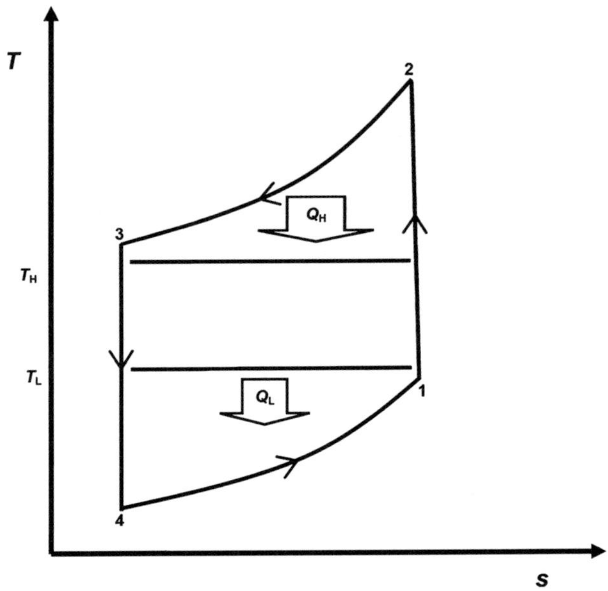

Fig. 10.2 *T–s* diagram of finite time Brayton refrigeration cycle

$$(\mathrm{LMTD})_{\mathrm{H}} = [(T_3 - T_{\mathrm{H}}) - (T_4 - T_{\mathrm{H}})]/\ln[(T_3 - T_{\mathrm{H}})/(T_4 - T_{\mathrm{H}})] \tag{10.9}$$

$$(\mathrm{LMTD})_{\mathrm{L}} = [(T_{\mathrm{L}} - T_1) - (T_{\mathrm{L}} - T_2)]/\ln[(T_{\mathrm{L}} - T_1)/(T_{\mathrm{L}} - T_4)] \tag{10.10}$$

From Eqs. (10.7), (10.8), (10.9), and (10.10), it yields:

$$Q_{\mathrm{H}} = \varepsilon_{\mathrm{H}} C_{\mathrm{W}}(T_3 - T_{\mathrm{H}}) = C_{\mathrm{w}}(T_3 - T_4) \tag{10.11}$$

$$Q_{\mathrm{L}} = \varepsilon_{\mathrm{L}} C_{\mathrm{W}}(T_{\mathrm{L}} - T_1) = C_{\mathrm{w}}(T_2 - T_1) \tag{10.12}$$

where ε_{H} and ε_{L} are the effectiveness of hot- and cold-side heat exchangers, respectively, and defined as (Tyagi et al. 2004a, b, c):

$$\varepsilon_{\mathrm{H}} = 1 - e^{-NTU_{\mathrm{H}}} \text{ and } \varepsilon_{\mathrm{L}} = 1 - e^{-NTU_{\mathrm{L}}} \tag{10.13}$$

where NTU_{H} and NTU_{L} are the number of heat transfer units on the sink- and source-side fluids, respectively, and are given as below (Tyagi et al. 2004a, b, c):

$$-NTU_{\mathrm{H}} = (\mathrm{UA})_{\mathrm{H}}/C_{\mathrm{W}} \quad \text{and} \quad -NTU_{\mathrm{L}} = (\mathrm{UA})_{\mathrm{L}}/C_{\mathrm{W}} \tag{10.14}$$

Using Eqs. (10.11) and (10.12) yields:

$$T_4 = (1 - \varepsilon_{\mathrm{H}})T_3 + \varepsilon_{\mathrm{H}} T_{\mathrm{H}} \tag{10.15}$$

$$T_2 = (1 - \varepsilon_{\mathrm{L}})T_1 + \varepsilon_{\mathrm{L}} T_{\mathrm{L}} \tag{10.16}$$

The second law of thermodynamics for Brayton refrigerator cycle yields:

$$T_3/T_2 = T_4/T_1 = X = \left(R_\mathrm{p}\right)^{(\gamma-1)/\gamma} \tag{10.17}$$

Again, using Eqs. (10.15), (10.16), and (10.17) yields:

$$T_3 = \frac{(1-\varepsilon_\mathrm{L})\varepsilon_\mathrm{H} T_\mathrm{H} + X\varepsilon_\mathrm{L} T_\mathrm{L}}{[1-(1-\varepsilon_\mathrm{H})(1-\varepsilon_\mathrm{L})]} \tag{10.18}$$

$$T_1 = \frac{X(1-\varepsilon_\mathrm{H})\varepsilon_\mathrm{L} T_\mathrm{L} + \varepsilon_\mathrm{H} T_\mathrm{H}}{[1-(1-\varepsilon_\mathrm{H})(1-\varepsilon_\mathrm{L})]X} \tag{10.19}$$

Using Eqs. (10.11), (10.12), (10.18), and (10.19), the cooling load, power input, and the COP of Brayton refrigeration cycle can be expressed as below:

$$Q_L = \frac{C_W \varepsilon_\mathrm{H}\varepsilon_\mathrm{L}(XT_\mathrm{L} - T_\mathrm{H})}{X[1-(1-\varepsilon_\mathrm{H})(1-\varepsilon_\mathrm{L})]} \tag{10.20}$$

$$P = Q_\mathrm{H} - Q_\mathrm{L} = \frac{C_W \varepsilon_\mathrm{H}\varepsilon_\mathrm{L}(XT_\mathrm{L} - T_\mathrm{H})(X-1)}{X[1-(1-\varepsilon_\mathrm{H})(1-\varepsilon_\mathrm{L})]} \tag{10.21}$$

$$\mathrm{COP} = \frac{Q_\mathrm{L}}{P} = \frac{1}{(X-1)} \tag{10.22}$$

From Eqs. (10.20), (10.21), and (10.22), it can be seen that the cooling load, power input, and the COP of an endoreversible Brayton refrigeration cycle are functions of a single variable (X) for a typical set of operating parameters. Thus, the optimum values of these performance parameters can be calculated for different set of operating conditions with respect to the cycle temperature ratio (X) using Lagrangian multiplier method (Blanchard 1980; Kaushik et al. 2002b, c; Tyagi et al. 2006a) as discussed in the previous chapter. Thus, the objective function for optimization is given as below:

$$L = P + \lambda Q_\mathrm{L} = \frac{C_W \varepsilon_\mathrm{H}\varepsilon_\mathrm{L}(XT_\mathrm{L} - T_\mathrm{H})(X-1+\lambda)}{X[1-(1-\varepsilon_\mathrm{H})(1-\varepsilon_\mathrm{L})]} \tag{10.23}$$

Thus, optimizing Eq. (10.23) with respect to x and λ yields the optimum values of different performance of an endoreversible Brayton refrigeration cycle, for a typical set of operating parameters, mentioned above.

10.3.2 Finite Heat Capacity

In this case of Brayton refrigeration system, the refrigerant enters the compressor at state point 1 and is compressed in the compressor to state point 2 in a compressor. The gas then cooled up to state point 3 by rejecting the heat to the heat sink of finite heat capacity whose temperature varies from $T_{\mathrm{H}1}$ to $T_{\mathrm{H}2}$. After releasing the heat to

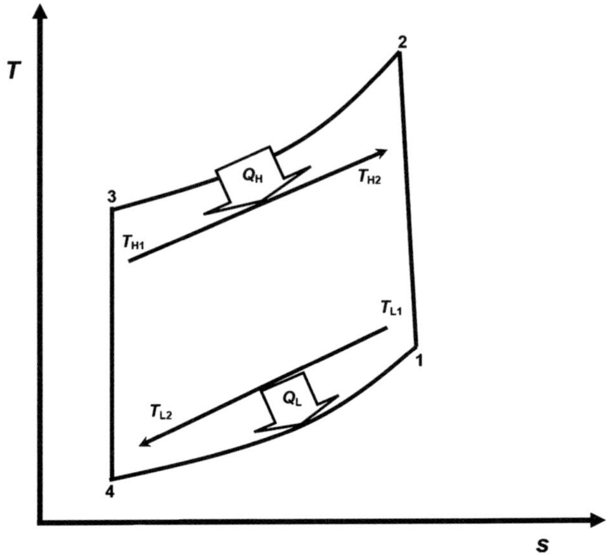

Fig. 10.3 *T–s* diagram of finite heat capacity Brayton refrigeration cycle

the sink/surrounding, the gas refrigerant is expanded in an expander up to the state point 4 in a real/ideal expander. At state point 4, the gas enters the heat exchanger and starts absorbing heat from the heat source (cold region) of finite heat capacity whose temperature varies from T_{L1} to T_{L2}, as it passes from state 4 to state 1, thereby, completing the cycle.

Thus, the real Brayton refrigeration cycle shown on the *T–s* diagram of Fig. 10.3 is denoted by 1–2–3–4–1 in which compression and expansion are assumed to be isentropic, and for simplicity, the irreversibility due to pressure drop has been ignored. The heat transfer rates (Q_L and Q_H) to and from the cycle can be written as (Tyagi et al. 2004a, b, c):

$$\begin{aligned}\dot{Q}_H &= C_{wf}(T_3 - T_4) = (UA)_H(LMTD)_H = C_H(T_{H2} - T_{H1}) \\ &= C_{H,min}\varepsilon_H(T_3 - T_{H1})\end{aligned} \tag{10.24}$$

$$\begin{aligned}\dot{Q}_L &= C_{wf}(T_2 - T_1) = (UA)_L(LMTD)_L = C_L(T_{L1} - T_{L2}) \\ &= C_{L,min}\varepsilon_L(T_{L1} - T_1)\end{aligned} \tag{10.25}$$

where $(LMTD)_H$ and $(LMTD)_L$ are the Log Mean Temperature Differences and ε_H and ε_L are the effectiveness of the source- and sink-side heat exchangers, respectively. Also $C_{H,min}$, $C_{L,min}$, $C_{H,max}$, and $C_{L,max}$ are the minimum and the maximum heat capacitance rates of the external fluids on the source and sink side, respectively, and for counterflow heat exchangers are given as below:

$$(\text{LMTD})_{\text{H}} = \frac{(T_3 - T_{\text{H2}}) - (T_4 - T_{\text{H1}})}{\ln\left(\frac{(T_3 - T_{\text{H2}})}{(T_4 - T_{\text{H1}})}\right)} \text{ and}$$
$$(\text{LMTD})_{\text{L}} = \frac{(T_{\text{L1}} - T_2) - (T_{\text{L2}} - T_1)}{\ln\left(\frac{(T_{\text{L1}} - T_2)}{(T_{\text{L1}} - T_1)}\right)} \tag{10.26}$$

$$\varepsilon_{\text{H}} = \frac{1 - e^{-NTU_{\text{H}}\left(1 - \frac{C_{\text{H,min}}}{C_{\text{H,max}}}\right)}}{1 - \frac{C_{\text{H,min}}}{C_{\text{H,max}}} e^{-NTU_{\text{H}}\left(1 - \frac{C_{\text{H,min}}}{C_{\text{H,max}}}\right)}}; \varepsilon_{\text{L}} = \frac{1 - e^{-NTU_{\text{L}}\left(1 - \frac{C_{\text{L,min}}}{C_{\text{L,max}}}\right)}}{1 - \frac{C_{\text{L,min}}}{C_{\text{L,max}}} e^{-NTU_{\text{L}}\left(1 - \frac{C_{\text{L,min}}}{C_{\text{L,max}}}\right)}} \tag{10.27}$$

$$\begin{pmatrix} C_{\text{H,min}} = \min(C_{\text{wf}}, C_{\text{H}}) \\ C_{\text{H,max}} = \max(C_{\text{wf}}, C_{\text{H}}) \end{pmatrix} \text{ and } \begin{pmatrix} C_{\text{L,min}} = \min(C_{\text{wf}}, C_{\text{L}}) \\ C_{\text{L,max}} = \max(C_{\text{wf}}, C_{\text{L}}) \end{pmatrix} \tag{10.28}$$

Using the second law of thermodynamics in this case of Brayton cycle yields:

$$\begin{aligned} &C_{\text{wf}} \ln (T_3/T_2) - C_{wf} \ln (T_4/T_1) = 0 \\ &\Rightarrow T_3/T_2 = T_4/T_1 = X = (R_{\text{p}})^{(\gamma-1)/\gamma} \end{aligned} \tag{10.29}$$

where R_{p} is the pressure ratio of the cycle and γ is the heat capacity ratio of the working fluid. Following Eqs. (10.24), (10.25), (10.26), (10.27), (10.28), and (10.29) for this case yields:

$$T_3 = \frac{(1 - \varepsilon_{\text{L}})\varepsilon_{\text{H}} T_{\text{H1}} + X \varepsilon_{\text{L}} T_{\text{L1}}}{[1 - (1 - \varepsilon_{\text{H}})(1 - \varepsilon_{\text{L}})]} \tag{10.30}$$

$$T_1 = \frac{X(1 - \varepsilon_{\text{H}})\varepsilon_{\text{L}} T_{L1} + \varepsilon_{\text{H}} T_{\text{H1}}}{[1 - (1 - \varepsilon_{\text{H}})(1 - \varepsilon_{\text{L}})]X} \tag{10.31}$$

Again, the cooling load, power input, and the COP in this case of Brayton refrigeration cycle can be expressed as below:

$$Q_{\text{L}} = \frac{C_{\text{W}} \varepsilon_{\text{H}} \varepsilon_{\text{L}} (X T_{\text{L1}} - T_{\text{H1}})}{X[1 - (1 - \varepsilon_{\text{H}})(1 - \varepsilon_{\text{L}})]} \tag{10.32}$$

$$P = Q_{\text{H}} - Q_{\text{L}} = \frac{C_{W} \varepsilon_{\text{H}} \varepsilon_{\text{L}} (X T_{\text{L1}} - T_{\text{H1}})(X - 1)}{X[1 - (1 - \varepsilon_{\text{H}})(1 - \varepsilon_{\text{L}})]} \tag{10.33}$$

$$\text{COP} = \frac{Q_{\text{L}}}{P} = \frac{1}{(X - 1)} \tag{10.34}$$

From Eqs. (10.32), (10.33), and (10.34), it can be seen that the cooling load, power input, and the COP of an endoreversible Brayton refrigeration cycle with finite heat capacity of external reservoirs are functions of a single variable (X) for a typical set of operating parameters. Thus, the optimum values of these performance parameters can be calculated for different set of operating conditions with respect to the cycle temperature ratio (X) using Lagrangian multiplier method (Blanchard 1980; Kaushik et al. 2002b, c; Tyagi et al. 2006a) as discussed in the previous chapter. Thus, the objective function for optimization is given as below:

$$L = P + \lambda Q_{\mathrm{L}} = \frac{C_W \varepsilon_{\mathrm{H}} \varepsilon_{\mathrm{L}} (X T_{\mathrm{L1}} - T_{\mathrm{H1}})(X - 1 + \lambda)}{X[1 - (1 - \varepsilon_{\mathrm{H}})(1 - \varepsilon_{\mathrm{L}})]} \tag{10.35}$$

Thus, optimizing Eq. (10.23) with respect to X and λ yields the optimum value of temperature ratio, and other performance of an endoreversible Brayton refrigeration cycle, for a typical set of operating parameters, mentioned above.

10.4 Irreversible Brayton Refrigeration Cycle

As it is well known that at the exit of the heat sink (high-temperature heat sink/reservoir), the temperature of the refrigerant is much higher than that of the inlet of the heat source (space to be cooled). In order to further reduce the exit temperature of the expander and/or to reduce the temperature at the entrance of the heat source, a regenerator may be included in the Brayton cycle as shown in Fig. 10.4a. This not only enhanced the cooling load but also reduced the compressor input power, thereby, enhancing the overall COP of the system. In this case of Brayton refrigeration system, the refrigerant enters the compressor at state point 1 and is compressed in the compressor to state point 2/2S in a real/ideal compressor. The gas then cooled up to state point 3 by rejecting the heat to the heat sink of finite heat capacity whose temperature remains constant at temperature T_{H}. After releasing the heat to the sink/surroundings, the gas refrigerant is expanded in an expander up to the state point 4S/4 in a real/ideal expander. At state point 4, the gas enters the heat exchanger and starts absorbing heat from the heat source (cold region) of infinite heat capacity whose temperature remains constant at temperature T_{L}, as it passes from state 4 to state 4R. Finally, the working fluid enters the regenerator where it gets heated to state point 1, thereby, completing the cycle.

Thus, the actual Brayton refrigeration cycle shown on the T–s diagram of Fig. 10.4b is denoted by 1–2–3–4–1 in which compression and expansion are assumed to be non-isentropic, due to the fact that in general, all real processes are nonadiabatic as there is always some loss of useful work during the actual processes. However, for simplicity the irreversibility due to pressure drop has been ignored. The heat transfer rates (Q_{L} and Q_{H}) to and from the cycle can be written as:

$$\begin{aligned} \dot{Q}_{\mathrm{H}} &= C_{\mathrm{wf}}(T_3 - T_{3\mathrm{R}}) = (\mathrm{UA})_{\mathrm{H}}(\mathrm{LMTD})_{\mathrm{H}} \\ &= C_{\mathrm{H}}(T_{\mathrm{H2}} - T_{\mathrm{H1}}) = C_{\mathrm{H,min}}\varepsilon_{\mathrm{H}}(T_3 - T_{\mathrm{H1}}) \end{aligned} \tag{10.36}$$

$$\begin{aligned} \dot{Q}_{\mathrm{L}} &= C_{wf}(T_{\mathrm{1R}} - T_1) = (\mathrm{UA})_{\mathrm{L}}(\mathrm{LMTD})_{\mathrm{L}} \\ &= C_{\mathrm{L}}(T_{\mathrm{L1}} - T_{\mathrm{L2}}) = C_{\mathrm{L,min}}\varepsilon_{\mathrm{L}}(T_{\mathrm{L1}} - T_1) \end{aligned} \tag{10.37}$$

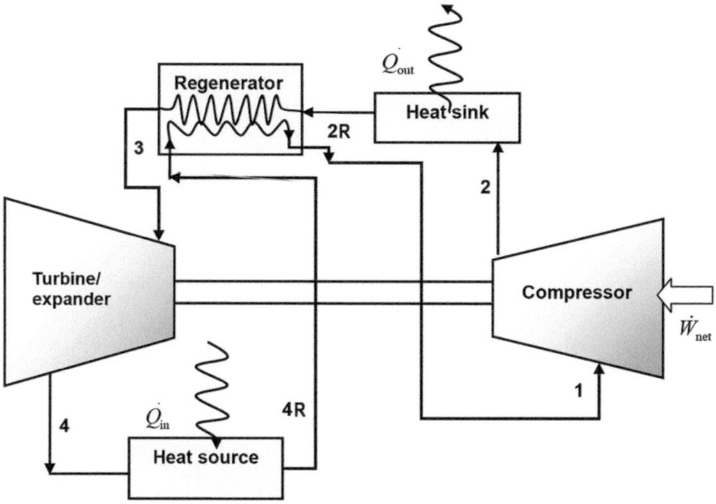

Fig. 10.4a Schematic of an irreversible regenerative Brayton refrigeration cycle

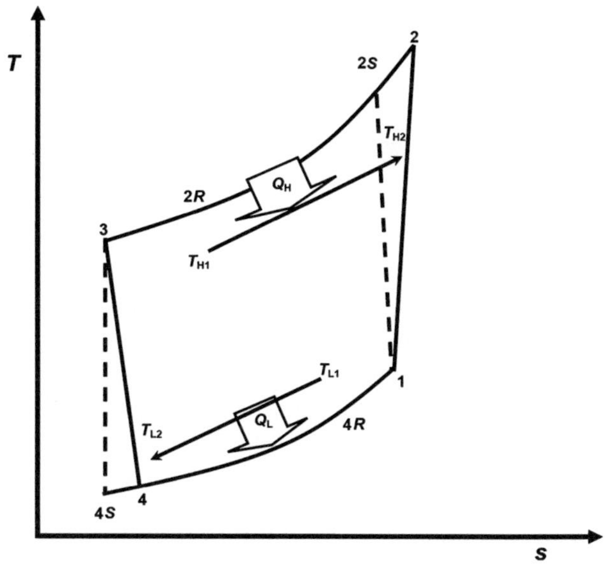

Fig. 10.4b *T–s* diagram of an irreversible regenerative Brayton refrigeration cycle

where $(\text{LMTD})_H$ and $(\text{LMTD})_L$ are the Log Mean Temperature Differences and ε_H and ε_L are the effectiveness of the source- and sink-side heat exchangers, respectively. Also $C_{H,min}$, $C_{L,min}$, $C_{H,max}$, and $C_{L,max}$ are the minimum and the maximum heat capacitance rates of the external fluids on the source and sink side, respectively, and for counterflow heat exchangers are given as below (Tyagi et al. 2006a):

$$\begin{aligned}(\mathrm{LMTD})_{\mathrm{H}} &= \frac{(T_{\mathrm{H1}}-T_2)-(T_{\mathrm{H2}}-T_3)}{\ln\left(\frac{(T_{\mathrm{H1}}-T_2)}{(T_{\mathrm{H2}}-T_3)}\right)}\\ (\mathrm{LMTD})_{\mathrm{L}} &= \frac{(T_4-T_{\mathrm{L2}})-(T_1-T_{\mathrm{L1}})}{\ln\left(\frac{(T_4-T_{\mathrm{L2}})}{(T_1-T_{\mathrm{L1}})}\right)}\end{aligned} \tag{10.38}$$

$$\varepsilon_{\mathrm{H}} = \frac{1-\mathrm{e}^{-NTU_{\mathrm{H}}\left(1-\frac{C_{\mathrm{H,min}}}{C_{\mathrm{H,max}}}\right)}}{1-\frac{C_{\mathrm{H,min}}}{C_{\mathrm{H,max}}}\mathrm{e}^{-NTU_{\mathrm{H}}\left(1-\frac{C_{\mathrm{H,min}}}{C_{\mathrm{H,max}}}\right)}};\ \varepsilon_{\mathrm{L}} = \frac{1-\mathrm{e}^{-NTU_{\mathrm{L}}\left(1-\frac{C_{\mathrm{L,min}}}{C_{\mathrm{L,max}}}\right)}}{1-\frac{C_{\mathrm{L,min}}}{C_{\mathrm{L,max}}}\mathrm{e}^{-NTU_{\mathrm{L}}\left(1-\frac{C_{\mathrm{L,min}}}{C_{\mathrm{L,max}}}\right)}};\ \varepsilon_R = \frac{NTU_{\mathrm{R}}}{1+NTU_{\mathrm{R}}} \tag{10.39}$$

$$\begin{pmatrix} C_{\mathrm{H,min}} = \min(C_{\mathrm{wf}}, C_{\mathrm{H}}) \\ C_{\mathrm{H,max}} = \max(C_{\mathrm{wf}}, C_{\mathrm{H}}) \end{pmatrix} \text{and} \begin{pmatrix} C_{\mathrm{L,min}} = \min(C_{\mathrm{wf}}, C_{\mathrm{L}}) \\ C_{\mathrm{L,max}} = \max(C_{\mathrm{wf}}, C_{\mathrm{L}}) \end{pmatrix} \tag{10.40}$$

Using T–s diagram, the expansion and compression efficiencies of the cycle can also be expressed, as below:

$$\eta_{\mathrm{e}} = \frac{C_{\mathrm{w}}(T_4-T_1)}{C_{\mathrm{w}}(T_4-T_1)} \text{ and } \eta_{\mathrm{c}} = \frac{C_{\mathrm{w}}(T_3-T_{2S})}{C_{\mathrm{w}}(T_3-T_2)} \tag{10.41}$$

Using the second law of thermodynamics in this case of Brayton cycle yields:

$$\begin{aligned}&C_{\mathrm{wf}}\ln\,(T_{3\mathrm{s}}/T_2) - C_{\mathrm{wf}}\ln\,(T_4/T_{1\mathrm{s}}) = 0\\ &\Rightarrow T_{3\mathrm{s}}/T_2 = T_4/T_{1\mathrm{s}} = X = (R_{\mathrm{p}})^{(\gamma-1)/\gamma}\end{aligned} \tag{10.42}$$

where X and R_{p} are, respectively, the temperature and pressure ratios of the cycle and γ is the heat capacity ratio of the working fluid. Following Eqs. (10.36) and (10.37) for this case yields:

$$T_4 = (1-\varepsilon_{\mathrm{R}})T_{3\mathrm{R}} + \varepsilon_{\mathrm{R}}T_{1\mathrm{R}} \quad \text{and} \quad T_2 = (1-\varepsilon_{\mathrm{R}})T_{1\mathrm{R}} + \varepsilon_{\mathrm{R}}T_{3\mathrm{R}} \tag{10.43a}$$

$$T_{1\mathrm{R}} = (1-a_2)T_1 + a_2T_{\mathrm{L1}} \quad \text{and} \quad T_{3\mathrm{R}} = (1-a_1)T_3 + a_1T_{\mathrm{H1}} \tag{10.43b}$$

where $a_1 = \varepsilon_{\mathrm{H}}C_{\mathrm{H,min}}/C_{\mathrm{w}}$ and $a_2 = \varepsilon_{\mathrm{H}}C_{\mathrm{L,min}}/C_{\mathrm{w}}$. Using Eqs. (10.42) and (10.43) yields:

$$T_1 = \frac{a_5a_8 + a_6(1-a_1)(1-\varepsilon_{\mathrm{R}})}{\left\{a_7a_8 - (1-a_1)(1-a_2)(1-\varepsilon_{\mathrm{R}})^2\right\}} \tag{10.44}$$

$$T_3 = \frac{a_6a_7 + a_5(1-a_2)(1-\varepsilon_R)}{\left\{a_7a_8 - (1-a_1)(1-a_2)(1-\varepsilon_R)^2\right\}} \tag{10.45}$$

where the different parameters are given as below:

$$a_8 = \frac{1}{a_4} - \varepsilon_R a_1; a_7 = \frac{1}{a_3} - \varepsilon_R a_2;$$
$$a_6 = (1-\varepsilon_R)a_2T_{L1} + \varepsilon_R a_1 T_{H1}; a_4 = \frac{(\eta_e - X + 1)}{X};$$
$$a_5 = (1-\varepsilon_R)a_1T_{H1} + \varepsilon_R a_2 T_{L1}; \quad \text{and} \quad a_3 = \frac{(X - 1 + \eta_c)}{\eta_c}$$

Again, the cooling load, power input, and the COP in this case of Brayton refrigeration cycle can be expressed as below:

$$Q_H = C_wC_1\left(\frac{a_6a_7 + a_5(1-a_2)(1-\varepsilon_R)}{\left\{a_7a_8 - (1-a_1)(1-a_2)(1-\varepsilon_R)^2\right\}} - T_{H1}\right) \tag{10.46}$$

$$Q_L = C_wC_2\left(T_{L1} - \frac{a_5a_8 + a_6(1-a_1)(1-\varepsilon_R)}{\left\{a_7a_8 - (1-a_1)(1-a_2)(1-\varepsilon_R)^2\right\}}\right) \tag{10.47}$$

$$P = C_w\left(\frac{\{a_6a_7 + a_5(1-a_2)\}C_1 + \{a_5a_8 + a_6(1-a_1)\}C_2}{\left\{a_7a_8 - (1-a_1)(1-a_2)(1-\varepsilon_R)^2\right\}}\right)(1-\varepsilon_R) - (C_1T_{H1} + C_2T_{L1}) \tag{10.48}$$

$$\text{COP} = \frac{Q_L}{P} \tag{10.49}$$

From Eqs. (10.46), (10.47), (10.48), and (10.49), it can be seen that the cooling load, power input, and the COP of an irreversible Brayton refrigeration cycle with finite heat capacity of external reservoirs are functions of a single variable (X) for a typical set of operating parameters. Thus, the optimum values of the performance parameter can be calculated for different set of operating conditions with respect to the cycle temperature ratio (X) using Lagrangian multiplier method (Blanchard 1980; Tyagi et al. 2004a, b, c) as discussed in the previous chapter. Thus, using Eqs. (10.46), (10.47), and (10.48), the objective function can be given as below:

$$L = P + \lambda Q_L \tag{10.50}$$

where λ is the Lagrangian undermined multiplier, which can be evaluated using boundary conditions. Equation (10.50) is a function of number of variables, but most of these variables are also functions of X only as mentioned earlier in Eqs. (10.44), (10.45), (10.46), (10.47), and (10.48).

It is also important to note that this particular case is more general, and most of the cases discussed here can be directly drawn from this particular model. For example, if the compression and expansion process are assumed to be ideal and if there is no regenerator, this cycle model will approach to that discussed in Sect. 10.3.1. Similarly, if there is no internal irreversibility and the external reservoirs are of infinite heat capacity, this cycle model will approach to that of given in Sect. 10.3.2.

10.5 Thermoeconomics of Brayton Refrigeration Cycle

An irreversible regenerative Brayton refrigerator cycle is shown on T–s diagrams of Figs. 10.4a and 10.4b. The heat transfer rates to and from the cycle (Q_L and Q_H) and the regenerative heat transfer rate (Q_R) can be written as (Tyagi et al. 2006a):

$$Q_H = (\mathrm{UA})_H(\mathrm{LMTD})_H = C_w(T_3 - T_{3R}) \tag{10.51}$$

$$Q_L = (\mathrm{UA})_L(\mathrm{LMTD})_L = C_w(T_{1R} - T_1) \tag{10.52}$$

$$Q_R = (\mathrm{UA})_R(\mathrm{LMTD})_R = C_w(T_{3R} - T_4) = C_w(T_2 - T_{1R}) \tag{10.53}$$

where ($\mathrm{UA_J}$) ($J = $ H, L, and R) are the heat transfer coefficient-area products on the respective-side heat exchangers, C_w is the heat capacitance rate of the working fluid, and LMTDs are the Log Mean Temperature Differences on the respective-side heat exchangers, defined as:

$$(\mathrm{LMTD})_H = \left[(T_3 - T_H) - (T_{3R} - T_H)\right] / \ln\left[(T_3 - T_H)/(T_{3R} - T_H)\right] \tag{10.54}$$

$$(\mathrm{LMTD})_L = \left[(T_L - T_1) - (T_L - T_{1R})\right] / \ln\left[(T_L - T_1)/(T_L - T_{1R})\right] \tag{10.55}$$

$$(\mathrm{LMTD})_R = \left[(T_{3R} - T_2) - (T_4 - T_{1R})\right] / \ln\left[(T_{3R} - T_2)/(T_4 - T_{1R})\right] \tag{10.56}$$

Using Eqs. (10.51), (10.52), (10.53), (10.54), (10.55), and (10.56), we have:

$$Q_H = \varepsilon_H C_W(T_3 - T_H) = C_W(T_3 - T_{3R}) \tag{10.57}$$

$$Q_L = \varepsilon_L C_W(T_L - T_1) = C_w(T_{1R} - T_1) \tag{10.58}$$

$$Q_R = \varepsilon_L C_W(T_{3R} - T_{1R}) = C_w(T_{3R} - T_4) = C_w(T_2 - T_{1R}) \tag{10.59}$$

where ε_H, ε_L, and ε_R are, respectively, the effectiveness on sink-, source-, and the regenerative-side heat exchanger, given as:

$$\varepsilon_{\mathrm{H}} = 1 - e^{\left(-(\mathrm{UA})_{\mathrm{H}}/C_{\mathrm{W}}\right)}, \quad \varepsilon_{\mathrm{L}} = 1 - e^{\left(-(\mathrm{UA})_{\mathrm{L}}/C_{\mathrm{W}}\right)} \quad \text{and} \quad \varepsilon_{\mathrm{R}} = \frac{(\mathrm{UA})_{\mathrm{R}}}{(\mathrm{UA})_{\mathrm{R}} + C_{\mathrm{W}}} \tag{10.60}$$

Using Fig. 10.4b, the compression and expansion efficiencies can also be expressed as below:

$$\eta_{\mathrm{c}} = (T_{3\mathrm{S}} - T_2)/(T_3 - T_2) \tag{10.61}$$

$$\eta_{\mathrm{e}} = (T_4 - T_1)/(T_4 - T_{1\mathrm{S}}) \tag{10.62}$$

Using Eqs. (10.57), (10.58), (10.59), (10.60), (10.61), and (10.62) yields (Tyagi et al. 2006a):

$$T_{3\mathrm{R}} = (1 - \varepsilon_{\mathrm{H}})T_3 + \varepsilon_{\mathrm{H}} T_{\mathrm{H}} \tag{10.63}$$

$$T_{1\mathrm{R}} = (1 - \varepsilon_{\mathrm{L}})T_1 + \varepsilon_{\mathrm{L}} T_{\mathrm{L}} \tag{10.64}$$

$$T_2 = (1 - \varepsilon_{\mathrm{R}})T_{1\mathrm{R}} + \varepsilon_{\mathrm{R}} T_{3\mathrm{R}} \tag{10.65}$$

$$T_4 = (1 - \varepsilon_{\mathrm{R}})T_{3\mathrm{R}} + \varepsilon_{\mathrm{R}} T_{1\mathrm{R}} \tag{10.66}$$

$$T_{3\mathrm{S}} = (1 - \eta_{\mathrm{c}})T_2 + \eta_{\mathrm{c}} T_3 \tag{10.67}$$

$$T_{1\mathrm{S}} = \left(1 - \eta_{\mathrm{e}}^{-1}\right)T_4 + \eta_{\mathrm{e}}^{-1} T_1 \tag{10.68}$$

From the second law of thermodynamics for a Brayton cycle, we have:

$$T_{3\mathrm{S}}/T_2 = T_4/T_{1\mathrm{S}} \Rightarrow T_{3\mathrm{S}} T_{1\mathrm{S}} = T_2 T_4 \tag{10.69}$$

Using Eqs. (10.61), (10.62), (10.63), (10.64), (10.65), (10.66), (10.67), (10.68), and (10.69), we obtained a quadratic equation in T_3 and T_1, for a given set of operating parameters, which on solving for T_3 (treating T_1 as constant) yields:

$$T_3 = \left(-B \pm \sqrt{B^2 - 4AC}\right)/2A \tag{10.70}$$

where the various parameters are given below:

$$A = a_2 b_2 - c_2 d_2, B = a_4 T_1 + a_5, C = a_6 T_1^2 + a_7 T_1 + a_8,$$
$$a_4 = a_1 b_2 + a_2 b_1 - c_1 d_2 - c_2 d_1,$$
$$a_5 = a_2 b_3 + a_3 b_2 - c_2 d_3 - c_3 d_2, a_6 = a_1 b_1 - c_1 d_1,$$
$$a_7 = a_1 b_3 + a_3 b_1 - c_1 d_3 - c_3 d_1, a_8 = a_3 b_3 - c_3 d_3$$
$$a_1 = (1 - \varepsilon_{\mathrm{R}})(1 - \varepsilon_{\mathrm{L}}), a_2 = \varepsilon_{\mathrm{R}}(1 - \varepsilon_{\mathrm{H}}), a_3 = (1 - \varepsilon_{\mathrm{R}})\varepsilon_{\mathrm{L}} T_{\mathrm{L}} + \varepsilon_{\mathrm{R}}\varepsilon_{\mathrm{H}} T_{\mathrm{H}},$$
$$b_1 = \varepsilon_{\mathrm{R}}(1 - \varepsilon_{\mathrm{L}}), b_2 = (1 - \varepsilon_{\mathrm{R}})(1 - \varepsilon_{\mathrm{H}}), b_3 = (1 - \varepsilon_{\mathrm{R}})\varepsilon_{\mathrm{H}} T_{\mathrm{H}} + \varepsilon_{\mathrm{R}}\varepsilon_{\mathrm{L}} T_{\mathrm{L}},$$
$$c_1 = h_{\mathrm{e}}^{-1} + b_1\left(1 - h_{\mathrm{e}}^{-1}\right), c_2 = \left(1 - h_{\mathrm{e}}^{-1}\right)b_2, c_3 = \left(1 - h_{\mathrm{e}}^{-1}\right)b_3,$$
$$d_1 = (1 - h_{\mathrm{c}})a_1, d_2 = h_{\mathrm{c}} + (1 - h_{\mathrm{c}})a_2 \text{ and } d_3 = (1 - h_{\mathrm{c}})a_3$$

The objective function for thermoeconomic optimization as proposed by Kodal et al. (2003) and Tyagi et al. (2004a) and modified by Tyagi et al. (2006a) is given as below:

$$F = \frac{Q_L}{C_i + C_e + C_m} \tag{10.71}$$

where C_i, C_e, and C_m refer to the annual investment, energy consumption, and maintenance costs, respectively. The investment cost was considered to be the costs of the main system components, such as evaporator, condenser, and the compression and expansion devices together. The investment cost of the heat evaporator and condenser is assumed to be proportional to the cooling and heating capacities, which is different from those of the earlier workers, Kodal et al. (2003) and Tyagi et al. (2004a, b). On the other hand, the investment cost of the compression and expansion devices is assumed to be proportional to their compression capacity or the power input of the cycle (Kodal et al. 2003; Tyagi et al. 2004a,b). Thus the investment cost of the system can be given by:

$$C_i = a_H Q_H + a_L Q_L + a_R Q_R + a_p P \tag{10.72}$$

where the proportionality constant for the investment cost of the condenser, evaporator, and regenerator, a_J (J = H, L, R), is equal to the annual cost per unit heat transfer rate and their dimension is ncu/year-kW. The proportionality constant for the investment cost for the compression and expansion devices together, a_p, is equal to the annual cost per unit power input capacity of the compressor, and its dimension is ncu/year-kW. The unit ncu stands for the 'national currency unit'. The average energy consumption and maintenance costs are, respectively, proportional to the power input and cooling rate (Kodal et al. 2003; Tyagi et al. 2004a, b), as given below:

$$C_e = a_e P \quad \text{and} \quad C_m = a_m \dot{Q}_L \tag{10.73}$$

where the coefficients a_e and a_m are equal to the annual cost per unit power input and energy input rate, respectively, and the dimensions of both the parameters are ncu/year-kW. Substituting Eqs. (10.72) and (10.73) into Eq. (10.71) yields (Tyagi et al. 2006a):

$$bF = \frac{Q_L}{a_H Q_H + a_L Q_L + a_R Q_R + b(Q_H - Q_L) + a_m Q_L} = \frac{(T_L - T_1)}{\gamma_1 T_3 + \gamma_2 T_1 + \gamma_5} \tag{10.74}$$

where $b = a_p + a_e$; $\gamma_1 = \{(a_H + b)e_H + a_R e_R(1 - e_H)\}/e_L$, $\gamma_2 = \{(b - a_m - a_L) e_L - a_R e_R(1 - e_L)\}/e_L$, $\gamma_5 = \gamma_3 + \gamma_4$, $\gamma_3 = \{(a_H + b)e_H + a_R e_R\}e_H T_H/e_L$, and $\gamma_4 = \{a_m + a_L - b - a_R e_R\}T_L$.

From Eq. (10.74), one can see that the objective function, cooling load, power input, and the COP an irreversible Brayton refrigerator cycle are functions of a single variable T_1 (as x and T_3 are also functions of T_1) for a typical set of operating parameters.

10.6 Discussion of Results

In order to verify the validity of this cycle model, we have plotted some characteristic curves for the thermoeconomic objective function, cooling load, power input, and COP with respect to state point temperature (T_1), while other parameters were chosen as $T_H = 300$ K, $T_L = 200$ K, $a_H = a_L = a_R = 0.15$, $a_p = 1.0$, $a_e = 0.1$, $a_m = 0.05$, $e_H = e_L = e_R = h_e = h_c = 0.85$, and $C_W = 1.0$ kW/K, and obtained the following results.

Effects of Compressor Inlet Temperature (T_1)

The variations of the objective function, cooling load, power input, and COP vs. expansion outlet temperature (T_1) for a typical set of operating parameters are shown in Fig. 10.5. It is seen from Fig. 10.5 that the power input and cooling load decrease while the COP increases as the temperature T_1 is increased. However, the objective function first increases and then decreases as T_1 is increased. Thus there is an optimal value of T_1 at which the objective function attains its maximum value for a given set of operating parameters. The above mentioned can be explained in different ways. For example, if the state point temperature T_1 tends to the source temperature, i.e. $T_1 \rightarrow T_L$, the heat transfer to the cycle tends to reversible. In other words, the heat transfer area or time tends to infinity, and hence, the cooling load

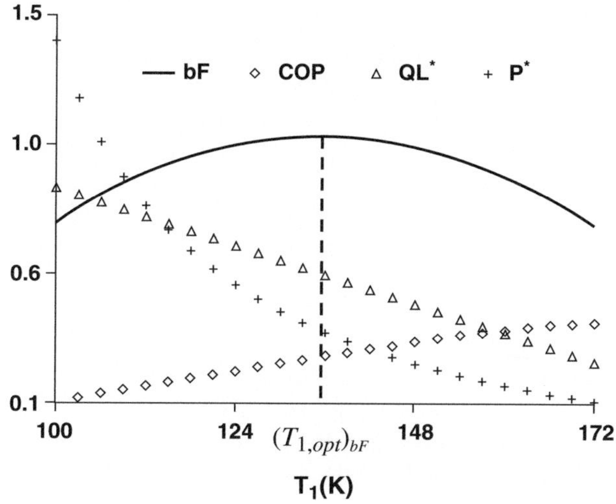

Fig. 10.5 Objective function, cooling load, COP, and power input versus T_1

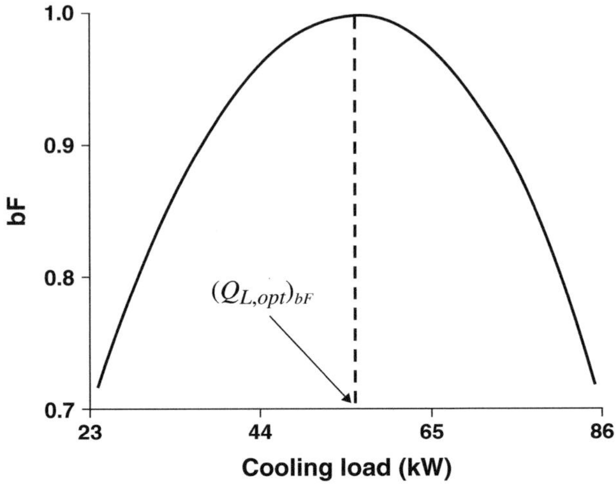

Fig. 10.6 Objective function versus cooling load

and power input tend to zero, while the COP tends to Carnot COP as can be seen from Fig. 10.5.

Again, as explained by Leff and Teeters (1978), there is no 'natural maxima' in the reverse heat engine cycles, which means we can optimize the power input for a given cooling/heating load or vice versa and the COP can be calculated for that condition as done by Blanchard (1980) for a Carnot heat pump. So the objective function which is cooling load per unit cost has a maximum value for a given set of operating parameters. Again, the maximum of the objective function will change as any of the operating parameter is changed.

Objective Function vs. Performance Parameters

The variation of the objective function with respect to the cooling load, power input, and COP of an irreversible regenerative Brayton refrigeration cycle is shown in Figs. 10.6, 10.7, and 10.8. It is seen from these figures that the objective function first increases and then decreases as any one of these parameter, viz. the power input, cooling load, and COP, is increased. Thus there are optimal values of all these parameters at which the Brayton refrigerator cycle attains its maximum objective function for a typical set of operating parameters. The results obtained in these figures can be explained in different ways. For example, if the cooling capacity is very small/low, the objective function, viz. the cooling load per unit cost, is low/small which means the cost is much higher for lower capacity. Again if the cooling capacity is much higher beyond a limit, the investment cost which includes the capital cost and running cost is much higher; hence, the objective function is again lower. Similar explanations can be given for the power input, and hence, there is an optimal range for the cooling load and power input for a given cycle at which it attains its maximum objective function.

The explanation for COP can be given using irreversible thermodynamic processes. For example, if the COP for a reversible cycle is much higher than a real

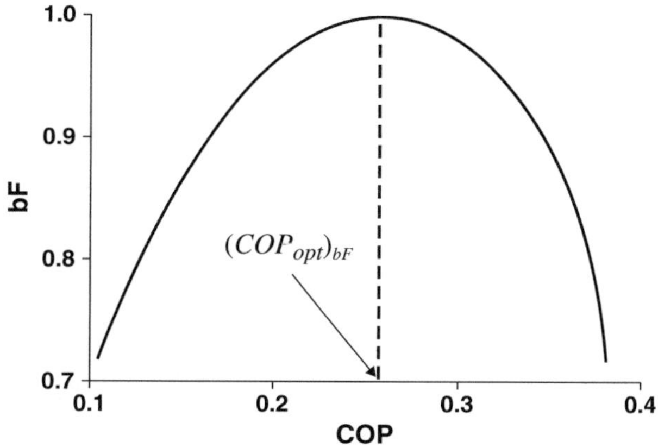

Fig. 10.7 Objective function versus COP

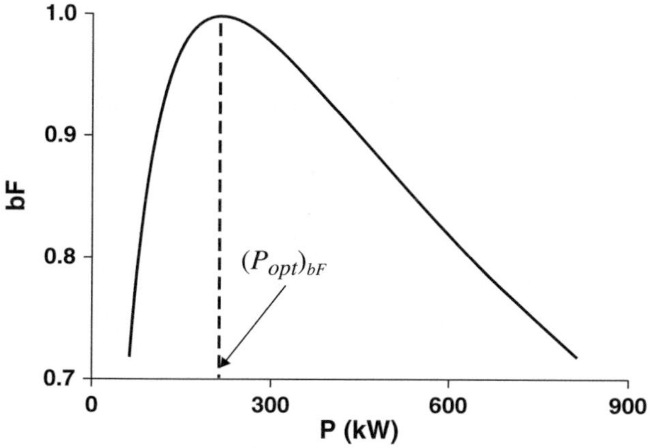

Fig. 10.8 Objective function versus power input

cycle, a reversible cycle needs much heat transfer area or time. But if the heat exchanger is much bigger, the system is very expensive, and if the heat transfer time is much higher, the cooling load is lower. Again if the COP is too low, the power input is much higher, and hence the cost again enhances. Thus, the objective function first increases and then decreases as the COP is increased.

So far we have given some characteristic curves based on the maximum objective function for a given set of operating parameters. We can further give the effects of some other cycle parameters such as component efficiency, effectiveness of different heat exchangers, economic parameters, etc. on the objective function and the corresponding cooling loads, power input, and COP, for a typical set of operating condition.

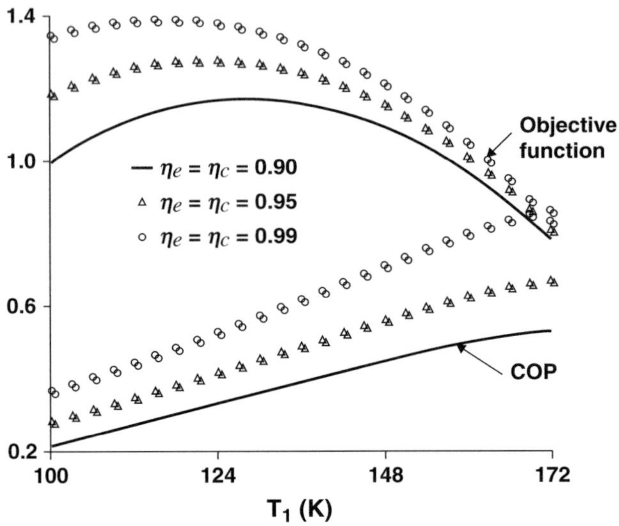

Fig. 10.9 Objective function and COP versus T_1 for component efficiencies

Effects of Component Efficiency (η_e and η_c)
The variation of the objective function and COP vs. T_1 for different values of component efficiencies of an irreversible regenerative reverse Brayton cycle is given in Fig. 10.9 for a typical set of operating parameters. It is seen from these figures that the COP increases and power input decreases as T_1 increases. However, the objective function first increases and then decreases as T_1 increases. Thus there is an optimal value of T_1 at which the objective function attains its maximum value which also can be given by Eq. (10.74). The objective function and COP further enhance, while the power input goes down as the component efficiencies are increased. The effects of the component efficiencies are significant for all the performance parameters, but the effect for power input is more significant.

This can be explained on the basis of internal irreversibility associated with the cycle. Since, the higher is the component's efficiency, the lower is the internal irreversibility associated with the cycle, and hence, the higher is the COP of the cycle, the lower will be the required power input, which also leads to lower cost of the cycle. As a result the objective function and the COP increase, while the power input decreases as the component's efficiencies are increased. Since the internal irreversibility affects the power input and COP, the objective function also enhances.

Effect of Economic Parameters
The variation of the objective function with respect to T_1 for different values of economic parameters is shown in Fig. 10.10, for a typical set of operating parameters. It is seen from Fig. 10.10 that the objective function first increases and then decreases as temperature T_1 increases. Thus there is an optimal value of T_1 at which the objective function attains its maximum value which is also given in Eq. (10.74).

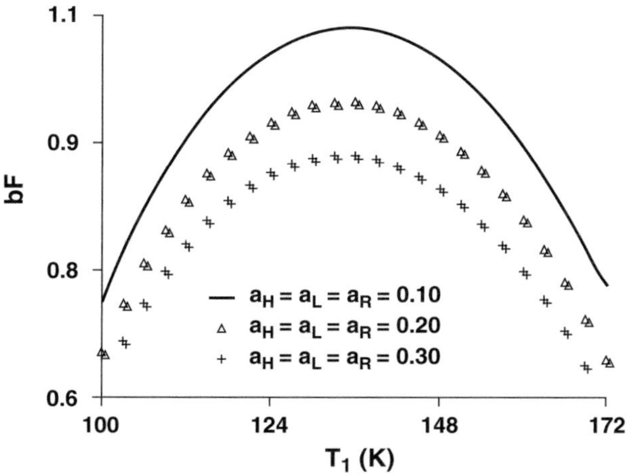

Fig. 10.10 Objective function versus T_1 for different values of economic parameter

The maxima of the objective function further go down as the economic parameters are increased, which show that the cost of each component has significant effect on the thermoeconomic performance of the cycle. The economic parameter consists of different sub-parameters such as the material technology, power cost, land cost, and other proxy parameters like availability of raw material, manpower, labour cost, etc. So the system operating/manufacturing in some Asian countries has a better thermoeconomic performance than those of the systems manufactured or operating in the developed countries.

So far we have given some characteristic curves based on different objective functions of an irreversible regenerative Brayton refrigerator cycle for a given set of operating parameters. Based on Fig. 10.5 and Eq. (10.74), one can also give some optimum criteria for the objective function and the corresponding cycle parameters such as cooling load, power input, and COP. Using Eq. (10.74) and its extremum condition, viz. $\frac{\partial bF}{\partial T_1} = 0$, one can prove that the optimal value of T_1 for the maximum objective function is given by:

$$A_1 T_1^2 + B_1 T_1 + C_1 = 0 \Rightarrow T_{1,\mathrm{opt}} = \frac{-B_1 \pm \sqrt{B_1^2 - A_1 C_1}}{A_1} \tag{10.75}$$

where the various parameters are:

$$\begin{aligned}
A_1 &= \left(a_9 - a_{12}^2/a_{11}^2\right), B_1 = \left(a_{10} - a_{12}a_{13}/a_{11}^2\right),\\
C_1 &= \left(a_5^2 - 4Aa_8 - a_{13}^2/a_{11}^2\right), a_9 = \left(a_4^2 - 4Aa_6\right),\\
a_{10} &= \left(a_4 a_5 - 2Aa_7\right), a_{11} = \left(a_4 T_\mathrm{L} + a_5 + 2A(\gamma_\mathrm{L} T_\mathrm{L} - \gamma_5)/\gamma_1\right),\\
a_{12} &= \left(a_{10} + a_9 T_\mathrm{L}\right) \text{and}\, a_{13} = \left(a_5^2 - 4Aa_8 + a_{10} T_\mathrm{L}\right)
\end{aligned}$$

Substituting the value from Eq. (10.59) into Eqs. (10.55), (10.56), (10.57), and (10.58), one can obtain the maximum value of the objective function and the corresponding power input, cooling load, and COP for a typical set of operating parameters.

10.7 Conclusion

The detailed parametric study of an irreversible regenerative Brayton refrigerator cycle using the new thermoeconomic approach is presented in this chapter. The thermoeconomic objective function defined as the cooling load per unit cost is optimized with respect to the state point temperatures for a typical set of operating conditions. On one hand, the power input and cooling load are found to be decreasing functions of the expansion outlet temperature (T_1), while it is reverse in the case of COP. On the other hand, there are optimal values of the temperature T_1, cooling load, power input, and COP at which the cycle attains the maximum objective function for a typical set of operating parameters. Again, the objective function, COP, and cooling load further enhance, while the power input goes down, as the various values of effectiveness and component efficiencies are increased. The maxima of the objective function further go down as the economic parameters are increased, which shows that the cost of each component has a significant effect on the thermoeconomic performance of a cycle. The economic parameters consist of various sub-parameters such as the material technology, power cost, land cost, and other proxy parameters like availability of raw material, manpower, labour cost, etc. So the system operating/manufacturing in some countries may have a better performance based on the thermoeconomics than those of the systems working/ manufactured in developed countries.

Chapter 11
Finite Time Thermodynamics of Stirling/Ericsson Refrigeration Cycles

11.1 Introduction

The Stirling and Ericsson cycles are among the important refrigeration systems for the production of very low temperature, especially in the cryogenic range. These cycles have been utilized by a number of engineering firms in the construction of practical systems and have promoted the development of new design of these cycles for different applications. The reversed Stirling and Ericsson cycles are also called gas refrigeration cycle as gas/air being the working fluid and are very similar to each other. The basic Stirling and Ericsson refrigeration cycles are very similar to each other. These refrigeration cycles also slightly deviate from the reverse Carnot cycle because the adiabatic processes of the latter replaced with the isochoric processes in the Stirling cycle and with the isobaric processes in the Ericsson cycle involving a regenerator for heat transfer during the operation of these cycles. Also the performance of these cycles approaches to the Carnot cycle, as the regenerator efficiency tends to unify, which seldom happens in real practice, and hence, the performance of Stirling and Ericsson cycles is always lesser than that of a Carnot cycle for the same set of operating parameters.

11.2 Ideal Stirling/Ericsson Refrigeration Cycle

As a general illustration, the ideal Stirling/Ericsson cycle consists of a compressor, a regenerator, and an expander as shown in the schematic diagram of Fig. 11.1a. The refrigerant, which may be air or a gas, enters the compressor at state point 1, where the temperature is somewhat below the temperature of the cold region, i.e. heat source temperature (T_L), and is compressed in the compressor to state point 2. The gas then cooled up to state point 3, where the temperature of the gas refrigerant approaches to the temperature of the sink/warm region (T_H). After releasing the

S.C. Kaushik et al., *Finite Time Thermodynamics of Power and Refrigeration Cycles*,
https://doi.org/10.1007/978-3-319-62812-7_11

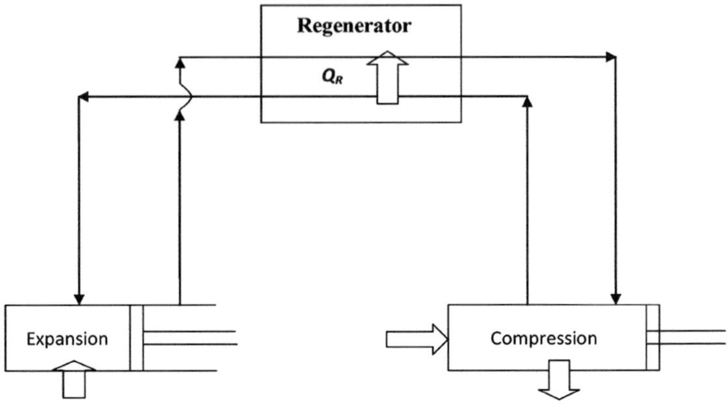

Fig. 11.1a Line diagram of Stirling/Ericsson refrigeration cycle

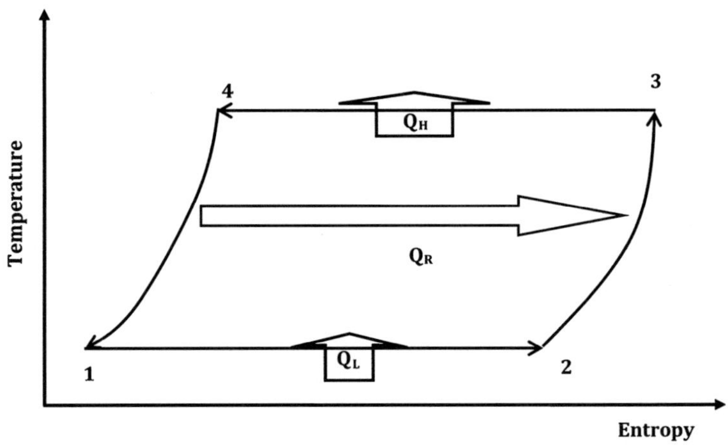

Fig. 11.1b *T–s* diagram of ideal Stirling refrigeration cycle

heat to the sink/surroundings, the gas refrigerant is expanded in an expander up to the state point 4, where the exit temperature is well below than that of the cold region/heat source (T_L) as can be seen from Fig. 11.1b. At state point 4, the gas enters the low-temperature heat exchanger and starts absorbing heat from the heat source (cold region) as it passes from state 4 to state 1, thereby completing the cycle. Thus, the ideal Stirling/Ericsson refrigeration cycle shown in Fig. 11.1b is denoted by 1–2–3–4–1. In these cycles, heat rejection processes (4–1 and 2–3) are isothermal, while the expansion (3–4) and compression (1–2) processes are isochoric in Stirling cycle, whereas they are isobaric in the Ericsson cycle, respectively (Fig. 11.1c).

For ideal cycle, the heat transferred to the regenerator during the compression process (1–2) is fully recovered during the expansion process (3–4), as can be seen

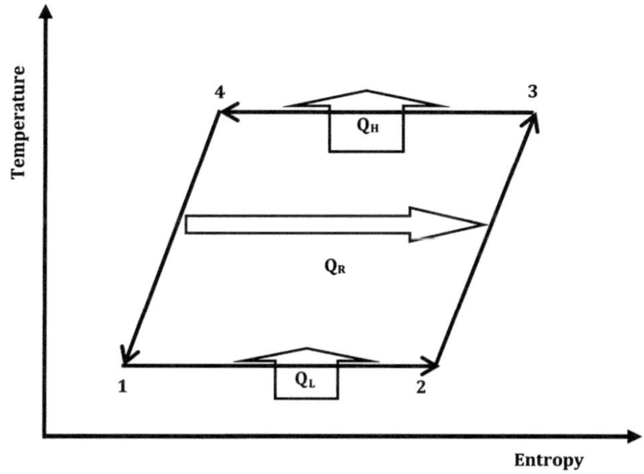

Fig. 11.1c *T–s* diagram of ideal Ericsson refrigeration cycle

from the *T–s* diagram of Fig. 11.1a. Again, for ideal gas refrigeration cycle, it is also assumed that all the processes are internally and externally reversible, and thus, the cycle executed is an ideal gas refrigeration cycle which is seldom approached in practice. Again, for steady-state operation, neglecting the kinetic and potential energy, the heat transfer to and from the cycle can be given as below:

$$Q_{\mathrm{L}} = nR\ln\lambda_{\mathrm{r}}T_{\mathrm{L}} = \Delta ST_{\mathrm{L}} \tag{11.1}$$

$$Q_{\mathrm{H}} = nR\ln\lambda_{\mathrm{r}}T_{\mathrm{H}} = \Delta ST_{\mathrm{H}} \tag{11.2}$$

where ΔS is the change in the entropy during isothermal heat addition and heat rejection processes of the cycle and λ_r is the volume ratio for Stirling and pressure ratio for the Ericsson cycle, respectively. Using the first law of thermodynamics, the work done on the cycle and the coefficient of performance (COP) are given as below:

$$W = Q_{\mathrm{H}} - Q_{\mathrm{L}} = \Delta S(T_{\mathrm{H}} - T_{\mathrm{L}}) \tag{11.3}$$

$$\mathrm{COP} = \frac{Q_{\mathrm{L}}}{W} = \frac{T_{\mathrm{L}}}{T_{\mathrm{H}} - T_{\mathrm{L}}} \tag{11.4}$$

It is clear from Eq. (11.4) that the performance of reversible Ericsson/Stirling refrigeration cycle is similar to that of a reversed Carnot for the same set of operating conditions. However, in the actual cycles, there are number of irreversible processes such as compression, expansion, regeneration, heat addition, heat rejections, etc. So in general, the real processes are not ideal because there is some loss of the useful energy during the actual processes leading to loss of potential to do useful work. Since no real cycle is as efficient as the ideal one, there is always a deviation in the real processes, and hence, the performance of a real cycle is always lower than that of an ideal one. Keeping this aspect in mind, Curzon and Ahlborn (1975) applied the concept of finite time thermodynamics on real cycle and gave a different direction in the area of thermodynamics followed by numerous works in this direction by number of authors later on.

11.3 Finite Time Stirling/Ericsson Refrigeration Cycle

It is well known that the practical refrigeration cycle is not as efficient as the ideal one, because there is always a deviation in the processes followed by a cycle, and hence, they are not reversible leading to loss of potential to do useful work. After the novel work by Curzon and Ahlborn (1975), the concept of finite time thermodynamics came into existence followed by numerous works in this direction. Leff and Teeters (1978) have noted that the straightforward C–A calculations will not work for reverse Carnot cycle because there is no "natural maximum" in the reverse cycles. Blanchard (1980) has applied the Lagrangian method of undetermined multiplier to find out the COP of endoreversible Carnot heat pump operated at minimum power input for a given heating load which was further extended by number of authors for the different cycles (Kaushik 1999; Kaushik and Kumar 2000b; Kaushik et al. 2002a; Tyagi et al. 2002).

According to the theory of finite time thermodynamics, in order to transfer a finite amount of heat in finite time, there should be a finite temperature difference between the cycle and its external reservoirs. There are two types of irreversibilities: One is the internal irreversibility which is due to entropy generation and friction pressure drop in different components of the cycle; second is the external which is due to finite temperature difference between the cycle and the external reservoirs as given by Curzon and Ahlborn (1975). The cycles which are internally reversible and externally irreversible are also called endoreversible cycles, while the cycles having both the internal and external irreversibilities are called irreversible cycles. The finite time Stirling and Ericsson refrigeration cycles with infinite and finite heat capacity of external reservoirs are evaluated and discussed as below.

11.3.1 Infinite Heat Capacity

The Stirling and Ericsson refrigeration cycle coupled with external reservoirs of infinite heat capacity are shown on the *T–s* diagram of Figs. 11.2a and 11.2b, respectively. These cycles approximate the expansion stroke of real cycle as an isothermal process 1–2 with irreversible heat addition (at temperature T_c) from a heat source of infinite heat capacity whose temperature remains constant at T_L. The heat addition to the working fluid from the regenerator is modelled as isobaric (in Ericsson cycle) or isochoric (in Stirling cycle) processes 2–3 in the ideal cycles. The compression stroke is modelled as an isothermal process 3–4 with irreversible

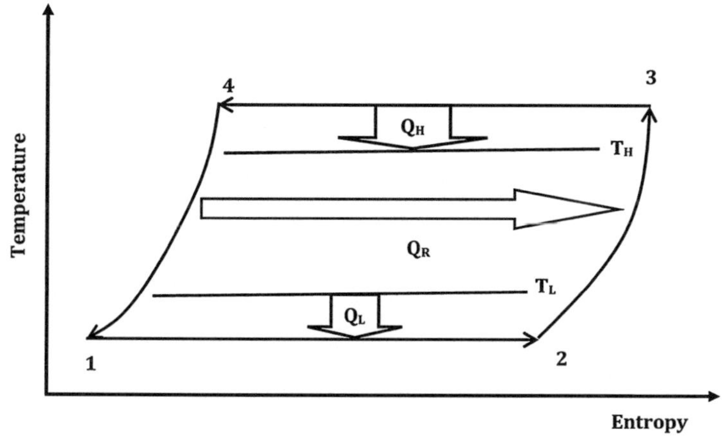

Fig. 11.2a *T–s* diagram of finite time Stirling refrigeration cycle

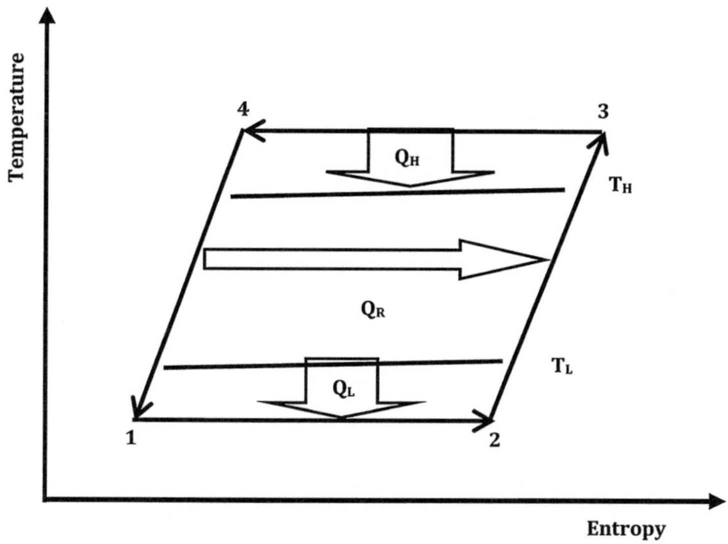

Fig. 11.2b *T–s* diagram of finite time Ericsson refrigeration cycle

heat rejection (at temperature T_h) to the heat sink of infinite heat capacity whose temperature remains constant at T_H. Finally, the heat rejection to the regenerator is modelled as isothermal/isochoric processes 4–1 an ideal Ericsson/Stirling cycle, thereby completing the cycle. As mentioned earlier, the heat transfer processes 1–2 and 3–4 in real cycles must occur in finite time. This requires that these heat processes must proceed through a finite temperature difference, therefore defined as endoreversible cycle, which means externally irreversible but internally reversible. There is also some heat loss through the regenerator, as an ideal regeneration requires infinite regeneration time or area, which is not the case in practice. Hence, it is desirable to consider a real regenerator. Let Q_c be the amount of heat absorbed from the source at temperature T_c and Q_h be the amount of heat release to the sink at temperature T_h; then (Kaushik 1999):

$$Q_h = k_1(T_w - T_H)t_H = \Delta S T_w \tag{11.5}$$

$$Q_c = k_2(T_L - T_c)t_L = \Delta S T_c \tag{11.6}$$

where k_1 and k_2 are the conductance between the cycle and the external reservoirs on the sink and source sides, respectively. ΔS is the change in the entropy during the heat addition and heat rejection processes and given as below:

$$\Delta S = nR \ln \lambda_r \tag{11.7}$$

where n is the number of moles of the working fluid, R is the universal gas constant, and λ_r is the volume ratio for Stirling cycle, while it is pressure ratio for the Ericsson cycle. It is also assumed that there is some loss in the regeneration process, as these cycles, in general, do not possess the condition of perfect regeneration. So it is desirable to consider the regenerator loss, which is proportional to the temperature difference of the two isothermal processes as assumed by earlier workers (Kaushik and Kumar 2000a, b; Tyagi et al. 2002; Kaushik et al. 2002a):

$$\Delta Q_R = nC_f(1 - \varepsilon_R)(T_h - T_c) \tag{11.8}$$

where C_f is the specific heat (at constant volume for Stirling cycle and at pressure for Ericssion cycle) of the working fluid. Owing to the influence of irreversibility of finite heat transfer, the regenerative time (t_R) should be finite and is assumed to be directly proportional to that of the two isothermal processes (Sahin and Kodal 1999; Kaushik 1999; Kaushik and Kumar 2000a, b; Kaushik et al. 2002a):

$$t_R \propto (t_H + t_L) = 2\alpha(t_H - T_L) \tag{11.9}$$

where t_H and t_L are, respectively, the heat rejection and heat addition time to and from the cycle and α is the proportionality constant which is independent of the temperature difference but depends on the property of the regenerative material. Thus, the total cycle time t_{cycle} will be:

$$t_{cycle} = (t_H + t_L + t_R) \tag{11.10}$$

When all the assumptions mentioned above are taken into account, the net amount of heat released to the sink and absorbed from the source is given by:

$$Q_H = Q_h - \Delta Q_R \tag{11.11}$$

$$Q_L = Q_c - \Delta Q_R \tag{11.12}$$

It is well known that the power input, cooling load, and the coefficient of performance (COP) are the important parameters of any refrigeration system. Using the above equations, one can obtain the expressions for these important performance parameters as below (Chen 1998; Kaushik 1999; Tyagi et al. 2002):

$$P = \frac{(Q_{\mathrm{H}} - Q_{\mathrm{L}})}{t_{\mathrm{cycle}}} = \frac{(Q_{\mathrm{h}} - Q_{\mathrm{c}})}{(t_{\mathrm{H}} + t_{\mathrm{L}} + t_{\mathrm{R}})} \tag{11.13}$$

$$R_{\mathrm{L}} = \frac{Q_{\mathrm{L}}}{t_{\mathrm{cycle}}} = \frac{(Q_{\mathrm{c}} - \Delta Q_{\mathrm{R}})}{(t_{\mathrm{H}} + t_{\mathrm{L}} + t_{\mathrm{R}})} \tag{11.14}$$

$$\mathrm{COP} = \frac{R_{\mathrm{L}}}{P} = \frac{Q_{\mathrm{c}} - \Delta Q_{\mathrm{R}}}{Q_{\mathrm{h}} - Q_{\mathrm{c}}} \tag{11.15}$$

Again, using the second law of thermodynamics for reversible cycle, one yields:

$$\frac{Q_{\mathrm{c}}}{T_{\mathrm{c}}} - \frac{Q_{\mathrm{h}}}{T_{\mathrm{h}}} = 0 \tag{11.16}$$

Thus from Eqs. (11.13), (11.14), (11.5), and (11.16), one can obtain the expression of important performance parameters, as below:

$$P = \frac{(x-1)}{\left[\frac{x}{k_1(xy - T_{\mathrm{H}})} + \frac{1}{k_2(T_L - y)} + b_1(x-1)\right]} \tag{11.17}$$

$$R_{\mathrm{L}} = \frac{[1 - a_1(x-1)]}{\left[\frac{x}{k_1(xy - T_{\mathrm{H}})} + \frac{1}{k_2(T_{\mathrm{L}} - y)} + b_1(x-1)\right]} \tag{11.18}$$

$$\mathrm{COP_L} = \frac{[1 - a_1(x-1)]}{(x-1)} \tag{11.19}$$

where $x = T_{\mathrm{h}}/T_{\mathrm{c}}$, $y = T_{\mathrm{c}}$, $b_1 = \frac{2\alpha}{n \ R \ln(\lambda_r)}$, and $a_1 = \frac{C_f(1-\varepsilon_R)}{R_0 \ln(\lambda_r)}$. Again, as pointed out by Leff and Teeters (1978), there is 'no natural maxima' in the reversed power cycle, so the straightforward calculations of Curzon and Ahlborn (1975) may not work for these cycles, because there is a constraint to extract as much as possible heat from the cold reservoir with the expenditure of minimum possible work or vice versa. This implies that one should try to minimize the power input force given cooling load or vice versa. For this purpose, one may introduce the Lagrangian (Chen 1998; Tyagi et al. 2002, 2008):

$$L = P + \lambda R_{\mathrm{L}} = \frac{x - 1 + \lambda[1 - a_1(x-1)]}{\frac{x}{k_1(xy - T_{\mathrm{H}})} + \frac{1}{k_2(T_{\mathrm{L}} - y)} + b_1(x-1)} \tag{11.20}$$

where λ is the Lagrangian multiplier. Again, using Eq. (11.20) and its extremal condition, one may obtain the optimum operating condition with the following relations:

$$x(T_L - y) = \sqrt{\frac{k_1}{k_2}}(xy - T_H) \quad \text{and} \quad y = \frac{\left(xT_L + \sqrt{\frac{k_1}{k_2}}T_H\right)}{x\left(1 + \sqrt{\frac{k_1}{k_2}}\right)} \tag{11.21}$$

Using Eqs. (11.21), (11.17), (11.18), and (11.19), one can obtain the optimum performance parameters for a typical set of operating conditions.

11.3.2 Finite Heat Capacity

In this case, these cycles approximate the expansion stroke of real cycle as an isothermal process 1–2 with irreversible heat addition (at temperature T_c) from a heat source of finite heat capacity whose temperature varies from T_{L1} to T_{L2}. The heat addition to the working fluid from the regenerator is modelled as isobaric (in Ericsson cycle) or isochoric (in Stirling cycle) processes 2–3 in the ideal cycles. The compression stroke is modelled as an isothermal process 3–4 with irreversible heat rejection (at temperature T_h) to the heat sink of finite heat capacity whose temperature varies from T_{H1} to T_{H2}. Finally, the heat rejection to the regenerator is modelled as isothermal/isochoric processes 4–1, thereby completing the cycle. Let Q_c be the amount of heat absorbed from the source at temperature T_c and Q_h be the amount of heat release to the sink at temperature T_h; then following Eqs. (11.5), (11.6), and (11.7), one yields:

$$Q_c = T_c \Delta S = C_L(T_{L1} - T_{L2})t_L \tag{11.22}$$

$$Q_h = T_h \Delta S = C_H(T_{H2} - T_{H1})t_H \tag{11.23}$$

where C_H and C_L are, respectively, the heat capacitance on the sink- and source-side reservoirs. From the heat transfer theory, the amount of heat Q_h and Q_c will be proportional to the Log Mean Temperature Difference (LMTD), i.e.:

$$Q_h = (\mathrm{UA})_H(\mathrm{LMTD})_H \tag{11.24}$$

$$Q_c = (\mathrm{UA})_L(\mathrm{LMTD})_L \tag{11.25}$$

where $(\mathrm{UA})_H$ and $(\mathrm{UA})_L$ are the overall heat transfer coefficient-area products and $(\mathrm{LMTD})_H$ and $(\mathrm{LMTD})_L$ are the Log Mean Temperature Difference on sink and source side, respectively, and defined as:

$$(\mathrm{LMTD})_H = \left[\frac{(T_h - T_{H1}) - (T_h - T_{H2})}{\left(\ln \frac{(T_h - T_{H1})}{(T_h - T_{H2})}\right)}\right] \tag{11.26}$$

$$(\mathrm{LMTD})_{\mathrm{L}} = \left[\frac{(T_{L1} - T_c) - (T_{L2} - T_c)}{\left(\ln \frac{(T_{L1} - T_c)}{(T_{L2} - T_c)} \right)} \right] \tag{11.27}$$

Using Eqs. (11.22), (11.23), (11.24), (11.25), (11.26), and (11.27), the heat transfer to and from the cycle is given by:

$$Q_{\mathrm{h}} = C_{\mathrm{H}}\varepsilon_{\mathrm{H}}(T_{\mathrm{h}} - T_{\mathrm{H1}})t_{\mathrm{H}} \tag{11.28}$$

$$Q_{\mathrm{c}} = C_{\mathrm{L}}\varepsilon_{\mathrm{L}}(T_{\mathrm{L1}} - T_{\mathrm{c}})t_{\mathrm{L}} \tag{11.29}$$

As these cycles, in general, do not possess the condition of perfect regeneration, there is some heat loss during the regeneration process as given in the earlier section. It is also assumed that the regenerative time should be finite as compared to the two isothermal processes. Thus considering the loss associated with these cycles mentioned above, the net amount of heat released to the sink and absorbed from the source may be given by Eqs. (11.11) and (11.12), while the performance parameters using the second law of thermodynamics are given as below (Sahin and Kodal 1999; Kaushik 1999; Kaushik and Kumar 2000a, b; Tyagi et al. 2002):

$$P = \frac{(x-1)}{\left[\frac{x}{k_1(xy - T_{\mathrm{H1}})} + \frac{1}{k_2(T_{\mathrm{L1}} - y)} + b_1(x-1) \right]} \tag{11.30}$$

$$R_{\mathrm{L}} = \frac{[1 - a_1(x-1)]}{\left[\frac{x}{k_1(xy - T_{\mathrm{H1}})} + \frac{1}{k_2(T_{\mathrm{L1}} - y)} + b_1(x-1) \right]} \tag{11.31}$$

$$\mathrm{COP}_{\mathrm{L}} = \frac{[1 - a_1(x-1)]}{(x-1)} \tag{11.32}$$

where $k_1 = C_{\mathrm{H}}\varepsilon_{\mathrm{H}}$, $k_2 = C_{\mathrm{L}}\varepsilon_{\mathrm{L}}$, $x = T_{\mathrm{h}}/T_{\mathrm{c}}$, $y = T_{\mathrm{c}}$, $b_1 = 2\alpha/nR\ln(\lambda_r)$, and $a_1 = C_{\mathrm{f}}(1 - \varepsilon_{\mathrm{R}})/R\ln(\lambda_{\mathrm{r}})$, while the objective function defined in the previous section is given as below (Tyagi et al. 2002):

$$L = P + \lambda R_L = \frac{x - 1 + \lambda[1 - a_1(x-1)]}{\frac{x}{k_1(xy - T_{\mathrm{H1}})} + \frac{1}{k_2(T_{\mathrm{L1}} - y)} + b_1(x-1)} \tag{11.33}$$

where λ is the Lagrangian multiplier. Again, using Eq. (11.33) and its extremal condition, one may obtain the optimum operating condition with the following relations:

$$x(T_{\mathrm{L1}} - y) = \sqrt{\frac{k_1}{k_2}}(xy - T_{\mathrm{H1}}) \quad \text{and} \quad y = \frac{\left(xT_{\mathrm{L1}} + \sqrt{\frac{k_1}{k_2}}T_{\mathrm{H1}}\right)}{x\left(1 + \sqrt{\frac{k_1}{k_2}}\right)} \tag{11.34}$$

Using Eqs. (11.34), (11.30), (11.31), and (11.32), the optimum performance can be obtained for a typical set of operating conditions in this case of Ericsson/Stirling refrigeration cycle (Tyagi et al. 2002):

$$R_{\mathrm{L}} = \frac{[1 - a_1(x-1)]}{\left[\frac{xk_3}{(xT_{\mathrm{L1}} - T_{\mathrm{H1}})} + b_1(x-1)\right]} \tag{11.35}$$

$$P = \frac{(x-1)}{\left[\frac{xk_3}{(xT_{\mathrm{L1}} - T_{\mathrm{H1}})} + b_1(x-1)\right]} \tag{11.36}$$

$$\mathrm{COP_L} = \frac{[1 - a_1(x-1)]}{(x-1)} \tag{11.37}$$

where $k_3 = \left[\sqrt{k_1} + \sqrt{k_2}\right]^2/(k_1 k_2)$ and the other parameters are the same as mentioned in the previous sections.

11.4 Irreversible Stirling/Ericsson Refrigeration Cycle

The practical refrigeration cycle is not only externally irreversible but also internally irreversible leading to losses of potential to do useful work. The internal irreversibility is due to entropy generation, friction, and pressure drop in different components, like turbine/expansion, compression, etc., while the external is due to finite temperature difference between the cycle and the external reservoirs. The mathematical modelling of these refrigeration cycles considering both the internal and external irreversibilities has been presented in the below sections.

11.4.1 Infinite Heat Capacity

In this particular case, these cycles approximate the expansion stroke of real cycle as an isothermal process 1–2 with irreversible heat addition (at temperature T_{c}) from a heat source of infinite heat capacity. The heat addition to the working fluid from the regenerator is modelled as isobaric (in Ericsson cycle) or isochoric (in Stirling cycle) processes 2–3 in the real cycles. The compression stroke is modelled as an isothermal process 3–4 with irreversible heat rejection (at temperature T_{h}) to the heat sink of infinite heat capacity. Finally, the heat rejection to the regenerator is modelled as isothermal/isochoric processes 4–1, thereby completing the cycle as shown on the *T–s* diagram of Fig. 11.3b. When the internal irreversibility is taken into account, the heat transfer to and from the cycle is given as below (Tyagi et al. 2002):

$$Q_c = T_c(S_2 - S_1) = T_c\,\Delta S_1 = k_1(T_L - T_c)t_L \tag{11.38}$$

$$Q_h = T_h(S_3 - S_4) = T_c\,\Delta S_2 = k_2(T_h - T_H)t_H \tag{11.39}$$

where $(S_2 - S_1)$ and $(S_3 - S_4)$ are, respectively, the entropy generation during the heat addition and heat rejection processes as can be seen in Figs. 11.3a, 11.3b, and 11.3c, while other parameters are the same as defined in the previous sections. It can also be seen from Figs. 11.3b, and 11.3c that there is net entropy generation per cycle, and the ratio of the entropy generation during heat rejection to that of heat addition is higher than unity and defined as the irreversibility parameter (Kaushik and Tyagi 2002) and also takes into consideration the other loss due to friction and pressure drop. Using the second law of thermodynamics for reversible cycle yields (Tyagi et al. 2002):

$$\frac{Q_\mathrm{c}}{T_\mathrm{c}} - \frac{Q_\mathrm{h}}{T_\mathrm{h}} \le 0 \quad \text{or} \quad \frac{R_{\mathrm{K-T}}Q_\mathrm{c}}{T_\mathrm{c}} = \frac{Q_\mathrm{h}}{T_\mathrm{h}} \tag{11.40}$$

where $R_{\mathrm{K-T}}$ is irreversibility parameter defined above, which is greater than unity for real cycles. Thus following Sect. 11.3.1 and Eq. (11.39), one can obtain the expressions of important performance parameters, as below (Tyagi et al. 2002):

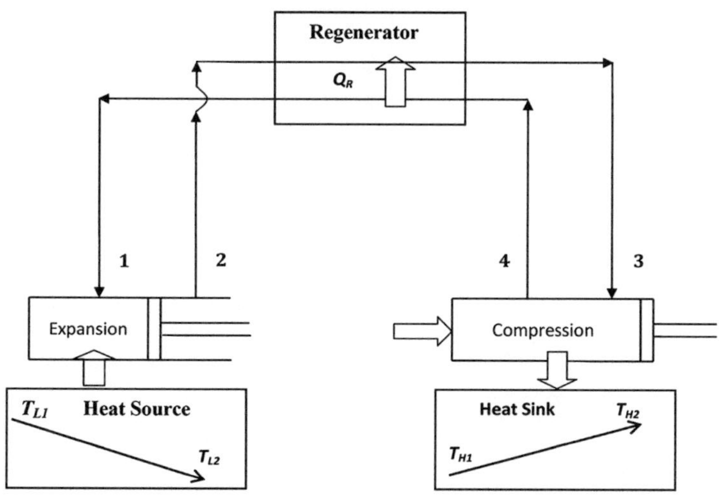

Fig. 11.3a Schematic of irreversible Stirling/Ericsson refrigeration cycle

Fig. 11.3b *T–s* diagram of irreversible Ericsson refrigeration cycle

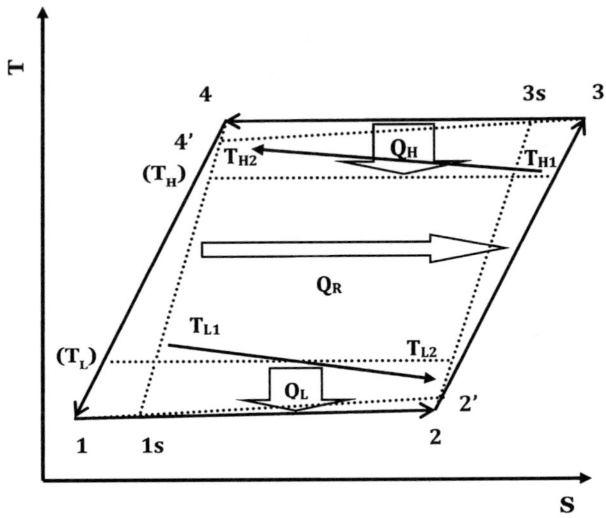

Fig. 11.3c *T–s* diagram of irreversible Stirling refrigeration cycle

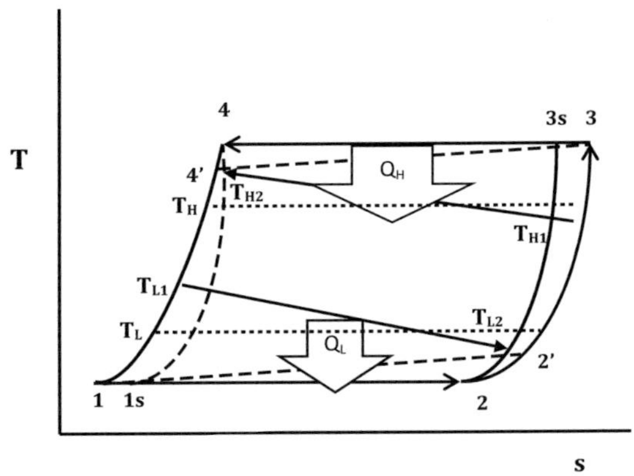

$$P = \frac{\left(x - R_{\mathrm{K-T}}^{-1}\right)}{\left[\frac{x}{k_1(xy - T_{\mathrm{H}})} + \frac{R_{\mathrm{K-T}}^{-1}}{k_2(T_{\mathrm{L}} - y)} + b_1(x-1)\right]} \tag{11.41}$$

$$R_{\mathrm{L}} = \frac{\left[R_{\mathrm{K-T}}^{-1} - a_1(x-1)\right]}{\left[\frac{x}{k_1(xy - T_{\mathrm{H}})} + \frac{R_{\mathrm{K-T}}^{-1}}{k_2(T_{\mathrm{L}} - y)} + b_1(x-1)\right]} \tag{11.42}$$

$$\mathrm{COP_L} = \frac{\left[R_{\mathrm{K-T}}^{-1} - a_1(x-1)\right]}{\left(x - R_{\mathrm{K-T}}^{-1}\right)} \tag{11.43}$$

In this case, introducing the Lagrangian (Blanchard 1980; Chen 1998; Tyagi et al. 2002), one yields:

$$L = P + \lambda R_{\mathrm{L}} = \frac{\left(x - R_{\mathrm{K-T}}^{-1}\right) + \lambda\left[R_{\mathrm{K-T}}^{-1} - a_1(x-1)\right]}{\left(\frac{x}{k_1(xy - T_{\mathrm{H}})} + \frac{R_{\mathrm{K-T}}^{-1}}{k_2(T_{\mathrm{L}} - y)} + b_1(x-1)\right)} \tag{11.44}$$

where λ is the Lagrangian multiplier. Again, using Eq. (11.44) and its extremal condition, one may obtain the optimum operating condition with the following relations:

$$y = \frac{\left(xT_{\mathrm{L}} + T_{\mathrm{H}}\sqrt{\frac{R_{\mathrm{K-T}}^{-1}k_1}{k_2}}\right)}{x\left(1 + \sqrt{\frac{R_{\mathrm{K-T}}^{-1}k_1}{k_2}}\right)} \tag{11.45}$$

Thus, using Eq. (11.45) and Eqs. (11.41), (11.42), and (11.43), one can obtain the optimum performance parameters for a typical set of operating conditions, as below (Tyagi et al. 2002):

$$R_{\mathrm{L}} = \frac{\left[R_{\mathrm{K-T}}^{-1} - a_1(x-1)\right]}{\left[\frac{xk_4}{(xT_{\mathrm{L}} - T_{\mathrm{H}})} + b_1(x-1)\right]} \tag{11.46}$$

$$P = \frac{\left(x - R_{\mathrm{K-T}}^{-1}\right)}{\left[\frac{xk_4}{(xT_{\mathrm{L}} - T_{\mathrm{H}})} + b_1(x-1)\right]} \tag{11.47}$$

$$\mathrm{COP_L} = \frac{\left[R_{\mathrm{K-T}}^{-1} - a_1(x-1)\right]}{\left(x - R_{\mathrm{K-T}}^{-1}\right)} \tag{11.48}$$

where $k_4 = \left[\sqrt{k_1 R_{\mathrm{K-T}}^{-1}} + \sqrt{k_2}\right]^2 / (k_1 k_2)$.

11.4.2 Finite Heat Capacity

Following earlier models discussed above, this particular case is more general in which the finite heat capacity of external and internal irreversibility due to friction, compression, and expansion is also taken into account similar to Sect. 11.4.1. The heat addition and heat rejection are modelled from the heat source and to the heat sink of finite heat capacities as shown on the *T–s* diagram of Fig. 11.3b. The heat addition to the working fluid from the regenerator is modelled as isobaric (in Ericsson cycle) or isochoric (in Stirling cycle) processes 2–3 in the real cycles, whereas the heat rejection to the regenerator is modelled as isothermal/isochoric processes 4–1, thereby completing the cycle.

Again, the objective of any refrigeration is to extract as much as heat from the low-temperature source with the expenditure of minimum possible work or vice versa. This implies that one should try to minimize the power input for the given cooling load or vice versa. For this purpose, one may introduce the Lagrangian, as given in the earlier section (Blanchard 1980; Chen 1998; Tyagi et al. 2002) (Figs. 11.3a, 11.3b, and 11.3c):

$$L = P + \lambda R_{\mathrm{L}} = \frac{x - R_{\mathrm{K-T}}^{-1} + \lambda\left[R_{\mathrm{K-T}}^{-1} - a_1(x-1)\right]}{\frac{x}{k_1(xy - T_{\mathrm{H1}})} + \frac{R_{\mathrm{K-T}}^{-1}}{k_2(T_{\mathrm{L1}} - y)} + b_1(x-1)} \tag{11.49}$$

Again, using Eq. (11.48) and Eqs. (11.45), (11.46), and (11.47) for finite heat capacity of external reservoirs, one can obtain the optimum performance parameters for a typical set of operating conditions, as below (Tyagi et al. 2002):

$$x(T_{\mathrm{L1}} - y) = \sqrt{\frac{k_1}{k_2}}(xy - T_{\mathrm{H1}}) \Rightarrow y = \frac{\left(xT_{\mathrm{L1}} + T_{\mathrm{H1}}\sqrt{\frac{R_{\mathrm{K-T}}^{-1}k_1}{k_2}}\right)}{x\left(1 + \sqrt{\frac{R_{\mathrm{K-T}}^{-1}k_1}{k_2}}\right)} \tag{11.50}$$

Similarly, using Eq. (11.50) and Eqs. (11.46), (11.47), and (11.48) for the finite heat capacity of external reservoirs, one can obtain the optimum performance parameters for a typical set of operating conditions, as below (Tyagi et al. 2002):

$$R_{\mathrm{L}} = \frac{\left[R_{\mathrm{K-T}}^{-1} - a_1(x-1)\right]}{\left[\frac{xk_4}{(xT_{\mathrm{L}} - T_{\mathrm{H1}})} + b_1(x-1)\right]} \tag{11.51}$$

$$P = \frac{\left(x - R_{\mathrm{K-T}}^{-1}\right)}{\left[\frac{xk_4}{(xT_{\mathrm{L1}} - T_{\mathrm{H1}})} + b_1(x-1)\right]} \tag{11.52}$$

$$\mathrm{COP_L} = \frac{\left[R_{\mathrm{K-T}}^{-1} - a_1(x-1)\right]}{\left(x - R_{\mathrm{K-T}}^{-1}\right)} \tag{11.53}$$

It is also clear from Eqs. (11.51), (11.52), and (11.53) that all the performance parameters are functions of a single variable 'x' for a typical set of operating parameters. Thus, using Eq. (11.50) and its extremal condition $\partial R_{\mathrm{L}}/\partial x = 0$, one may prove that the optimal value of 'x' corresponding to the maximum cooling load, for a given set of operating parameters, can be obtained by the relation as given below:

$$x_{\text{opt}} = \frac{-B \pm \sqrt{B^2 - 4AC}}{2A} \tag{11.54}$$

where the different parameters are given as below:

$$A = k_4 a_1 T_{\text{L1}} + b_1 R_{\text{K-T}}^{-1} T_{\text{L1}}^2; \quad B = -2R_{\text{K-T}}^{-1} b_1 T_{\text{H1}} T_{\text{L1}} \quad \text{and}$$
$$C = -k_4\left(a_1 + R_{\text{K-T}}^{-1}\right) T_{\text{H1}} + b_1 T_{\text{H1}}^2.$$

Again, substituting Eq. (11.54) into Eqs. (11.51), (11.52), and (11.53), one may find the optimal values of the cooling load, power input, and cooling coefficient of performance of these refrigerators at a typical set of operating parameters. It is also important to note that the case given in this particular section is more general, and the results obtained and presented in other sections may be directly obtained from this cycle mode. Besides, the work presented by earlier researchers can be directly obtained from the present cycle model as a special case, for different performance parameters, and is presented as special cases given below.

11.4.2.1 Some Special Cases

Case 1 When $\varepsilon_{\text{R}} = 1.0$, this indicates that these cycles possess the condition of perfect regeneration. The performance of these cycles obtained is also given by Eqs. (11.13), (11.14), and (11.15) by substituting $a_1 = 0$ which indicates that the performance characteristics of Ericsson/Stirling refrigeration cycle with perfect regeneration are similar to that of the reversed Carnot cycle operating at the same condition, while the time of two adiabatic processes is given by Eq. (11.9). However, physically for finite regeneration time, ε_{R} should be less than unity. This shows that in the investigation of these cycles, it would be difficult to obtain new conclusion if the regenerative losses were not considered in the analysis.

Case 2 When $t_{\text{R}} = \gamma(t_{\text{H}} + t_{\text{L}})$, this means that the time of regeneration processes is directly proportional to that of the two isothermal processes, and hence, the performance characteristics in such a case are given by (Tyagi et al. 2002):

$$P = \frac{\left(x - R_{\text{K-T}}^{-1}\right)\left(xT_{\text{L1}} - T_{\text{H1}}\right)}{(1+\gamma)x_1 k_4} \tag{11.55}$$

$$R_{\text{L}} = \frac{\left[R_{\text{K-T}}^{-1} - a_1(x-1)\right]\left(x_1 T_{\text{L1}} - T_{\text{H1}}\right)}{(1+\gamma)x_1 k_4} \tag{11.56}$$

$$\text{COP} = \frac{\left[R_{\text{KT}}^{-1} - a_1(x_1 - 1)\right]}{\left(x_1 - R_{\text{KT}}^{-1}\right)} \tag{11.57}$$

where

$$x_1 = \frac{-B_1 \pm \sqrt{B_1^2 - 4A_1 C_1}}{2A_1} \tag{11.58}$$

and $A_1 = a_1 T_{L1};\ B_1 = -2a_1 T_{H1}$ and $C_1 = \left(a_1 + R_{K-T}^{-1}\right) T_{H1}$.

11.5 Discussion of Results

In order to have the numerical appreciation of the results for the cryogenic refrigerators, the sink and source temperatures were considered to be in the range of $T_{H1} = 290$–325 K and $T_{L1} = 180$–260 K, respectively, while the effectiveness of the heat exchangers in the range 0.60–1.00, the pressure/volume ratio of $\lambda_r = 2.50$, whereas the irreversibility parameter of $R_{K-T} = 1.25$. In this study, the effect of each of these parameters (while keeping the others as constant) on the power input, the cooling load, the COP, and the refrigeration cycle and the discussion of results are given below:

Tables 11.1a, 11.1b, and 11.1c show the effect of the various heat exchangers (ε_H, ε_L, and ε_R) on the heat transfer to and from the refrigerators, the regenerative heat transfer, the power input, the cooling load, the cooling COP, and the working fluid temperatures.

Effect of $\boldsymbol{\varepsilon_H}$ It can be seen from Table 11.1a that as the effectiveness of the sink-side heat exchanger increases, the heat transfer to the refrigerators, the power input, the cooling load, and the cooling COP increase while the heat transfer from the refrigerators, the regenerative heat transfer, and the working fluid temperatures decrease. The effect of ε_H is more pronounced for cooling load and less pronounced for the heat transfer to the regenerator.

Effect of $\boldsymbol{\varepsilon_L}$ Table 11.1b shows that as effectiveness of the source-side heat exchanger increases, the heat transfer to and from the refrigerators, the power input, the cooling load, the cooling COP, and the working fluid temperatures increase while the

Table 11.1a Effect of ε_H on the performance of Stirling/Ericsson refrigerators ($\varepsilon_L = 0.80$, $\varepsilon_R = 0.90$, $T_{L1} = 250$ K, $T_{H1} = 300$ K, $C_H = C_L = 1.0$, and $R_{K-T} = 1.25$)

ε_H	Q_H	Q_L	Q_R	P_m	R_L	COP_L	T_h	T_c
0.50	77.81	57.91	48.79	0.90	0.83	0.92	316.50	240.83
0.60	77.56	57.95	48.10	0.92	0.87	0.94	315.27	240.68
0.70	77.37	57.98	47.55	0.95	0.90	0.95	314.29	240.55
0.75	77.28	57.99	47.31	0.96	0.92	0.96	313.87	240.50
0.80	77.20	58.00	47.09	0.97	0.93	0.96	313.48	240.45
0.85	77.13	58.01	46.90	0.97	0.94	0.97	313.13	240.40
0.90	77.07	58.02	46.71	0.98	0.96	0.97	312.80	240.36
0.95	77.01	58.03	46.54	0.99	0.97	0.98	312.50	240.32
1.00	76.95	58.04	46.38	1.00	0.98	0.98	312.22	240.29

Table 11.1b Effect of ε_L on the performance of Stirling/Ericsson refrigerators ($\varepsilon_H = 0.80$, $\varepsilon_R = 0.90$, $T_{L1} = 250$ K, $T_{H1} = 300$ K, $C_H = C_L = 1.0$, and $R_{K-T} = 1.25$)

ε_L	Q_H	Q_L	Q_R	P_m	R_L	COP_L	T_h	T_c
0.50	77.09	57.42	48.22	0.90	0.84	0.93	313.51	238.73
0.60	77.18	57.69	47.80	0.93	0.87	0.94	313.68	239.55
0.70	77.25	57.90	47.46	0.95	0.90	0.95	313.81	240.21
0.75	77.28	57.99	47.31	0.96	0.92	0.96	313.87	240.50
0.80	77.31	58.07	47.18	0.97	0.93	0.96	313.92	240.76
0.85	77.34	58.15	47.05	0.97	0.94	0.97	313.96	241.00
0.90	77.36	58.22	46.94	0.98	0.95	0.97	314.01	241.22
0.95	77.38	58.28	46.83	0.99	0.96	0.98	314.05	241.42
1.00	77.40	58.34	46.73	0.99	0.97	0.98	314.08	241.61

Table 11.1c Effect of ε_R on the performance of Stirling/Ericsson refrigerators ($\varepsilon_L = \varepsilon_H = 0.80$, $T_{L1} = 250$ K, $T_{H1} = 300$ K, $C_H = C_L = 1.0$, and $R_{K-T} = 1.25$)

ε_R	Q_H	Q_L	Q_R	P_m	R_L	COP_L	T_h	T_c
0.70	67.78	53.13	27.95	0.37	0.09	0.24	303.31	247.58
0.75	69.68	54.01	32.02	0.56	0.22	0.39	305.58	245.98
0.80	71.87	55.08	36.59	0.72	0.40	0.56	308.10	244.26
0.85	74.39	56.39	41.68	0.85	0.64	0.75	310.87	242.42
0.90	77.28	57.99	47.31	0.96	0.92	0.96	313.87	240.50
0.95	80.57	59.91	53.49	1.04	1.23	1.19	317.09	238.51
1.00	83.56	62.61	57.08	1.16	1.74	1.50	317.76	238.09

regenerative heat transfer decreases. The effect of the ε_L is more pronounced for the heating load and less pronounced for the sink-side working fluid temperature.

Effect of ε_R Table 11.1c shows that as the effectiveness of the regenerator increases, the heat transfer to and from the refrigerators, the regenerative heat transfer, the power input, the cooling load, the cooling COP, and the sink-side working fluid temperature increase while the source-side working fluid temperature decreases. The effect of ε_R is more pronounced for the cooling load and less pronounced for the source-side working fluid temperature.

Tables 11.2a and 11.2b show the effect of source and sink inlet temperatures (T_{L1} and T_{H1}) on the heat transfers, the power input, the cooling load, the cooling COP, and the working fluid temperatures of these cycles.

Effect of T_{H1} It can be seen from Table 11.2a that as the inlet temperature of the sink fluid increases, the heat transfer from the refrigerators, the regenerative heat transfer, and the working fluid temperatures increase, while the heat transfer to the refrigerators, the power input, the cooling load, and the cooling COP decrease. The effect of T_{H1} is more pronounced for the cooling load and less pronounced for the source-side working fluid temperature. Thus, it is desirable to have lower T_{H1} for better performance of both the refrigerators.

Table 11.2a Effect of T_{L1} on the performance of Stirling/Ericsson refrigerators ($\varepsilon_H = \varepsilon_L = 0.80$, $\varepsilon_R = 0.90$, $T_{H1} = 300$ K, $C_H = C_L = 1.0$, and $R_{K\text{-}T} = 1.25$)

T_{L1}	Q_H	Q_L	Q_R	P_m	COP_L	R_L	T_h	T_c
180	70.87	37.81	81.06	0.50	0.04	0.08	303.74	178.04
190	71.94	40.53	77.03	0.65	0.11	0.16	306.12	186.66
200	72.93	43.34	72.56	0.74	0.19	0.26	307.99	195.47
210	73.86	46.21	67.81	0.80	0.30	0.37	309.53	204.37
220	74.76	49.12	62.87	0.85	0.42	0.50	310.84	213.35
230	75.62	52.06	57.78	0.89	0.56	0.63	311.97	222.37
240	76.46	55.01	52.59	0.92	0.73	0.79	312.97	231.42
250	77.28	57.99	47.31	0.96	0.92	0.96	313.87	240.50
260	78.09	60.98	41.96	0.99	1.14	1.15	314.67	249.60

Table 11.2b Effect of T_{H1} on the performance of Stirling/Ericsson refrigerators ($\varepsilon_H = \varepsilon_L = 0.80$, $\varepsilon_R = 0.90$, $T_{L1} = 250$ K, $C_H = C_L = 1.0$, and $R_{K\text{-}T} = 1.25$)

T_{H1}	Q_H	Q_L	Q_R	P_m	R_L	COP_L	T_h	T_c
290	75.40	58.49	41.47	0.98	1.09	1.12	304.26	239.94
295	76.34	58.24	44.40	0.97	1.00	1.03	309.07	240.22
300	77.28	57.99	47.31	0.96	0.92	0.96	313.87	240.50
305	78.22	57.74	50.22	0.95	0.84	0.89	318.65	240.77
310	79.15	57.49	53.12	0.94	0.77	0.82	323.43	241.05
315	80.08	57.24	56.02	0.93	0.71	0.76	328.19	241.32
320	81.01	56.99	58.91	0.92	0.65	0.71	332.95	241.60
325	81.94	56.74	61.79	0.91	0.60	0.65	337.69	241.87

Effect of T_{L1} Table 11.2b shows that as T_{L1} increases, the heat transfer to and from the refrigerators, the power input, the cooling load, the cooling COP, and the working fluid temperatures increase, while the regenerative heat transfer decreases. The effect of the T_{L1} is more pronounced for the cooling load and less pronounced for the sink-side working fluid temperature.

Tables 11.3a, 11.3b, and 11.3c show the effect of heat capacitance rates (C_H and C_L) and the internal irreversibility parameter (R_{DS}) on the heat transfers, the power input, the cooling load, the cooling COP, and the working fluid temperatures of these refrigerators.

Effect of C_H Table 11.3a shows that as the heat capacitance rates of the sink-side fluid increase, the heat transfers (Q_H and Q_R) and the working fluid temperatures decrease, but the heat transfer to the refrigerators, the power input, the cooling load, and the cooling COP increase. The effect of C_H is more pronounced for the cooling load and less pronounced for the heat transfer to the refrigerators.

Effect of C_L Table 11.3b shows that as the heat capacitance rates of the source-side fluid increase, the heat transfers to and from the refrigerators (Q_H and Q_L), the power input, cooling load, the cooling COP, and the working fluid temperatures

Table 11.3a Effect of C_H on the performance of Stirling/Ericsson refrigerators ($\varepsilon_H = \varepsilon_L = 0.80$, $\varepsilon_R = 0.90$, $T_{L1} = 250$ K, $T_{H1} = 300$ K, $= C_L = 1.0$, and $R_{K\text{–}T} = 1.25$)

C_H	Q_H	Q_L	Q_R	P_m	R_L	COP_L	T_h	T_c
0.50	78.23	57.85	49.97	0.85	0.76	0.90	318.59	241.10
0.60	77.96	57.89	49.21	0.88	0.80	0.91	317.24	240.93
0.70	77.74	57.92	48.60	0.90	0.84	0.93	316.16	240.79
0.80	77.56	57.95	48.10	0.92	0.87	0.94	315.27	240.68
0.90	77.41	57.97	47.68	0.94	0.89	0.95	314.51	240.58
1.00	77.28	57.99	47.31	0.96	0.92	0.96	313.87	240.50
1.10	77.17	58.00	46.99	0.97	0.94	0.97	313.30	240.43
1.20	77.07	58.02	46.71	0.98	0.96	0.97	312.80	240.36
1.30	76.98	58.03	46.46	0.99	0.97	0.98	312.36	240.31
1.40	76.90	58.04	46.24	1.00	0.99	0.98	311.96	240.26
1.50	76.83	58.05	46.03	1.01	1.00	0.99	311.59	240.21

Table 11.3b Effect of C_L on the performance of Stirling/Ericsson refrigerators ($\varepsilon_H = \varepsilon_L = 0.80$, $\varepsilon_R = 0.90$, $T_{L1} = 250$ K, $T_{H1} = 300$ K,$= C_H = 1.0$, and $R_{K\text{–}T} = 1.25$)

C_L	Q_H	Q_L	Q_R	P_m	R_L	COP_L	T_h	T_c
0.50	76.93	56.97	48.94	0.86	0.77	0.90	313.22	237.33
0.60	77.03	57.26	48.48	0.89	0 81	0.92	313.41	238.23
0.70	77.11	57.49	48.11	0.91	0.85	0.93	313.56	238.96
0.80	77.18	57.69	47.80	0.93	0.87	0.94	313.68	239.55
0.90	77.23	57.85	47.54	0.94	0.90	0.95	313.78	240.06
1.00	77.28	57.99	47.31	0.96	0.92	0.96	313.87	240.50
1.10	77.32	58.11	47.11	0.97	0.94	0.97	313.94	240.88
1.20	77.36	58.22	46.94	0.98	0.95	0.97	314.01	241.22
1.30	77.39	58.31	46.78	0.99	0.97	0.98	314.06	241.52
1.40	77.42	58.40	46.64	1.00	0.98	0.98	314.12	241.79
1.50	77.45	58.48	46.51	1.01	0.99	0.99	314.16	242.03

Table 11.3c Effect of $R_{K\text{–}T}$ on the performance of Stirling/Ericsson refrigerators ($\varepsilon_H = \varepsilon_L = 0.80$, $\varepsilon_R = 0.90$, $T_{L1} = 250$ K, $T_{H1} = 300$ K, and $C_H = C_L = 1.0$)

$R^{-1}{}_{K\text{–}T}$	Q_H	Q_L	Q_R	P_m	R_L	COP_L	T_h	T_c
0.50	76.70	59.79	41.46	1.07	0.32	0.30	309.17	244.87
0.60	76.93	59.16	43.56	1.08	0.50	0.46	310.92	243.38
0.70	77.12	58.56	45.49	1.04	0.70	0.67	312.47	241.92
0.75	77.20	58.27	46.42	1.00	0.80	0.80	313.18	241.21
0.80	77.28	57.99	47.31	0.96	0.92	0.96	313.87	240.50
0.85	77.36	57.71	48.19	0.90	1.03	1.14	314.52	239.79
0.90	77.43	57.43	49.04	0.84	1.15	1.36	315.15	239.10
0.95	77.51	57.15	51.01	0.78	1.28	1.64	315.78	238.41
1.00	78.15	55.27	53.62	0.72	1.39	1.92	320.47	335.29

increase, while the regenerative heat transfer (Q_R) decreases. The effect of C_L is more pronounced for the cooling load and less pronounced for the sink-side working fluid temperature.

Since higher value of C_H decreased the heat transfers and increased the performance, while higher value of C_L increased the heat transfers but slightly decreased the performance, thus it is desirable to have higher C_H rather than higher C_L.

Effect of R_{K-T} Table 11.3c shows that as the internal irreversibility parameter increases, the heat transfers (Q_H and Q_R), the cooling load, the cooling COP, and the sink-side working fluid temperature increase, while heat transfer to the refrigerators and the power input increase, and the source-side working fluid temperature decease. The effect of R_{K-T} is more pronounced for the power input and less pronounced for the sink-side working fluid temperature; also the effect of R_{K-T} is found to be more than the other parameters mentioned above, on the performance of this refrigeration cycle. Thus, it is desirable to have lesser internal irreversibility as compared to the external irreversibility for better performance of these cycles.

11.6 Conclusion

The performance characteristics of irreversible Ericsson/Stirling refrigeration cycle have been evaluated including external as well as internal irreversibilities. The external irreversibility is due to finite temperature difference, and internal irreversibilities are due to regenerative loss and entropy generation in the cycle. It is found that the effectiveness of each heat exchanger, the heat capacitance rate on sink-side fluid, and the irreversibility parameter should be higher enough while the sink-side inlet temperature of the fluid and the heat capacitance rates on source side should be lower for better performance of both the cycles. Since the larger heat capacitance allows the fluid to reject the heat at lower temperature, it is desirable to have higher capacitance rate on sink side in comparison to that on source side (i.e. $C_H > C_L$) for better performance of both the refrigerators. Hence, the present analysis will be useful and more general for evaluating the performance of these cycles and other regenerative cycles as well.

Chapter 12
Finite Time Thermodynamics of Vapour Absorption Airconditioning and Heat Pump Cycles

12.1 Introduction

Chen (1995) has analysed an endoreversible absorption refrigeration system coupled with thermal reservoirs of infinite heat capacity and affected by the irreversibility of finite-rate heat transfer between the working fluid and the heat reservoirs. He proposed that an absorption refrigeration system may be modelled as a combined cycle, which consists of two subsystems – an endoreversible heat engine and an endoreversible refrigerator. He developed the optimal performance parameters for these systems, but study was limited to endoreversible systems coupled with thermal reservoirs of infinite heat capacity.

The main objective of this chapter is to analyse vapour absorption airconditioning/heat pump systems coupled with external heat reservoirs of finite heat capacity and including both external irreversibility due to the finite-rate heat transfer between the working fluid and heat reservoirs and internal irreversibility due to the working fluid dissipation. Internal irreversibility is included in the analysis by introducing an internal irreversibility parameter, defined by second law of thermodynamics (Goktun and Ozkaynak 1997; Goktun 1997). The absorption airconditioning/heat pump system is considered as combined cycle, consisting of a heat engine and a refrigerator/heat pump having non-uniform source/sink temperatures but uniform component temperatures following Chen (1995) and Yan and Chen (1989). This assumption is made to simplify the analysis for studying the external irreversibility of finite heat transfer with surroundings. The optimal coefficient of performance is determined at maximum cooling/heating load. The effect of inlet temperature of thermal reservoirs (heat source/sink), heat capacitance rates of external fluids, effectiveness of various heat exchangers, and internal irreversibility parameter on heat transfer rates, internal working fluid temperatures, and coefficient of performance of the system are investigated. For a given quantity of heat supplied in generator, refrigeration/heating effect has been maximized, and the primary performance parameters (such as coefficient of performance, heat

S.C. Kaushik et al., *Finite Time Thermodynamics of Power and Refrigeration Cycles*,
https://doi.org/10.1007/978-3-319-62812-7_12

transfer areas of heat exchangers, and temperatures of the working fluid in heat exchanger) are optimized. Analytical expressions for optimal COP are derived and typical numerical results are presented and discussed.

12.2 Vapour Absorption Cycle

Figure 12.1a shows a schematic diagram of an absorption airconditioning/heat pump system, which mainly consists of four components – a generator, an absorber, an evaporator, and a condenser. The output power of generator–absorber assembly (which can be considered as a virtual heat engine) is used as input power to the evaporator–condenser assembly of an airconditioning/heat pump cycle. The equivalent cycle model of an absorption airconditioning/heat pump system is shown in Fig. 12.1b. Since thermal reservoirs (heat source/sink) coupled with the system are of finite heat capacity, temperature of the external fluids will vary during heat transfer to and from the system. In the generator–absorber assembly, let initial temperature of the heat source T_{G1} go down to T_{G2} as a result of heat transfer Q_G to the generator and temperature of the heat sink warms from T_{A1} to T_{A2} as a result of receiving heat Q_A from the absorber. Similarly, in the evaporator–condenser assembly, temperature of the fluid or space, which is to be cooled, decreases from T_{E1} to T_{E2} as a result of extracting heat from it to the evaporator, and temperature of the heat sink increases from T_{C1} to T_{C2} as a result of rejecting heat from the condenser to it.

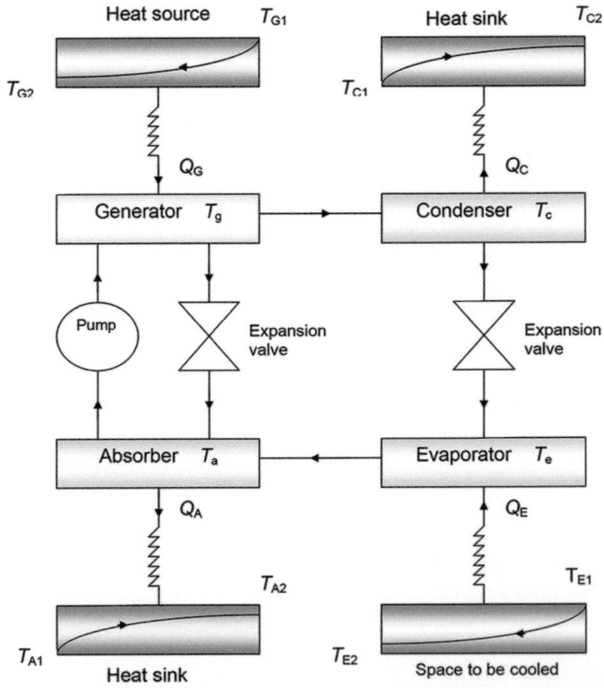

Fig. 12.1a A schematic diagram of vapour absorption airconditioning/ heat pump system

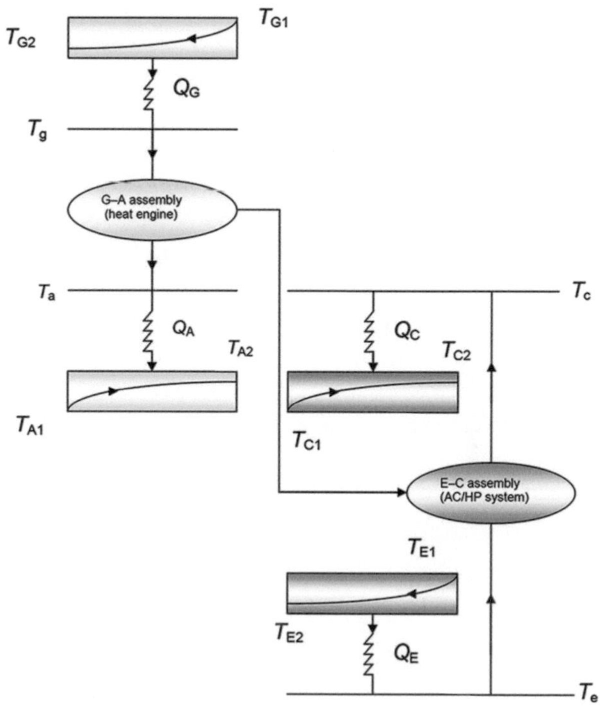

Fig. 12.1b The equivalent cycle of absorption airconditioning/heat pump system

12.3 Finite Time Vapour Absorption Cycle

By employing the Lagrangian multiplier method, for given heat transfer rate in the generator, output power of the heat engine is maximized, and then for the same output power used as input to the evaporator–condenser assembly, cooling/heating load has been maximized (Bhardwaj et al. 2003b, c, 2005).

12.3.1 Vapour Absorption Airconditioning Cycle

12.3.1.1 Generator–Absorber Assembly Analysis

The rate of heat input from heat source to the generator is given by:

$$\dot{Q}_G = U_G A_G (\mathrm{LMTD})_G = \dot{m}_G C_{PG} (T_{G1} - T_{G2}) \tag{12.1}$$

Similarly, the rate of heat output from absorber to the heat sink:

$$\dot{Q}_A = U_A A_A (\mathrm{LMTD})_A = \dot{m}_A C_{PA}(T_{A2} - T_{A1}) \tag{12.2}$$

where

$$(\mathrm{LMTD})_G = \frac{(T_{G1} - T_g) - (T_{G2} - T_g)}{\ln \frac{(T_{G1} - T_g)}{(T_{G2} - T_g)}} \quad \text{and}$$

$$(\mathrm{LMTD})_A = \frac{(T_a - T_{A1}) - (T_a - T_{A2})}{\ln \frac{(T_a - T_{A1})}{(T_a - T_{A2})}}$$

Using LMTD expressions in Eqs. (12.1) and (12.2), we have:

$$T_{G2} = T_g + (T_{G1} - T_g)\ e^{-U_G A_G/\dot{m}_G C_{PG}} \tag{12.3}$$

$$T_{A2} = T_a - (T_a - T_{A1})\ e^{-U_A A_A/\dot{m}_A C_{PA}} \tag{12.4}$$

Using Eqs. (12.1), (12.2), (12.3), and (12.4), we have:

$$\dot{Q}_G = \mathrm{GE}(T_{G1} - T_g) \tag{12.5}$$

$$\dot{Q}_A = \mathrm{AE}(T_a - T_{A1}) \tag{12.6}$$

where

$$\mathrm{GE} = C_G \varepsilon_G, \quad \mathrm{AE} = C_A \varepsilon_A, \quad C_G = \dot{m}_G C_{PG}, \quad C_A = \dot{m}_A C_{PA},$$
$$\varepsilon_G = 1 - e^{-U_G A_G/\dot{m}_G C_{PG}}, \quad \text{and} \quad \varepsilon_A = 1 - e^{-U_A A_A/\dot{m}_A C_{PA}}$$

Output power of the generator–absorber assembly is given by:

$$P = \mathrm{GE}(T_{G1} - T_g) - \mathrm{AE}(T_a - T_{A1}) \tag{12.7}$$

and Clausius inequality gives:

$$\int \frac{dQ}{T} = \frac{Q_G}{T_g} - \frac{Q_A}{T_a} \le 0.$$

In this inequality, equality sign holds only for reversible system, and less than is for irreversible case. So for internally irreversible system, an internal irreversibility parameter is introduced as:

$$\Delta S = \frac{Q_g}{T_g} - R_{\Delta s}\frac{Q_a}{T_a} = 0 \tag{12.8}$$

$R_{\Delta S}$ is internal irreversibility parameter of generator–absorber assembly. $R_{\Delta S} = 1$ for endoreversible system and $R_{\Delta S} < 1$ for internally irreversible (real) system.

For given heat transfer rate in the generator, output power of the generator–absorber assembly can be maximized by defining the modified Lagrangian operator as given by:

$$L = P - \lambda \dot{Q}_G - \mu \Delta S \tag{12.9}$$

where μ and λ are undetermined Lagrangian multipliers.

$\partial L/\partial T_g = 0$ and $\partial L/\partial T_a = 0$ give the optimal values of T_g and T_a, and substituting these values in Eqs. (12.5), (12.6), (12.7), and (12.8), we have the optimal working fluid temperatures as given by:

$$T_{go} = T_{G1} - \frac{\dot{Q}_G}{GE} \tag{12.10}$$

$$T_{ao} = \frac{T_{A1}\left(T_{G1} - \frac{\dot{Q}_G}{GE}\right)}{T_{G1} - \frac{\dot{Q}_G}{GE} - \frac{\dot{Q}_G}{R_{\Delta S}AE}} \tag{12.11}$$

Substituting the optimal value of T_a from Eq. (12.11) into Eq. (12.6), we have the thermal efficiency at maximum power output of generator–absorber assembly:

$$\eta = 1 - \frac{T_{A1}/R_{\Delta S}}{\left(T_{G1} - \frac{\dot{Q}_G}{K}\right)} \tag{12.12}$$

where

$$K = \frac{\mathrm{GE}R_{\Delta S}\mathrm{AE}}{\mathrm{GE} + R_{\Delta S}\mathrm{AE}}.$$

and maximum power output of generator–absorber assembly is given by:

$$P = \dot{Q}_G\left[1 - \frac{T_{A1}/R_{\Delta S}}{\left(T_{G1} - \frac{\dot{Q}_G}{K}\right)}\right] \tag{12.13}$$

12.3.1.2 Evaporator–Condenser Assembly Analysis

The rate of heat extracted from the fluid or space to be cooled to the evaporator is given by:

$$\dot{Q}_{\rm E} = U_{\rm E}A_{\rm E}(LMTD)_{\rm E} = \dot{m}_{\rm E}C_{\rm PE}(T_{\rm E1} - T_{\rm E2}) \tag{12.14}$$

Similarly, the rate of heat rejected from condenser to the heat sink:

$$\dot{Q}_{\rm C} = U_{\rm C}A_{\rm C}(\text{LMTD})_{\rm C} = \dot{m}_{\rm C}C_{\rm PC}(T_{\rm C2} - T_{\rm C1}) \tag{12.15}$$

where,

$$(\text{LMTD})_{\rm E} = \frac{(T_{\rm E1} - T_{\rm e}) - (T_{\rm E2} - T_{\rm e})}{\ln\dfrac{(T_{\rm E1} - T_{\rm e})}{(T_{\rm E2} - T_{\rm e})}} \text{ and}$$

$$(\text{LMTD})_{\rm C} = \frac{(T_{\rm c} - T_{\rm C1}) - (T_{\rm c} - T_{\rm C2})}{\ln\dfrac{(T_{\rm c} - T_{\rm C1})}{(T_{\rm c} - T_{\rm C2})}}$$

Using LMTD expressions in Eqs. (12.14) and (12.15), we have:

$$T_{\rm E2} = T_{\rm e} + (T_{\rm E1} - T_{\rm e})\mathrm{e}^{-U_{\rm E}A_{\rm E}/\dot{m}_{\rm E}C_{\rm PE}} \tag{12.16}$$

$$T_{\rm C2} = T_{\rm c} + (T_{\rm c} - T_{\rm C1})\mathrm{e}^{-U_{\rm C}A_{\rm C}/\dot{m}_{\rm C}C_{\rm PC}} \tag{12.17}$$

Using Eqs. (12.14), (12.15), (12.16), and (12.17), we have:

$$\dot{Q}_{\rm E} = EE(T_{\rm E1} - T_{\rm e}) \tag{12.18}$$

$$\dot{Q}_{\rm C} = CE(T_{\rm c} - T_{\rm C1}) \tag{12.19}$$

where

$$EE = C_{\rm E}\varepsilon_{\rm E},\ \text{CE} = C_{\rm C}\varepsilon_{\rm C},\ C_{\rm E} = \dot{\rm m}_{\rm E}C_{\rm PE},\ C_{\rm C} = \dot{\rm m}_{\rm C}C_{\rm PC},$$
$$\varepsilon_{\rm E} = 1 - \mathrm{e}^{-U_{\rm E}A_{\rm E}/\dot{m}_{\rm E}C_{\rm PE}},\ \text{and } \varepsilon_{\rm C} = 1 - \mathrm{e}^{-U_{\rm C}A_{\rm C}/\dot{m}_{\rm C}C_{\rm PC}}.$$

Clausius inequality requires:

$$\int\frac{\mathrm{d}Q}{T} = \frac{Q_{\rm E}}{T_{\rm e}} - \frac{Q_{\rm C}}{T_{\rm c}} \le 0.$$

In this inequality, equality sign holds for reversible system, and less than is for irreversible case. Introducing an internal irreversibility parameter as in the previous section:

$$\Delta S = \frac{Q_{\rm E}}{T_{\rm e}} - R'_{\Delta S}\frac{Q_{\rm C}}{T_{\rm c}} = 0 \tag{12.20}$$

$R'_{\Delta S}$ is internal irreversibility parameter of evaporator–condenser assembly, $R'_{\Delta S} = 1$ for endoreversible system, and $R'_{\Delta S} < 1$ for irreversible (real) system. Now, output power of the generator–absorber assembly is used as the input power for evaporator–condenser assembly, and cooling load is maximized. Modified Lagrangian is defined as:

$$L = \dot{Q}_{\mathrm{E}} - \lambda P - \mu \Delta S. \tag{12.21}$$

$\partial L/\partial T_{\mathrm{e}} = 0$ and $\partial L/\partial T_{\mathrm{c}} = 0$ give the optimal values of T_{e} and T_{c}, and substituting these values in Eqs. (12.18), (12.19), and (12.20), we have the optimal working fluid temperatures as given by:

$$T_{\mathrm{eo}} = T_{\mathrm{C1}} + \frac{P + EE(T_{\mathrm{E1}} - T_{\mathrm{eo}})}{CE} \tag{12.22}$$

$$T_{\mathrm{eo}} = \frac{-b \pm \sqrt{b^2 - 4ac}}{2a} \tag{12.23}$$

where

$$a = EE\left(EE + CER'_{\Delta S}\right)$$

$$b = -\left[\left(EE + CER'_{\Delta S}\right)\left(P + EET_{\mathrm{E1}}\right) + EE^2 T_{\mathrm{E1}} + \mathrm{CE}EE\ T_{\mathrm{C1}}\right]$$

$$c = EET_{\mathrm{E1}}\left(P + EET_{\mathrm{E1}}\right) + CEEET_{\mathrm{C1}}T_{\mathrm{E1}}$$

Now, the optimal overall coefficient of performance of an irreversible absorption airconditioning system is given by:

$$\mathrm{COP_o} = \frac{LE(T_{\mathrm{E1}} - T_{\mathrm{eo}})\eta}{P} \tag{12.24}$$

12.3.2 Vapour Absorption Heat Pump Cycle

12.3.2.1 Generator–Absorber Assembly Cycle

This analysis is the same as for Sect. 12.3.1; therefore we have the thermal efficiency at maximum power output of generator–absorber assembly:

$$\eta = 1 - \frac{T_{\mathrm{A1}}/R_{\Delta\mathrm{S}}}{\left(T_{\mathrm{G1}} - \frac{\dot{Q}_{\mathrm{G}}}{K}\right)} \tag{12.25}$$

where $K = (GE\,R_{\Delta S}AE)/(GE + R_{\Delta S}AE)$ and

$$P = \dot{Q}_G\left[1 - \frac{T_{A1}/R_{\Delta S}}{\left(T_{G1} - \frac{\dot{Q}_G}{K}\right)}\right] \tag{12.26}$$

12.3.2.2 Evaporator–Condenser Assembly Analysis

Output power of the generator–absorber assembly is used as the input power for evaporator–condenser assembly, and heating load is maximized. Modified Lagrangian operator is defined as:

$$L = \dot{Q}_C - \lambda P - \mu \Delta S \tag{12.27}$$

$\partial L/\partial T_e = 0$ and $\partial L/\partial T_c = 0$ give the optimal values of T_e and T_c. After substituting these values in Eqs. (12.18), (12.19), and (12.20), we have the optimal working fluid temperatures as given by:

$$T_{co} = T_{C1} + \frac{P + EE(T_{E1} - T_{eo})}{CE} \tag{12.28}$$

$$T_{eo} = \frac{-b \pm \sqrt{b^2 - 4ac}}{2a} \tag{12.29}$$

where

$$a = EE\left(EE + CE\,R'_{\Delta S}\right)$$
$$b = -\left[\left(EE + CE\,R'_{\Delta S}\right)(P + EE\,T_{E1}) + EE^2 T_{E1} + CE\,EE\ T_{C1}\right]$$
$$c = EE\,T_{E1}(P + EE\,T_{E1}) + CE\,EE\ T_{C1}T_{E1}$$

Now, the optimal overall coefficient of performance of an irreversible vapour absorption heat pump system is given by:

$$\mathrm{COP_o} = \frac{CE(T_{co} - T_{C1})\eta}{P} \tag{12.30}$$

12.4 Results and Discussion

12.4.1 Vapour Absorption Airconditioning Cycle

For a given heat input rate in the generator ($\dot{Q}_G = 3.0$ kW), power output of the generator–absorber assembly is maximized, and for the same output power used as input power for evaporator–condenser assembly, cooling load ($\dot{Q}_G$) is maximized. In order to have numerical appreciation of the theoretical analysis of vapour absorption airconditioning cycle, we have studied the effect of various input parameters on the optimal working fluid temperatures, various heat transfer rates to and from the system, cooling load, and optimal coefficient of performance of the vapour absorption airconditioning cycle. The results are shown in Tables 12.1a, 12.1b, 12.2a, 12.2b, 12.3a, 12.3b, 12.4, 12.5a, 12.5b, 12.6a, 12.6b, 12.7a, 12.7b, and 12.8. During the variation of any one parameter, all other parameters are assumed to be constant as given below:

Table 12.1a Effect of T_{G1} on various working fluid temperatures, heat transfer rates, and coefficient of performance of vapour absorption airconditioning cycle

T_{G1}	T_{go}	T_{ao}	η	P	$\dot{Q}_A$	T_{eo}	T_{co}	COP	$\dot{Q}_E$	$\dot{Q}_C$	COP_o
K	K	K	–	kW	kW	K	K	–	kW	kW	–
373	369.0	321.48	0.13	0.39	2.61	284.17	322.35	7.44	2.88	3.26	0.96
375	371.0	321.47	0.13	0.40	2.60	284.05	322.48	7.39	2.96	3.36	0.99
377	373.0	321.45	0.14	0.41	2.59	283.94	322.61	7.34	3.04	3.46	1.01
379	375.0	321.43	0.14	0.43	2.57	283.83	322.74	7.30	3.13	3.56	1.04
381	377.0	321.41	0.15	0.44	2.56	283.72	322.87	7.25	3.21	3.65	1.07
383	379.0	321.39	0.15	0.46	2.54	283.62	322.99	7.20	3.29	3.74	1.10
385	381.0	321.37	0.16	0.47	2.53	283.52	323.11	7.16	3.36	3.83	1.12
387	383.0	321.36	0.16	0.48	2.52	283.42	323.23	7.12	3.44	3.92	1.15
389	385.0	321.34	0.17	0.50	2.50	283.32	323.34	7.08	3.51	4.01	1.17
391	387.0	321.32	0.17	0.51	2.49	283.22	323.46	7.04	3.58	4.09	1.19

Table 12.1b Effect of T_{A1} on various working fluid temperatures, heat transfer rates, and coefficient of performance of vapour absorption airconditioning cycle

T_{A1}	T_{go}	T_{ao}	η	P	$\dot{Q}_A$	T_{eo}	T_{co}	COP	$\dot{Q}_E$	$\dot{Q}_C$	COP_o
K	K	K	–	kW	kW	K	K	–	kW	kW	–
308	371.00	311.36	0.16	0.48	2.52	283.42	323.22	7.12	3.43	3.92	1.14
310	371.00	313.38	0.16	0.47	2.53	283.54	323.08	7.17	3.34	3.81	1.11
312	371.00	315.40	0.15	0.45	2.55	283.67	322.93	7.23	3.25	3.70	1.08
314	371.00	317.42	0.14	0.43	2.57	283.80	322.78	7.28	3.15	3.59	1.05
316	371.00	319.44	0.14	0.42	2.58	283.92	322.63	7.34	3.06	3.47	1.02
318	371.00	321.47	0.13	0.40	2.60	284.05	322.48	7.39	2.96	3.36	0.99
320	371.00	323.49	0.13	0.38	2.62	284.18	322.33	7.45	2.86	3.25	0.95
322	371.00	325.51	0.12	0.37	2.63	284.32	322.17	7.51	2.76	3.13	0.92
324	371.00	327.53	0.12	0.35	2.65	284.45	322.02	7.57	2.66	3.01	0.89
326	371.00	329.55	0.11	0.34	2.66	284.59	321.86	7.64	2.56	2.89	0.85

Table 12.2a Effect of C_G on various working fluid temperatures, heat transfer rates, and coefficient of performance of vapour absorption airconditioning cycle

C_G	T_{go}	T_{ao}	η	P	$\dot{Q}_A$	T_{eo}	T_{co}	COP	$\dot{Q}_E$	$\dot{Q}_C$	COP_o
kW/K	K	K	–	kW	kW	K	K	–	kW	kW	–
1.1	371.36	321.46	0.13	0.40	2.60	284.03	322.51	7.38	2.98	3.38	0.99
1.2	371.67	321.46	0.14	0.41	2.59	284.01	322.53	7.38	2.99	3.39	1.00
1.3	371.92	321.46	0.14	0.41	2.59	284.00	322.54	7.37	3.00	3.41	1.00
1.4	372.14	321.46	0.14	0.41	2.59	283.99	322.56	7.36	3.01	3.42	1.00
1.5	372.33	321.45	0.14	0.41	2.59	283.98	322.57	7.36	3.02	3.43	1.01
1.6	372.50	321.45	0.14	0.41	2.59	283.97	322.58	7.35	3.02	3.43	1.01
1.7	372.65	321.45	0.14	0.41	2.59	283.96	322.59	7.35	3.03	3.44	1.01
1.8	372.78	321.45	0.14	0.41	2.59	283.95	322.60	7.35	3.04	3.45	1.01
1.9	372.89	321.45	0.14	0.41	2.59	283.95	322.61	7.35	3.04	3.45	1.01
2.0	373.00	321.45	0.14	0.41	2.59	283.94	322.61	7.34	3.04	3.46	1.01

Table 12.2b Effect of C_A on various working fluid temperatures, heat transfer rates, and coefficient of performance of vapour absorption airconditioning cycle

C_A	T_{go}	T_{ao}	η	P	$\dot{Q}_A$	T_{eo}	T_{co}	COP	$\dot{Q}_E$	$\dot{Q}_C$	COP_o
kW/K	K	K	–	kW	kW	K	K	–	kW	kW	–
1.1	371.00	321.15	0.13	0.40	2.60	284.03	322.51	7.38	2.98	3.38	0.99
1.2	371.00	320.88	0.14	0.41	2.59	284.01	322.53	7.38	2.99	3.39	1.00
1.3	371.00	320.66	0.14	0.41	2.59	284.00	322.54	7.37	3.00	3.41	1.00
1.4	371.00	320.47	0.14	0.41	2.59	283.99	322.56	7.36	3.01	3.42	1.00
1.5	371.00	320.30	0.14	0.41	2.59	283.98	322.57	7.36	3.02	3.43	1.01
1.6	371.00	320.16	0.14	0.41	2.59	283.97	322.58	7.35	3.02	3.43	1.01
1.7	371.00	320.03	0.14	0.41	2.59	283.96	322.59	7.35	3.03	3.44	1.01
1.8	371.00	319.92	0.14	0.41	2.59	283.95	322.60	7.35	3.04	3.45	1.01
1.9	371.00	319.81	0.14	0.41	2.59	283.95	322.61	7.35	3.04	3.45	1.01
2.0	371.00	319.72	0.14	0.41	2.59	283.94	322.61	7.34	3.04	3.46	1.01

Table 12.3a Effect of ε_G on various working fluid temperatures, heat transfer rates, and coefficient of performance of vapour absorption airconditioning cycle

ε_G	T_{go}	T_{ao}	η	P	$\dot{Q}_A$	T_{eo}	T_{co}	COP	$\dot{Q}_E$	$\dot{Q}_A$	COP_o
–	K	K	–	kW	kW	K	K	–	kW	kW	–
0.20	360.00	321.57	0.11	0.32	2.68	284.71	321.71	7.70	2.46	2.78	0.82
0.30	365.00	321.52	0.12	0.36	2.64	284.40	322.07	7.55	2.70	3.06	0.90
0.40	367.50	321.50	0.13	0.38	2.62	284.25	322.25	7.48	2.81	3.19	0.94
0.50	369.00	321.48	0.13	0.39	2.61	284.17	322.35	7.44	2.88	3.26	0.96
0.60	370.00	321.48	0.13	0.39	2.61	284.11	322.42	7.42	2.92	3.31	0.97
0.70	370.71	321.47	0.13	0.40	2.60	284.07	322.46	7.40	2.95	3.35	0.98
0.75	371.00	321.47	0.13	0.40	2.60	284.05	322.48	7.39	2.96	3.36	0.99
0.80	371.25	321.46	0.13	0.40	2.60	284.04	322.50	7.39	2.97	3.37	0.99
0.90	371.67	321.46	0.14	0.41	2.59	284.01	322.53	7.38	2.99	3.39	1.00
1.00	372.00	321.46	0.14	0.41	2.59	284.00	322.55	7.37	3.00	3.41	1.00

Table 12.3b Effect of ε_A on various working fluid temperatures, heat transfer rates, and coefficient of performance of vapour absorption airconditioning cycle

ε_A	T_{go}	T_{ao}	η	P	$\dot{Q}_A$	T_{eo}	T_{co}	COP	$\dot{Q}_E$	$\dot{Q}_C$	COP_o
–	K	K	–	kW	kW	K	K	–	kW	kW	–
0.20	371.00	331.40	0.11	0.32	2.68	284.71	321.71	7.70	2.46	2.78	0.82
0.30	371.00	326.81	0.12	0.36	2.64	284.40	322.07	7.55	2.70	3.06	0.90
0.40	371.00	324.56	0.13	0.38	2.62	284.25	322.25	7.48	2.81	3.19	0.94
0.50	371.00	323.23	0.13	0.39	2.61	284.17	322.35	7.44	2.88	3.26	0.96
0.60	371.00	322.34	0.13	0.39	2.61	284.11	322.42	7.42	2.92	3.31	0.97
0.70	371.00	321.72	0.13	0.40	2.60	284.07	322.46	7.40	2.95	3.35	0.98
0.75	371.00	321.47	0.13	0.40	2.60	284.05	322.48	7.39	2.96	3.36	0.99
0.80	371.00	321.25	0.13	0.40	2.60	284.04	322.50	7.39	2.97	3.37	0.99
0.90	371.00	320.88	0.14	0.41	2.59	284.01	322.53	7.38	2.99	3.39	1.00
1.00	371.00	320.59	0.14	0.41	2.59	284.00	322.55	7.37	3.00	3.41	1.00

Table 12.4 Effect of $R_{\Delta S}$ on various working fluid temperatures, heat transfer rates, and coefficient of performance of vapour absorption airconditioning cycle

$R_{\Delta S}$	T_{go}	T_{ao}	η	P	$\dot{Q}_A$	T_{eo}	T_{co}	COP	$\dot{Q}_E$	$\dot{Q}_C$	COP_o
–	K	K	–	kW	kW	K	K	–	kW	kW	–
0.90	371.00	321.86	0.04	0.11	2.89	286.73	319.41	8.77	0.95	1.06	0.32
0.91	371.00	321.81	0.05	0.14	2.86	286.39	319.79	8.58	1.20	1.34	0.40
0.92	371.00	321.77	0.06	0.17	2.83	286.08	320.15	8.40	1.44	1.61	0.48
0.93	371.00	321.73	0.07	0.20	2.80	285.78	320.49	8.23	1.67	1.87	0.56
0.94	371.00	321.69	0.08	0.23	2.77	285.49	320.82	8.08	1.88	2.11	0.63
0.95	371.00	321.65	0.09	0.26	2.74	285.22	321.13	7.94	2.08	2.34	0.69
0.96	371.00	321.61	0.10	0.29	2.71	284.97	321.42	7.82	2.27	2.57	0.76
0.97	371.00	321.57	0.11	0.32	2.68	284.72	321.70	7.70	2.46	2.78	0.82
0.98	371.00	321.54	0.12	0.35	2.65	284.49	321.97	7.59	2.63	2.98	0.88
1.00	371.00	321.47	0.13	0.40	2.60	284.05	322.48	7.39	2.96	3.36	0.99

Table 12.5a Effect of T_{E1} on various working fluid temperatures, heat transfer rates, and coefficient of performance of vapour absorption airconditioning cycle

T_{E1}	T_{eo}	T_{co}	COP	$\dot{Q}_E$	$\dot{Q}_C$	COP_o
K	K	K	–	kW	kW	–
278	274.87	321.67	5.87	2.35	2.75	0.76
280	276.73	321.81	6.14	2.46	2.86	0.80
282	278.58	321.96	6.42	2.57	2.97	0.83
284	280.41	322.12	6.72	2.69	3.09	0.87
286	282.24	322.29	7.05	2.82	3.22	0.92
288	284.06	322.48	7.39	2.96	3.36	0.96
290	285.86	322.67	7.76	3.11	3.51	1.01
292	287.65	322.89	8.16	3.27	3.67	1.06
294	289.42	323.11	8.59	3.44	3.84	1.12
296	291.17	323.36	9.05	3.62	4.02	1.18

Table 12.5b Effect of T_{C1} on various working fluid temperatures, heat transfer rates, and coefficient of performance of vapour absorption airconditioning cycle

T_{C1}	T_{eo}	T_{co}	COP	$\dot{Q}_E$	$\dot{Q}_C$	COP_o
K	K	K	–	kW	kW	–
313	283.60	317.94	8.26	3.30	3.70	1.07
315	283.79	319.74	7.89	3.16	3.56	1.03
317	283.97	321.56	7.55	3.02	3.42	0.98
319	284.14	323.39	7.24	2.90	3.30	0.94
321	284.30	325.24	6.94	2.78	3.18	0.90
323	284.44	327.09	6.67	2.67	3.07	0.87
325	284.58	328.95	6.41	2.57	2.97	0.83
327	284.71	330.83	6.17	2.47	2.87	0.80
329	284.83	332.71	5.95	2.38	2.78	0.77
331	284.94	334.59	5.74	2.30	2.70	0.75

Table 12.6a Effect of C_E on various working fluid temperatures, heat transfer rates, and coefficient of performance of vapour absorption airconditioning cycle

C_E	T_{eo}	T_{co}	COP	$\dot{Q}_E$	$\dot{Q}_C$	COP_o
kW/K	K	K	–	kW	kW	–
1.1	284.38	322.51	7.46	2.98	3.38	0.97
1.2	284.66	322.54	7.51	3.01	3.41	0.98
1.3	284.90	322.57	7.56	3.03	3.43	0.98
1.4	285.10	322.59	7.61	3.04	3.44	0.99
1.5	285.28	322.61	7.64	3.06	3.46	0.99
1.6	285.44	322.63	7.68	3.07	3.47	1.00
1.7	285.58	322.64	7.71	3.08	3.48	1.00
1.8	285.71	322.66	7.73	3.09	3.49	1.01
1.9	285.82	322.67	7.76	3.10	3.50	1.01
2.0	285.93	322.68	7.78	3.11	3.51	1.01

Table 12.6b Effect of C_C on various working fluid temperatures, heat transfer rates, and coefficient of performance of vapour absorption airconditioning cycle

C_C	T_{eo}	T_{co}	COP	$\dot{Q}_E$	$\dot{Q}_C$	COP_o
kW/K	K	K	–	kW	kW	–
1.1	284.02	322.10	7.46	2.98	3.38	0.97
1.2	283.99	321.78	7.51	3.01	3.41	0.98
1.3	283.97	321.51	7.56	3.03	3.43	0.98
1.4	283.94	321.28	7.61	3.04	3.44	0.99
1.5	283.92	321.07	7.64	3.06	3.46	0.99
1.6	283.91	320.89	7.68	3.07	3.47	1.00
1.7	283.89	320.73	7.71	3.08	3.48	1.00
1.8	283.88	320.59	7.73	3.09	3.49	1.01
1.9	283.86	320.46	7.76	3.10	3.50	1.01
2.0	283.85	320.34	7.78	3.11	3.51	1.01

Table 12.7a Effect of ε_E on various working fluid temperatures, heat transfer rates, and coefficient of performance of vapour absorption airconditioning cycle

ε_E	T_{eo}	T_{co}	COP	$\dot{Q}_E$	$\dot{Q}_C$	COP_o
–	K	K	–	kW	kW	–
0.20	275.95	321.75	6.03	2.41	2.81	0.78
0.30	279.29	322.02	6.54	2.61	3.01	0.85
0.40	281.15	322.19	6.85	2.74	3.14	0.89
0.50	282.35	322.30	7.07	2.83	3.23	0.92
0.60	283.18	322.39	7.22	2.89	3.29	0.94
0.70	283.80	322.45	7.34	2.94	3.34	0.95
0.75	284.06	322.48	7.39	2.96	3.36	0.96
0.80	284.28	322.50	7.44	2.98	3.38	0.97
0.90	284.66	322.54	7.51	3.01	3.41	0.98
1.00	284.97	322.57	7.58	3.03	3.43	0.99

Table 12.7b Effect of ε_C on various working fluid temperatures, heat transfer rates, and coefficient of performance of vapour absorption airconditioning cycle

ε_C	T_{eo}	T_{co}	COP	$\dot{Q}_E$	$\dot{Q}_C$	COP_o
–	*K*	*K*	–	*kW*	*kW*	–
0.20	284.79	332.05	6.03	2.41	2.81	0.78
0.30	284.51	328.05	6.54	2.61	3.01	0.85
0.40	284.35	325.85	6.85	2.74	3.14	0.89
0.50	284.23	324.45	7.07	2.83	3.23	0.92
0.60	284.15	323.48	7.22	2.89	3.29	0.94
0.70	284.08	322.77	7.34	2.94	3.34	0.95
0.75	284.06	322.48	7.39	2.96	3.36	0.96
0.80	284.03	322.22	7.44	2.98	3.38	0.97
0.90	283.99	321.78	7.51	3.01	3.41	0.98
1.00	283.96	321.43	7.58	3.03	3.43	0.99

Table 12.8 Effect of $R'_{\Delta S}$ on various working fluid temperatures, heat transfer rates, and coefficient of performance of vapour absorption airconditioning cycle

$R'_{\Delta S}$	T_{eo}	T_{co}	COP	$\dot{Q}_E$	$\dot{Q}_C$	COP_o
–	K	K	–	kW	kW	–
0.90	285.84	320.70	4.05	1.62	2.02	0.53
0.91	285.72	320.81	4.28	1.71	2.11	0.56
0.92	285.59	320.94	4.52	1.81	2.21	0.59
0.93	285.45	321.08	4.77	1.91	2.31	0.62
0.94	285.30	321.23	5.06	2.02	2.42	0.66
0.95	285.14	321.39	5.36	2.15	2.55	0.70
0.96	284.96	321.57	5.70	2.28	2.68	0.74
0.97	284.77	321.77	6.06	2.43	2.83	0.79
0.98	284.55	321.98	6.47	2.59	2.99	0.84
1.00	284.06	322.48	7.39	2.96	3.36	0.96

$T_{\mathrm{G1}} = 375\,\mathrm{K}, T_{\mathrm{A1}} = 318\,\mathrm{K}, C_{\mathrm{G}} = C_{\mathrm{A}} = 1.0\,\mathrm{kW/K}, \varepsilon_{\mathrm{G}} = \varepsilon_{\mathrm{A}} = 0.75, \dot{Q}_{\mathrm{G}} = 3.0\,\mathrm{kW}, R_{\Delta\mathrm{S}} = 1.00,$
$T_{\mathrm{C1}} = 318\,\mathrm{K}, T_{\mathrm{E1}} = 288\mathrm{K}, C_{\mathrm{C}} = C_{\mathrm{E}} = 1.0\,\mathrm{kW/K}, \varepsilon_{\mathrm{C}} = \varepsilon_{\mathrm{E}} = 0.75, R'_{\Delta\mathrm{S}} = 1.00.$

For the above typical operating conditions, the COP of a completely reversible system is found as 1.46. The COP of a completely irreversible vapour absorption airconditioning cycle is calculated for $R_{\mathrm{DS}} = R'_{\Delta S} = 0.97$, and it is found $\mathrm{COP_o} = 0.66$, whereas for endoreversible cycle, it is 0.99.

Table 12.1a shows that by increasing generator heat source inlet temperature (T_{G1}), optimal working fluid temperature on generator side increases, and thermal efficiency h of the generator–absorber assembly increases which is expected. Further due to higher optimal value of absorber temperature, COP of evaporator–condenser assembly decreases while heat rejection rate on the absorber side slowly decreases. The optimal overall coefficient of performance and cooling load of the cycle increase by increasing the generator heat source inlet temperature. Therefore, it is desirable to have higher temperature of generator heat source for higher optimal overall coefficient of performance as well as higher cooling load of absorption airconditioning cycle.

Table 12.1b shows that by increasing absorber heat sink inlet temperature (T_{A1}), optimal working fluid temperatures on the absorber and evaporator side increase, whereas on condenser side it decreases and generator side remains constant. Heat rejection rate on the absorber side increases, while thermal efficiency and power output of generator–absorber assembly decrease, and optimal overall coefficient of performance and cooling load of the cycle decrease by increasing the absorber heat sink inlet temperature. Therefore, it is desirable to have lower temperature of absorber heat sink for higher optimal overall coefficient of performance and cooling load of the absorption airconditioning cycle.

It can be seen from Table 12.2a that heat capacitance rate of the generator heat source has the same effect as generator heat source inlet temperature. The optimal overall coefficient of performance and cooling load of the absorption airconditioning cycle increase by increasing the heat capacitance rate of external fluid in generator heat source.

Table 12.2b shows the effect of heat capacitance rate of absorber heat sink. By increasing it, working fluid temperature on absorber and evaporator side decreases, whereas it slowly increases on condenser side and remains constant on generator side. Heat rejection rate on the absorber side decreases slowly while thermal efficiency and power output of generator–absorber assembly increase by increasing it. The optimal overall coefficient of performance and cooling load of the absorption airconditioning cycle increase by increasing the heat capacitance rate of external fluid of absorber heat sink.

It can be seen from Tables 12.3a and 12.3b that effectiveness of the generator heat exchanger has the same effect as heat capacitance rate of external fluid in generator heat source. Similarly, the effectiveness of the absorber heat exchanger has the same effect as the heat capacitance rate of external fluid in absorber heat sink. The optimal overall coefficient of performance and cooling load of the absorption airconditioning cycle increase by increasing any one of the effectiveness of these two heat exchangers.

Table 12.4 shows that if internal irreversibility of generator–absorber assembly increases ($R_{\Delta S}$ decreases, say from 1.00 to 0.90), thermal efficiency and power output of the generator–absorber assembly decrease drastically. COP of the evaporator–condenser assembly increases, but as a result, the optimal overall coefficient of performance and cooling load of the absorption airconditioning cycle decrease rapidly with decreasing this internal irreversibility parameter. It is seen that internal irreversibility parameter is more responsible for performance and cooling load reduction than other external irreversibility parameters.

Tables 12.5a, 12.5b, 12.6a, 12.6b, 12.7a, 12.7b, and 12.8 show the effect of input parameters of evaporator–condenser assembly. During this parameter variation study, power output (P = 0.40 kW) of the generator–absorber assembly remains the same and is used as input power for the evaporator–condenser assembly.

Table 12.5a shows that by increasing inlet temperature of external fluid in evaporator thermal reservoir, optimal working fluid temperatures on evaporator and condenser side, heat rejection rate on condenser side, cooling load, and COP of evaporator–condenser assembly increase, and hence optimal overall coefficient of performance of the absorption airconditioning cycle increases. Table 12.5b shows the effect of inlet temperature of external fluid in condenser thermal reservoir. It is seen that by increasing this temperature, the optimal working fluid temperature on evaporator and condenser side increases, and heat rejection rate on condenser side, cooling load, and COP of the evaporator–condenser assembly decrease, and hence the optimal overall coefficient of performance of the vapour absorption airconditioning cycle decreases. Obviously, it is desirable to have low condenser thermal reservoir inlet temperature and high evaporator thermal reservoir inlet temperature, to have higher performance and cooling load of absorption airconditioning cycle.

Table 12.6a shows that the effect of heat capacitance rate of external fluid in evaporator thermal reservoir is similar to the effect of inlet temperature of external fluid in evaporator thermal reservoir. Table 12.6b shows that by increasing the heat capacitance rate of the external fluid in condenser thermal reservoir, working fluid temperature on the evaporator and condenser side decreases, while heat rejection rate on the condenser side, cooling load, and COP of the evaporator–condenser assembly increase, and hence the optimal overall coefficient of performance of the absorption airconditioning cycle increases.

Table 12.7a shows that effect of the effectiveness of evaporator heat exchanger is similar to the effect of the heat capacitance rate of external fluid in evaporator thermal reservoir. Table 12.7b shows that effect of the effectiveness of the condenser heat exchanger is similar to the effect of the heat capacitance rate of external fluid in the condenser thermal reservoir. Cooling load and optimal overall coefficient of performance of absorption airconditioning cycle increase with increasing effectiveness of heat exchanger on either evaporator or condenser side.

Table 12.8 shows that by increasing internal irreversibility of evaporator–condenser assembly (decreasing, say from 1.00 to 0.9), working fluid temperature on condenser side, heat rejection rate on the condenser side, cooling load, and COP of the evaporator–condenser assembly decrease, and hence the optimal overall coefficient of performance of the absorption airconditioning cycle decreases drastically. Thus, internal irreversibility has drastic effect on the coefficient of performance and cooling load of the absorption airconditioning cycle than other external irreversibility parameters.

12.4.2 Vapour Absorption Heat Pump Cycle

For a given heat input rate in the generator ($\dot{Q}_{\mathrm{G}} = 3.0\ \mathrm{kW}$), heating load ($\dot{Q}_{\mathrm{G}}$) is maximized. In order to have numerical appreciation of the theoretical analysis of vapour absorption heat pump cycle, we have studied the effect of various input parameters on the optimal working fluid temperatures, heat transfer rates to and from the cycle, heating load, and optimal overall coefficient of performance of the vapour absorption heat pump cycle, and results are shown in Tables 12.9a, 12.9b, 12.10a, 12.10b, 12.11a, 12.11b, 12.12, 12.13a, 12.13b, 12.14a, 12.14b, 12.15a, 12.15b, and 12.16. During the variation of any one parameter, all other parameters are assumed to be constant as given below:

$$T_{\mathrm{G1}} = 393K, T_{\mathrm{A1}} = 318K, C_{\mathrm{G}} = C_{\mathrm{A}} = 1.0\mathrm{kW}/K, \varepsilon_{\mathrm{G}} = \varepsilon_{\mathrm{A}} = 0.75, \dot{Q}_{\mathrm{G}} = 3.0\mathrm{kW}, R_{\Delta\mathrm{S}} = 0.9$$
$$T_{\mathrm{C1}} = 318K, T_{\mathrm{E1}} = 288K, C_{\mathrm{C}} = C_{\mathrm{E}} = 1.0\mathrm{kW}/K, \varepsilon_{\mathrm{C}} = \varepsilon_{\mathrm{E}} = 0.75, R'_{\Delta\mathrm{S}} = 0.9$$

Tables 12.9a and 12.9b show the effect of heat source and sink inlet temperature of external fluid on generator and absorber side. Table 12.9a shows that by increasing generator heat source inlet temperature (T_{G1}), optimal working fluid temperature on generator and condenser side increases, and on the absorber and evaporator side, it decreases slowly; thermal efficiency and power output of the generator–absorber assembly increase, while heat rejection rate on the absorber side slowly decreases. The optimal overall coefficient of performance and heating load of the cycle increase by increasing the generator heat source inlet temperature. Therefore, it is desirable to have higher temperature of generator heat source for higher COP as well as higher heating load of absorption heat pump cycle.

Table 12.9b shows that by increasing absorber heat sink inlet temperature (T_{A1}), working fluid temperature on the absorber and evaporator side increases, whereas on condenser side, it decreases and generator side remains constant. Heat rejection rate on the absorber side increases, while thermal efficiency and power output of generator–absorber assembly decrease, and optimal overall coefficient of performance and heating load of the system decrease by increasing the absorber heat sink inlet temperature. Therefore, it is desirable to have lower absorber temperature for higher performance and heating load of the absorption heat pump cycle.

Tables 12.10a and 12.10b show the effect of the heat capacitance rate of external fluid in the generator heat source and absorber heat sink. It can be seen from Table 12.10a that heat capacitance rate of external fluid in generator heat source

Table 12.9a Effect of T_{G1} on various working fluid temperatures, heat transfer rates, and coefficient of performance of vapour absorption heat pump cycle

T_{G1}	T_{go}	T_{ao}	h	P	$\dot{Q}_A$	T_{eo}	T_{co}	COP	$\dot{Q}_E$	$\dot{Q}_C$	COP_o
K	K	K	–	kW	kW	K	K	–	kW	kW	–
393	389.0	321.68	0.08	0.24	2.76	286.64	319.68	5.18	1.02	1.26	0.42
395	391.0	321.66	0.09	0.26	2.74	286.57	319.78	5.17	1.07	1.33	0.44
397	393.0	321.64	0.09	0.27	2.73	286.49	319.87	5.16	1.13	1.40	0.47
399	395.0	321.62	0.10	0.29	2.71	286.42	319.96	5.14	1.19	1.47	0.49
401	397.0	321.60	0.10	0.30	2.70	286.35	320.05	5.13	1.24	1.54	0.51
403	399.0	321.58	0.10	0.31	2.69	286.28	320.14	5.12	1.29	1.61	0.54
405	401.0	321.56	0.11	0.33	2.67	286.21	320.23	5.11	1.34	1.67	0.56
407	403.0	321.55	0.11	0.34	2.66	286.14	320.32	5.10	1.40	1.74	0.58
409	405.0	321.53	0.12	0.35	2.65	286.07	320.40	5.09	1.45	1.80	0.60
411	407.0	321.51	0.12	0.37	2.63	286.00	320.48	5.08	1.50	1.86	0.62

Table 12.9b Effect of T_{A1} on various working fluid temperatures, heat transfer rates, and coefficient of performance of vapour absorption heat pump cycle

T_{A1}	T_{go}	T_{ao}	η	P	$\dot{Q}_A$	T_{eo}	T_{co}	COP	$\dot{Q}_E$	$\dot{Q}_C$	COP_o
K	K	K	–	kW	kW	K	K	–	kW	kW	–
308	389.0	311.56	0.11	0.33	2.67	286.19	320.25	5.11	1.36	1.69	0.56
310	389.0	313.58	0.10	0.31	2.69	286.28	320.14	5.12	1.29	1.60	0.53
312	389.0	315.61	0.10	0.30	2.70	286.37	320.02	5.14	1.22	1.52	0.51
314	389.0	317.63	0.09	0.28	2.72	286.46	319.91	5.15	1.16	1.43	0.48
316	389.0	319.65	0.09	0.26	2.74	286.55	319.80	5.17	1.09	1.35	0.45
318	389.0	321.68	0.08	0.24	2.76	286.64	319.68	5.18	1.02	1.26	0.42
320	389.0	323.70	0.08	0.23	2.77	286.73	319.57	5.20	0.95	1.18	0.39
322	389.0	325.72	0.07	0.21	2.79	286.83	319.45	5.21	0.88	1.09	0.36
324	389.0	327.74	0.06	0.19	2.81	286.92	319.33	5.22	0.81	1.00	0.33
326	389.0	329.77	0.06	0.17	2.83	287.01	319.22	5.24	0.74	0.91	0.30

has the same effect as inlet temperature of this external fluid in generator heat source. The optimal overall coefficient of performance and heating load of the absorption heat pump system increase by increasing the heat capacitance of external fluid in the generator heat source.

Table 12.10b shows the effect of heat capacitance rate of external fluid in absorber heat sink. By increasing it, optimal working fluid temperature on absorber and evaporator side decreases, whereas temperature increases on condenser side and remains constant on generator side. Heat rejection rate on the absorber side decreases, while thermal efficiency and power output of generator–absorber assembly increase by increasing it. The optimal overall coefficient of performance and heating load of the absorption heat pump cycle increase by increasing the heat capacitance rate of external fluid in absorber heat sink.

Table 12.10a Effect of C_G on various working fluid temperatures, heat transfer rates, and coefficient of performance of vapour absorption heat pump cycle

C_G	T_{go}	T_{ao}	η	P		T_{eo}	T_{co}	COP	$\dot{Q}_E$	$\dot{Q}_C$	COP_o
kW/ K	K	K	–	kW	kW	K	K	–	kW	kW	–
1.1	389.36	321.67	0.08	0.25	2.75	286.63	319.70	5.18	1.03	1.28	0.43
1.2	389.67	321.67	0.08	0.25	2.75	286.62	319.71	5.18	1.04	1.29	0.43
1.3	389.92	321.67	0.08	0.25	2.75	286.61	319.73	5.18	1.04	1.29	0.43
1.4	390.14	321.66	0.08	0.25	2.75	286.60	319.74	5.17	1.05	1.30	0.43
1.5	390.33	321.66	0.08	0.25	2.75	286.59	319.75	5.17	1.06	1.31	0.44
1.6	390.50	321.66	0.08	0.25	2.75	286.59	319.75	5.17	1.06	1.32	0.44
1.7	390.65	321.66	0.09	0.26	2.74	286.58	319.76	5.17	1.06	1.32	0.44
1.8	390.78	321.66	0.09	0.26	2.74	286.58	319.77	5.17	1.07	1.32	0.44
1.9	390.89	321.66	0.09	0.26	2.74	286.57	319.77	5.17	1.07	1.33	0.44
2.0	391.00	321.66	0.09	0.26	2.74	286.57	319.78	5.17	1.07	1.33	‘0.44

Table 12.10b Effect of C_A on various working fluid temperatures, heat transfer rates, and coefficient of performance of vapour absorption heat pump cycle

C_A	T_{go}	T_{ao}	η	P	$\dot{Q}_A$	T_{eo}	T_{co}	COP	$\dot{Q}_E$	$\dot{Q}_C$	COP_o
kW/K	K	K	–	kW	kW	K	K	–	kW	kW	–
1.1	389.0	321.34	0.08	0.25	2.75	286.63	319.70	5.18	1.03	1.28	0.43
1.2	389.0	321.06	0.08	0.25	2.75	286.61	319.72	5.18	1.04	1.29	0.43
1.3	389.0	320.82	0.08	0.25	2.75	286.60	319.73	5.17	1.05	1.30	0.43
1.4	389.0	320.62	0.08	0.25	2.75	286.59	319.74	5.17	1.05	1.31	0.44
1.5	389.0	320.44	0.08	0.25	2.75	286.59	319.75	5.17	1.06	1.31	0.44
1.6	389.0	320.29	0.09	0.26	2.74	286.58	319.76	5.17	1.07	1.32	0.44
1.7	389.0	320.15	0.09	0.26	2.74	286.57	319.77	5.17	1.07	1.33	0.44
1.8	389.0	320.03	0.09	0.26	2.74	286.57	319.78	5.17	1.07	1.33	0.44
1.9	389.0	319.92	0.09	0.26	2.74	286.56	319.78	5.17	1.08	1.34	0.45
2.0	389.0	319.83	0.09	0.26	2.74	286.56	319.79	5.17	1.08	1.34	0.45

Tables 12.11a and 12.11b show the effect of effectiveness of the heat exchangers on generator and absorber side. It can be seen from these tables that effectiveness of the generator heat exchanger has the same effect as heat capacitance rate of external fluid in generator heat source. Similarly, the effectiveness of the absorber heat exchanger has the same effect as the heat capacitance rate of external fluid in the absorber heat sink. The optimal overall coefficient of performance and heating load of the absorption heat pump system increase by increasing any one of the effectiveness of these two heat exchangers.

Table 12.12 shows that if internal irreversibility increases (value of $R_{\Delta S}$ decreases, say from 1.00 to 0.90), thermal efficiency and power output of the generator–absorber assembly decrease drastically. COP of the evaporator–condenser assembly increases, but as a result, the optimal overall coefficient of performance and heating load of the vapour absorption heat pump cycle decrease

Table 12.11a Effect of ε_G on various working fluid temperatures, heat transfer rates, and coefficient of performance of vapour absorption heat pump cycle

ε_G	T_{go}	T_{ao}	η	P	$\dot{Q}_A$	T_{eo}	T_{co}	COP	$\dot{Q}_E$	$\dot{Q}_C$	COP_o
–	K	K	–	kW	kW	K	K	–	kW	kW	–
0.20	378.00	321.78	0.05	0.16	2.84	287.08	319.14	5.25	0.69	0.85	0.28
0.30	383.00	321.73	0.07	0.20	2.80	286.88	319.39	5.22	0.84	1.04	0.35
0.40	385.50	321.71	0.07	0.22	2.78	286.78	319.51	5.20	0.92	1.14	0.38
0.50	387.00	321.69	0.08	0.23	2.77	286.72	319.59	5.19	0.96	1.19	0.40
0.60	388.00	321.68	0.08	0.24	2.76	286.68	319.63	5.19	0.99	1.23	0.41
0.70	388.71	321.68	0.08	0.24	2.76	286.65	319.67	5.18	1.01	1.25	0.42
0.75	389.00	321.68	0.08	0.24	2.76	286.64	319.68	5.18	1.02	1.26	0.42
0.80	389.25	321.67	0.08	0.25	2.75	286.63	319.69	5.18	1.03	1.27	0.42
0.90	389.67	321.67	0.08	0.25	2.75	286.62	319.71	5.18	1.04	1.29	0.43
1.00	390.00	321.67	0.08	0.25	2.75	286.60	319.73	5.18	1.05	1.30	0.43

Table 12.11b Effect of ε_A on various working fluid temperatures, heat transfer rates, and coefficient of performance of vapour absorption heat pump cycle

ε_A	T_{go}	T_{ao}	η	P	$\dot{Q}_A$	T_{eo}	T_{co}	COP	$\dot{Q}_E$	$\dot{Q}_C$	COP_o
–	K	K	–	kW	kW	K	K	–	kW	kW	–
0.20	389.0	332.23	0.05	0.15	2.85	287.13	319.07	5.26	0.65	0.81	0.27
0.30	389.0	327.35	0.06	0.19	2.81	286.90	319.36	5.22	0.82	1.02	0.34
0.40	389.0	324.96	0.07	0.22	2.78	286.79	319.49	5.20	0.91	1.12	0.37
0.50	389.0	323.54	0.08	0.23	2.77	286.73	319.58	5.19	0.95	1.18	0.39
0.60	389.0	322.61	0.08	0.24	2.76	286.68	319.63	5.19	0.99	1.22	0.41
0.70	389.0	321.94	0.08	0.24	2.76	286.65	319.67	5.18	1.01	1.25	0.42
0.75	389.0	321.68	0.08	0.24	2.76	286.64	319.68	5.18	1.02	1.26	0.42
0.80	389.0	321.44	0.08	0.25	2.75	286.63	319.70	5.18	1.03	1.27	0.42
0.90	389.0	321.06	0.08	0.25	2.75	286.61	319.72	5.18	1.04	1.29	0.43
1.00	389.0	320.75	0.08	0.25	2.75	286.60	319.74	5.17	1.05	1.30	0.43

rapidly with decreasing this internal irreversibility parameter. It is seen that internal irreversibility parameter is more responsible for overall coefficient of performance and heating load reduction than other external irreversibility parameters.

Tables 12.13a, 12.13b, 12.14a, 12.14b, 12.15a, 12.15b, and 12.16 show the effect of input parameters of evaporator–condenser assembly. During this parametric study, power output ($P = 0.24$ kW) of the generator–absorber assembly is assumed to be constant and used as input power for the evaporator–condenser assembly.

Table 12.13a shows that by increasing inlet temperature of external fluid in evaporator thermal reservoir, optimal working fluid temperatures on evaporator and condenser side, heat rejection rate on evaporator side, heating load, and COP of evaporator–condenser assembly increase, and hence the optimal overall coefficient of performance of the absorption heat pump cycle increases.

Table 12.12 Effect of $R_{\Delta S}$ on various working fluid temperatures, heat transfer rates, and coefficient of performance of vapour absorption heat pump cycle

$R_{\Delta S}$	T_{go}	T_{ao}	η	P	$\dot{Q}_A$	T_{eo}	T_{co}	COP	$\dot{Q}_E$	$\dot{Q}_C$	COP_o
–	K	K	–	kW	kW	K	K	–	kW	kW	–
0.90	389.0	321.68	0.08	0.24	2.76	286.64	319.68	5.18	1.02	1.26	0.42
0.91	389.0	321.63	0.09	0.27	2.73	286.48	319.88	5.16	1.14	1.41	0.47
0.92	389.0	321.59	0.10	0.30	2.70	286.32	320.08	5.13	1.26	1.56	0.52
0.93	389.0	321.56	0.11	0.33	2.67	286.17	320.27	5.11	1.37	1.70	0.57
0.94	389.0	321.52	0.12	0.36	2.64	286.03	320.46	5.08	1.48	1.84	0.61
0.95	389.0	321.48	0.13	0.39	2.61	285.89	320.63	5.06	1.59	1.98	0.66
0.96	389.0	321.44	0.14	0.42	2.58	285.75	320.81	5.04	1.69	2.11	0.70
0.97	389.0	321.41	0.15	0.44	2.56	285.62	320.98	5.02	1.79	2.23	0.74
0.98	389.0	321.37	0.16	0.47	2.53	285.49	321.14	5.00	1.88	2.36	0.79
1.00	389.0	321.30	0.17	0.52	2.48	285.24	321.46	4.97	2.07	2.59	0.86

Table 12.13a Effect of T_{E1} on various working fluid temperatures, heat transfer rates, and coefficient of performance of vapour absorption heat pump cycle

T_{E1}	T_{eo}	T_{co}	COP	$\dot{Q}_E$	$\dot{Q}_C$	COP_o
K	K	K	–	kW	kW	–
278	276.87	319.45	4.55	0.85	1.09	0.36
280	278.83	319.49	4.66	0.88	1.12	0.37
282	280.79	319.53	4.78	0.91	1.15	0.38
284	282.75	319.57	4.91	0.94	1.18	0.39
286	284.71	319.61	5.04	0.97	1.21	0.40
288	286.66	319.66	5.18	1.00	1.24	0.41
290	288.61	319.71	5.33	1.04	1.28	0.43
292	290.56	319.76	5.49	1.08	1.32	0.44
294	292.51	319.81	5.66	1.12	1.36	0.45
296	294.45	319.87	5.83	1.16	1.40	0.47

Table 12.13b Effect of T_{C1} on various working fluid temperatures, heat transfer rates, and coefficient of performance of vapour absorption heat pump cycle

T_{C1}	T_{eo}	T_{co}	COP	$\dot{Q}_E$	$\dot{Q}_C$	COP_o
K	K	K	–	kW	kW	–
313	286.55	314.77	5.53	1.09	1.33	0.44
315	286.60	316.72	5.39	1.05	1.29	0.43
317	286.64	318.68	5.25	1.02	1.26	0.42
319	286.68	320.64	5.12	0.99	1.23	0.41
321	286.72	322.60	5.00	0.96	1.20	0.40
323	286.76	324.56	4.88	0.93	1.17	0.39
325	286.79	326.53	4.77	0.91	1.15	0.38
327	286.83	328.49	4.67	0.88	1.12	0.37
329	286.86	330.46	4.57	0.86	1.10	0.37
331	286.89	332.43	4.48	0.83	1.07	0.36

Table 12.14a Effect of C_E on various working fluid temperatures, heat transfer rates, and coefficient of performance of vapour absorption heat pump cycle

C_E	T_{eo}	T_{co}	COP	$\dot{Q}_E$	$\dot{Q}_C$	COP_o
kW/K	K	K	–	kW	kW	–
1.1	286.78	319.66	5.19	1.01	1.25	0.42
1.2	286.88	319.66	5.20	1.01	1.25	0.42
1.3	286.96	319.67	5.21	1.01	1.25	0.42
1.4	287.04	319.67	5.21	1.01	1.25	0.42
1.5	287.10	319.67	5.22	1.01	1.25	0.42
1.6	287.16	319.67	5.22	1.01	1.25	0.42
1.7	287.20	319.67	5.23	1.01	1.25	0.42
1.8	287.25	319.67	5.23	1.01	1.25	0.42
1.9	287.29	319.67	5.23	1.02	1.26	0.42
2.0	287.32	319.67	5.23	1.02	1.26	0.42

Table 12.14b Effect of C_C on various working fluid temperatures, heat transfer rates, and coefficient of performance of vapour absorption heat pump cycle

C_C	T_{eo}	T_{co}	COP	$\dot{Q}_E$	$\dot{Q}_C$	COP_o
kW/K	K	K	–	kW	kW	–
1.1	286.66	319.51	5.19	1.01	1.25	0.42
1.2	286.66	319.39	5.20	1.01	1.25	0.42
1.3	286.65	319.28	5.21	1.01	1.25	0.42
1.4	286.65	319.19	5.21	1.01	1.25	0.42
1.5	286.65	319.11	5.22	1.01	1.25	0.42
1.6	286.65	319.05	5.22	1.01	1.25	0.42
1.7	286.65	318.98	5.23	1.01	1.25	0.42
1.8	286.65	318.93	5.23	1.02	1.26	0.42
1.9	286.64	318.88	5.24	1.02	1.26	0.42
2.0	286.64	318.84	5.24	1.02	1.26	0.42

Table 12.13b shows that by increasing inlet temperature of external fluid in condenser thermal reservoir, optimal working fluid temperatures on evaporator and condenser side increase, and heat rejection rate on evaporator side, heating load, and COP of the evaporator–condenser assembly decrease, and hence the optimal overall coefficient of performance of absorption heat pump cycle decreases. Therefore, it is desirable to have lower inlet temperature of external fluid in condenser thermal reservoir for having higher performance and heating load of absorption heat pump cycle.

Table 12.14a shows that the effect of heat capacitance rate of external fluid in evaporator thermal reservoir is similar to the effect of inlet temperature of external fluid in evaporator thermal reservoir. Table 12.14b shows that by increasing the heat capacitance rate of the external fluid in condenser thermal reservoir, working fluid temperature on the evaporator and condenser side decreases, while heat rejection rate on the evaporator side, heating load, and COP of the evaporator–condenser assembly increase slowly, and hence the overall coefficient of performance of the absorption heat pump cycle increases.

Table 12.15a Effect of ε_E on various working fluid temperatures, heat transfer rates, and coefficient of performance of vapour absorption heat pump cycle

ε_E	T_{eo}	T_{co}	COP	$\dot{Q}_E$	$\dot{Q}_C$	COP_o
–	K	K	–	kW	kW	–
0.20	283.27	319.58	4.94	0.95	1.19	0.40
0.30	284.76	319.62	5.05	0.97	1.21	0.40
0.40	285.54	319.63	5.10	0.98	1.22	0.41
0.50	286.01	319.64	5.14	0.99	1.23	0.41
0.60	286.34	319.65	5.16	1.00	1.24	0.41
0.70	286.57	319.66	5.18	1.00	1.24	0.41
0.75	286.66	319.66	5.18	1.00	1.24	0.41
0.80	286.74	319.66	5.19	1.01	1.25	0.42
0.90	286.88	319.66	5.20	1.01	1.25	0.42
1.00	286.99	319.67	5.21	1.01	1.25	0.42

Table 12.15b Effect of ε_C on various working fluid temperatures, heat transfer rates, and coefficient of performance of vapour absorption heat pump cycle

ε_C	T_{eo}	T_{co}	COP	$\dot{Q}_E$	$\dot{Q}_C$	COP_o
–	K	K	–	kW	kW	–
0.20	286.75	323.90	4.92	0.94	1.18	0.39
0.30	286.71	322.03	5.03	0.97	1.21	0.40
0.40	286.69	321.06	5.09	0.98	1.22	0.41
0.50	286.68	320.46	5.13	0.99	1.23	0.41
0.60	286.67	320.06	5.16	1.00	1.24	0.41
0.70	286.66	319.77	5.18	1.00	1.24	0.41
0.75	286.66	319.66	5.18	1.00	1.24	0.41
0.80	286.66	319.56	5.19	1.01	1.25	0.42
0.90	286.66	319.39	5.20	1.01	1.25	0.42
1.00	286.65	319.25	5.21	1.01	1.25	0.42

Table 12.16 Effect of $R'_{\Delta S}$ on various working fluid temperatures, heat transfer rates, and coefficient of performance of vapour absorption heat pump cycle

$R'_{\Delta S}$	T_{eo}	T_{co}	COP	$\dot{Q}_E$	$\dot{Q}_C$	COP_o
–	K	K	–	kW	kW	–
0.90	286.66	319.66	5.18	1.00	1.24	0.41
0.91	286.58	319.74	5.42	1.06	1.30	0.43
0.92	286.50	319.82	5.69	1.12	1.36	0.45
0.93	286.41	319.91	5.97	1.19	1.43	0.48
0.94	286.31	320.01	6.29	1.27	1.51	0.50
0.95	286.20	320.12	6.64	1.35	1.59	0.53
0.96	286.07	320.25	7.02	1.44	1.68	0.56
0.97	285.94	320.38	7.45	1.55	1.79	0.60
0.98	285.79	320.53	7.92	1.66	1.90	0.63
1.00	285.43	320.89	9.05	1.93	2.17	0.72

Table 12.15a shows that effect of the effectiveness of evaporator heat exchanger is similar to the effect of the heat capacitance rate of external fluid in evaporator thermal reservoir. Similarly, Table 12.15b shows that effect of the effectiveness of the condenser heat exchanger is similar to the effect of the heat capacitance rate of external fluid in the condenser thermal reservoir. The optimal overall coefficient of performance and heating load increase with increasing effectiveness of either heat exchanger on evaporator or condenser side.

Table 12.16 shows that by increasing the internal irreversibility of evaporator–condenser assembly (decreasing value, say from 1.00 to 0.9), optimal working fluid temperatures on condenser side, heat rejection rate on the evaporator side, heating load, and COP of the evaporator–condenser assembly decrease, and hence the optimal overall coefficient of performance of absorption heat pump cycle decreases drastically. Thus, internal irreversibility has more drastic effect on performance and heating load of the absorption heat pump system than other external irreversibility parameters.

12.5 Conclusion

Finite time heat transfer and thermodynamic optimization of endoreversible and irreversible vapour absorption airconditioning/heat pump cycles have been carried out with external thermal reservoirs of finite heat capacitance and considering both external as well as internal irreversibilities often existing in real absorption airconditioning/heat pump cycles. Analytical expressions are obtained for the optimal COP by maximizing the cooling/heating load with given heat input rate in the generator. It is found that overall performance is more sensitive to internal irreversibility parameter of generator–absorber assembly than internal irreversibility parameter of evaporator–condenser assembly. The optimum formulae obtained are useful for further research work needed in explaining the deviation of the actual performance from the ideal performance of these cycles.

Correction to: Finite Time Thermodynamic Analysis of Modified Brayton Cycle

Correction to:
Chapter 4 in: S.C. Kaushik et al., *Finite Time Thermodynamics of Power and Refrigeration Cycles*,
https://doi.org/10.1007/978-3-319-62812-7_4

The original version of this chapter was inadvertently published with incorrect Tables 4.1 and 4.2. The original chapter has been corrected and updated.

The updated online version of the chapter can be found at
https://doi.org/10.1007/978-3-319-62812-7_4

S. C. Kaushik et al., *Finite Time Thermodynamics of Power and Refrigeration Cycles*,
https://doi.org/10.1007/978-3-319-62812-7_13

Appendices

Appendix 1: Natural Maximum Analysis for RAC and HP Systems

The rate of heat input from low-temperature heat source to the heat pump is given by:

$$\dot{Q}_{\mathrm{E}} = \frac{Q_{\mathrm{E}}}{t_{\mathrm{E}}} = U_{\mathrm{E}}A_{\mathrm{E}}(\mathrm{LMTD})_{\mathrm{E}} = \dot{m}_{\mathrm{E}}C_{\mathrm{PE}}(T_{\mathrm{E1}} - T_{\mathrm{E2}}) \tag{A1.1}$$

Similarly, the rate of heat output from the heat pump to the heat sink is given by:

$$\dot{Q}_{\mathrm{C}} = \frac{Q_{\mathrm{C}}}{t_{\mathrm{C}}} = U_{\mathrm{C}}A_{\mathrm{C}}(\mathrm{LMTD})_{\mathrm{C}} = \dot{m}_{\mathrm{C}}C_{\mathrm{PC}}(T_{\mathrm{C2}} - T_{\mathrm{C1}}) \tag{A1.2}$$

where

$$(\mathrm{LMTD})_{\mathrm{E}} = \frac{(T_{\mathrm{E1}} - T_{\mathrm{e}}) - (T_{\mathrm{E2}} - T_{\mathrm{e}})}{\ln \dfrac{(T_{\mathrm{E1}} - T_{\mathrm{e}})}{(T_{\mathrm{E2}} - T_{\mathrm{e}})}} \quad \text{and}$$

$$(\mathrm{LMTD})_{\mathrm{C}} = \frac{(T_{\mathrm{c}} - T_{\mathrm{C1}}) - (T_{\mathrm{c}} - T_{\mathrm{C2}})}{\ln \dfrac{(T_{\mathrm{c}} - T_{\mathrm{C1}})}{(T_{\mathrm{c}} - T_{\mathrm{C2}})}}$$

Using LMTD expressions in Eqs. (A1.1) and (A1.2), we have:

$$T_{\mathrm{E2}} = T_{\mathrm{e}} + (T_{\mathrm{E1}} - T_{\mathrm{e}})\ \mathrm{e}^{-U_{\mathrm{E}}A_{\mathrm{E}}/\dot{m}_{\mathrm{E}}C_{\mathrm{PE}}} \tag{A1.3}$$

$$T_{\mathrm{C2}} = T_{\mathrm{c}} + (T_{\mathrm{c}} - T_{\mathrm{C1}})\ \mathrm{e}^{-U_{\mathrm{C}}A_{\mathrm{C}}/\dot{m}_{\mathrm{C}}C_{\mathrm{PC}}} \tag{A1.4}$$

Using Eqs. (A1.1), (A1.2), (A1.3), and (A1.4), we have:

S.C. Kaushik et al., *Finite Time Thermodynamics of Power and Refrigeration Cycles*,
https://doi.org/10.1007/978-3-319-62812-7

$$Q_{\mathrm{E}} = EE(T_{\mathrm{E1}} - T_{\mathrm{e}})t_{\mathrm{E}} \tag{A1.5}$$

$$Q_{\mathrm{C}} = CE(T_{\mathrm{c}} - T_{\mathrm{C1}})t_{\mathrm{C}} \tag{A1.6}$$

where

$$CE = C_{\mathrm{E}}\varepsilon_{\mathrm{E}}, \quad C_{\mathrm{E}} = \dot{m}_{\mathrm{E}} C_{\mathrm{PE}}, \quad C_{\mathrm{C}} = \dot{m}_{\mathrm{C}} C_{\mathrm{PC}},$$
$$\varepsilon_{\mathrm{E}} = 1 - \mathrm{e}^{-U_{\mathrm{E}} A_{\mathrm{E}}/\dot{m}_{\mathrm{E}} C_{\mathrm{PE}}}, \quad \text{and} \quad \varepsilon_{\mathrm{C}} = 1 - \mathrm{e}^{-U_{\mathrm{C}} A_{\mathrm{C}}/\dot{m}_{\mathrm{C}} C_{\mathrm{PC}}}$$

Now the total cycle time becomes:

$$t = \frac{Q_{\mathrm{E}}}{EE(T_{\mathrm{E1}} - T_{\mathrm{e}})} + \frac{Q_{\mathrm{C}}}{CE(T_{\mathrm{c}} - T_{\mathrm{C1}})} \tag{A1.7}$$

According to second law of thermodynamics:

$$\frac{\dot{Q}_{\mathrm{E}}}{T_{\mathrm{e}}} = \frac{\dot{Q}_{\mathrm{C}}}{T_{\mathrm{c}}} \tag{A1.8}$$

Equation (A1.8) gives:

$$T_{\mathrm{e}} = \frac{EET_{\mathrm{E1}}T_{\mathrm{c}}}{EET_{\mathrm{c}} + CET_{\mathrm{c}} - CET_{\mathrm{C1}}} \tag{A1.9}$$

Power input to heat pump system:

$$P = CE(T_{\mathrm{c}} - T_{\mathrm{C1}}) - \frac{CE\,EE\,(T_{\mathrm{c}} - T_{\mathrm{C1}})T_{\mathrm{E1}}}{(CE + EE)T_{\mathrm{c}} - CET_{\mathrm{C1}}} \tag{A1.10}$$

For fixed T_{C1} and T_{E1}, power input is the function of refrigerant temperature T_{c} only. Minimizing power with respect to T_{c}, $\frac{\mathrm{d}P}{\mathrm{d}T_{\mathrm{c}}} = 0$, gives:

$$(T_{\mathrm{c}})_{\mathrm{O}} = \frac{\left(CE\sqrt{T_{\mathrm{C1}}} + EE\sqrt{T_{\mathrm{E1}}}\right)}{(CE + EE)}\sqrt{T_{\mathrm{C1}}} \tag{A1.11}$$

$$(T_{\mathrm{e}})_{\mathrm{O}} = \frac{\left(CE\sqrt{T_{\mathrm{C1}}} + EE\sqrt{T_{\mathrm{E1}}}\right)}{(CE + EE)}\sqrt{T_{\mathrm{E1}}} \tag{A1.12}$$

For numerical appreciation, let us assume:

$$T_{\mathrm{E1}} = 5^{\circ}\mathrm{C} = 278\mathrm{K}, T_{\mathrm{C1}} = 25^{\circ}\mathrm{C} = 298\mathrm{K}, C_{\mathrm{C}} = C_{\mathrm{E}} = 1.00\,\mathrm{kW/K},$$
$$\varepsilon_{\mathrm{C}} = \varepsilon_{\mathrm{E}} = 0.75 \ \ \mathrm{kW/K}$$

Using these above typical values, Eqs. (A1.11) and (A1.12) give:

$$(T_{\mathrm{c}})_{\mathrm{O}} = 292.91 \ \ \mathrm{K}, \quad (T_{\mathrm{e}})_{\mathrm{O}} = 282.91\,\mathrm{K}$$

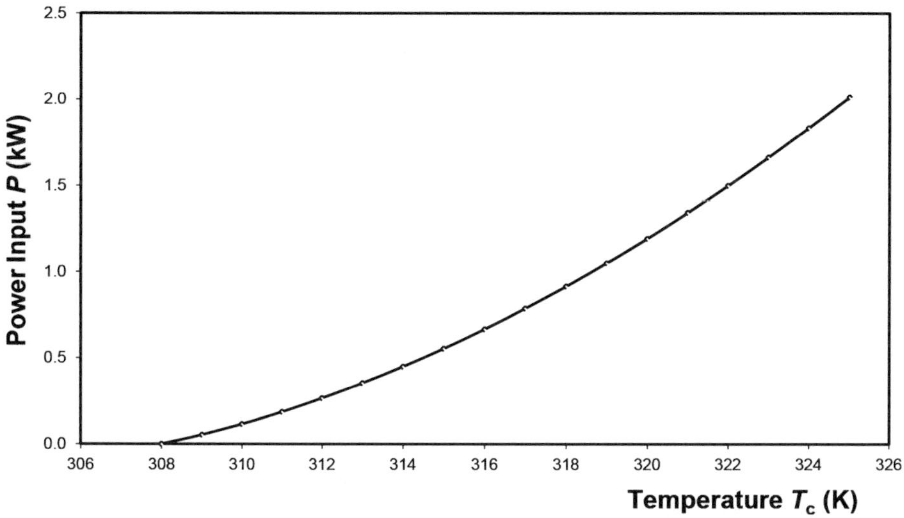

Fig. A1.1 Power input versus refrigerant temperature in condenser

These typical calculations show that $(T_e)_O$ is greater than T_{E1} and $(T_c)_O$ is less than T_{C1}, which is unrealistic. This result shows that these straightforward calculations do not work for a heat pump. A graph between power input (P) to the heat pump system and refrigerant temperature (T_c) is shown in Fig. A1.1, which indicates that power input is a monotonic increasing function of refrigerant temperature (T_c).

Similar results can be obtained for refrigeration/airconditioning systems. Thus, it is seen that there is no 'natural maximum' in case of R/AC and HP systems. Therefore, we have to use alternative method (Lagrangian multiplier method) for seeking the minimum input power (P) for a given desired output (viz. cooling load, heating load) for these systems.

Appendix 2: Optimization Method: Lagrangian Multiplier Method

To extremalize a function ψ_K subject to certain conditions or constraints ϕ_K, Lagrangian's undetermined multiplier method can be used in which each conditional function ϕ_K is multiplied by a multiplier λ_K, and the $\lambda_K\phi_K$ is added to the objective function ψ_K to define Lagrangian operator. Partial differentiation of the Lagrangian operator with respect to the variable parameters is taken and the extremal value of the variable parameters equated to zero. Now after substituting these extremal values in objective function, we can find optimal objective function (Band et al. 1982; Tolle 1975):

1. The method of Lagrangian multipliers gives a set of necessary conditions to identify optimal points of equality constrained optimization problems.
2. This is done by converting a constrained problem to an equivalent unconstrained problem with the help of certain unspecified parameters known as Lagrangian multipliers.

For example, the classical problem formulation:

Minimize $f(x_1, x_2, x_3, \ldots\ldots x_n)$
Subject to $h_1(x_1, x_2, x_3, \ldots\ldots x_n) = 0$

can be converted to minimize $L(x, v) = f(x) - v\, h_1(x)$
where $L(x, v)$ is the Lagrangian function and 'v' is an unspecified positive or negative constant called the Lagrangian multiplier.

Finding an Optimum Using Lagrangian Multipliers

1. Suppose that we fix $v = v_0$ and the unconstrained minimum of $L(x, v)$ occurs at $x = x^*$ and x^* satisfies $h_1(x^*) = 0$, then x^* minimizes $f(x)$ subject to $h_1(x) = 0$.
2. Trick is to find appropriate value for Lagrangian multiplier v.
3. This can be done by treating v as a variable, finding the unconstrained minimum of $L(x, v)$, and adjusting v so that $h_1(x) = 0$ is satisfied.

Typical Approach

1. Take derivative of $L(x, v)$ with respect to x_i and set them equal to zero.
2. If there are n variables, you will get n equations with $n + 1$ unknowns.
3. Express all x_i in terms of Lagrangian multiplier v.
4. Plug x in terms of v in constraint $h_1(x) = 0$ and solve v.
5. Calculate x by using the just found value for v.
6. Note that the n derivatives and one constraint equation result in $n + 1$ equations.

Appendix 3: Comparative Study of RAC and HP Systems Alternatively and Continuously Connected to Thermal Reservoirs

The refrigeration/airconditioning and heat pump systems alternatively as well as continuously connected to the thermal reservoirs (through external heating and cooling fluids) are analysed, in Chap. 7, and found the analytical expressions of optimal performance for both cases (Kumar, 2002).

The coefficient of performance for alternatively connected case is having similar form of the COP as that of continuously connected case but slightly differs because of different form of '*K*' which depends on effectiveness, heat capacitance rates, and internal irreversibility. In order to have a numerical appreciation of the theoretical analysis for alternatively and continuously connected case, we have studied the effect of various input parameters on optimal working fluid temperatures, power input, and optimal coefficient of performance. The optimal coefficient of performance of refrigeration/airconditioning and heat pump systems and results are shown in Tables A3.1, A3.2, A3.3, A3.4, A3.5, A3.6, A3.7, A3.8, A3.9, A3.10, A3.11, A3.12, A3.13, and A3.14. For refrigeration/airconditioning system, during the variation of any one parameter, all other parameters are assumed to be constant as given below:

$$T_{\mathrm{C1}} = 318\,\mathrm{K} \quad T_{\mathrm{E1}} = 275\,\mathrm{K} \quad C_{\mathrm{C}} = C_{\mathrm{E}} = 1.00\,\mathrm{kW/K}$$
$$\varepsilon_{\mathrm{C}} = \varepsilon_{\mathrm{E}} = 0.75 \quad R_{\Delta S} = 1.0 \quad \text{and} \quad \mathrm{P_L} = 3.0\,\mathrm{kW}$$

For heat pump system, during the variation of any one parameter, all other parameters are assumed to be constant as given below:

$$T_{\mathrm{C1}} = 310\,\mathrm{K} \quad T_{\mathrm{E1}} = 283\,\mathrm{K} \quad C_{\mathrm{C}} = C_{\mathrm{E}} = 1.00\,\mathrm{kW/K}$$
$$\varepsilon_{\mathrm{C}} = \varepsilon_{\mathrm{E}} = 0.75 \quad R_{\Delta S} = 1.0 \quad \text{and} \quad \mathrm{P_H} = 3.0\,\mathrm{kW}$$

It can be seen from Tables A3.1, A3.2, A3.3, A3.4, A3.5, A3.6, A3.7, A3.8, A3.9, A3.10, A3.11, A3.12, A3.13, and A3.14 that coefficient of performance for continuously connected case is slightly higher than the corresponding COP for alternatively connected case. However, Tables A3.7 and A3.14 show that the difference in COP decreases as the internal irreversibility increases (value of $R_{\Delta S}$ decreases).

Table A3.1 Effect of heat source inlet fluid temperature on working fluid temperatures, power input, and performance of heat pump system in alternatively and continuously connected cases

T_{E1}	Alternatively connected				Continuously connected			
K	$(T_e)_o$ K	$(T_c)_o$ K	$\mathrm{COP_o}$	P_{in} kW	$(T_e)_o$ K	$(T_c)_o$ K	$\mathrm{COP_o}$	P_{in} kW
278	271.18	318.00	6.79	0.44	274.50	314.00	7.95	0.38
279	272.15	318.00	6.94	0.43	275.49	314.00	8.15	0.37
280	273.13	318.00	7.09	0.42	276.48	314.00	8.37	0.36
281	274.10	318.00	7.24	0.41	277.47	314.00	8.59	0.35
282	275.08	318.00	7.41	0.40	278.45	314.00	8.83	0.34
283	276.06	318.00	7.58	0.40	279.44	314.00	9.09	0.33
284	277.03	318.00	7.76	0.39	280.43	314.00	9.35	0.32
285	278.01	318.00	7.95	0.38	281.42	314.00	9.64	0.31
286	278.98	318.00	8.15	0.37	282.40	314.00	9.94	0.30
287	279.96	318.00	8.36	0.36	283.39	314.00	10.26	0.29

Table A3.2 Effect of heat sink inlet fluid temperature on working fluid temperatures, power input, and performance of heat pump system in alternatively and continuously connected cases

T_{C1}	Alternatively connected				Continuously connected			
K	$(T_e)_o$ K	$(T_c)_o$ K	COP_o	P_{in} kW	$(T_e)_o$ K	$(T_c)_o$ K	COP_o	P_{in} kW
300	275.84	308	9.58	0.31	279.32	304	12.32	0.24
302	275.88	310	9.09	0.33	279.35	306	11.48	0.26
304	275.93	312	8.65	0.35	279.37	308	10.76	0.28
306	275.97	314	8.26	0.36	279.39	310	10.13	0.3
308	276.01	316	7.90	0.38	279.42	312	9.58	0.31
310	276.06	318	7.58	0.40	279.44	314	9.09	0.33
312	276.10	320	7.29	0.41	279.46	316	8.65	0.35
314	276.14	322	7.02	0.43	279.48	318	8.26	0.36
316	276.18	324	6.78	0.44	279.51	320	7.90	0.38
318	276.22	326	6.55	0.46	279.53	322	7.58	0.40

Table A3.3 Effect of heat capacitance of heat source fluid on working fluid temperatures, power input, and performance of heat pump system in alternatively and continuously connected cases

C_C	Alternatively connected				Continuously connected			
kW/K	$(T_e)_o$ K	$(T_c)_o$ K	COP_o	P_{in} kW	$(T_e)_o$ K	$(T_c)_o$ K	COP_o	P_{in} kW
0.60	275.16	321.83	6.90	0.44	279.47	316.67	8.51	0.35
0.70	275.45	320.50	7.12	0.42	279.46	315.71	8.71	0.34
0.80	275.69	319.47	7.30	0.41	279.45	315.00	8.86	0.34
0.90	275.89	318.66	7.45	0.40	279.45	314.44	8.98	0.33
1.00	276.06	318.00	7.58	0.40	279.44	314.00	9.09	0.33
1.10	276.20	317.45	7.70	0.39	279.44	313.64	9.17	0.33
1.20	276.33	316.98	7.80	0.38	279.43	313.33	9.24	0.32
1.30	276.44	316.59	7.89	0.38	279.43	313.08	9.30	0.32
1.40	276.55	316.24	7.97	0.38	279.43	312.86	9.36	0.32
1.50	276.64	315.93	8.04	0.37	279.43	312.67	9.41	0.32

Table A3.4 Effect of heat capacitance of heat sink fluid on working fluid temperatures, power input, and performance of heat pump system in alternatively and continuously connected cases

C_E	Alternatively connected				Continuously connected			
kW/K	$(T_e)_o$ K	$(T_c)_o$ K	COP_o	P_{in} kW	$(T_e)_o$ K	$(T_c)_o$ K	COP_o	P_{in} kW
0.60	272.88	319.16	6.90	0.44	277.12	314.00	8.51	0.35
0.70	273.98	318.78	7.12	0.42	277.94	314.00	8.71	0.34
0.80	274.83	318.47	7.30	0.41	278.56	314.00	8.86	0.34
0.90	275.50	318.22	7.45	0.40	279.05	314.00	8.98	0.33
1.00	276.06	318.00	7.58	0.40	279.44	314.00	9.09	0.33
1.10	276.52	317.81	7.70	0.39	279.76	314.00	9.17	0.33
1.20	276.91	317.65	7.80	0.38	280.03	314.00	9.24	0.32
1.30	277.25	317.51	7.89	0.38	280.25	314.00	9.30	0.32
1.40	277.55	317.38	7.97	0.38	280.45	314.00	9.36	0.32
1.50	277.81	317.27	8.04	0.37	280.62	314.00	9.41	0.32

Table A3.5 Effect of effectiveness of sink side heat exchanger on working fluid temperatures, power input, and performance of heat pump system in alternatively and continuously connected cases

ε_C	Alternatively connected				Continuously connected			
–	$(T_e)_o$ K	$(T_c)_o$ K	COP_o	P_{in} kW	$(T_e)_o$ K	$(T_c)_o$ K	COP_o	P_{in} kW
0.20	273.35	332.75	5.60	0.54	279.56	325.00	7.15	0.42
0.30	274.32	326.32	6.28	0.48	279.51	320.00	7.90	0.38
0.40	274.93	322.98	6.72	0.45	279.48	317.50	8.35	0.36
0.50	275.36	320.90	7.05	0.43	279.46	316.00	8.65	0.35
0.60	275.69	319.47	7.30	0.41	279.45	315.00	8.86	0.34
0.70	275.95	318.43	7.50	0.40	279.44	314.29	9.02	0.33
0.75	276.06	318.00	7.58	0.40	279.44	314.00	9.09	0.33
0.80	276.15	317.62	7.66	0.39	279.44	313.75	9.14	0.33
0.90	276.33	316.98	7.80	0.38	279.43	313.33	9.24	0.32
1.00	276.48	316.46	7.91	0.38	279.43	313.00	9.32	0.32

Table A3.6 Effect of effectiveness of source side heat exchanger on working fluid temperatures, power input, and performance of heat pump system in alternatively and continuously connected cases

ε_E	Alternatively connected				Continuously connected			
–	$(T_e)_o$ K	$(T_c)_o$ K	COP_o	P_{in} kW	$(T_e)_o$ K	$(T_c)_o$ K	COP_o	P_{in} kW
0.20	264.31	321.75	5.60	0.54	270.10	314.00	7.15	0.42
0.30	269.28	320.32	6.28	0.48	274.27	314.00	7.90	0.38
0.40	271.95	319.48	6.72	0.45	276.40	314.00	8.35	0.36
0.50	273.65	318.90	7.05	0.43	277.69	314.00	8.65	0.35
0.60	274.83	318.47	7.30	0.41	278.56	314.00	8.86	0.34
0.70	275.70	318.14	7.50	0.40	279.19	314.00	9.02	0.33
0.75	276.06	318.00	7.58	0.40	279.44	314.00	9.09	0.33
0.80	276.37	317.87	7.66	0.39	279.66	314.00	9.14	0.33
0.90	276.91	317.65	7.80	0.38	280.03	314.00	9.24	0.32
1.00	277.35	317.46	7.91	0.38	280.32	314.00	9.32	0.32

Appendix 4: Entropic Average Temperature and Internal Irreversibility Parameter for VCR Cycle

Figure A4.1 is a *T–s* diagram of a simple ideal vapour compression cycle, and its equivalent Carnot cycle is shown in Fig. A4.2. A simple ideal vapour compression cycle involves four sequential processes, namely, isobaric evaporation, isentropic compression, isobaric condensation, and isentropic expansion. In equivalent Carnot cycle, the above-mentioned isobaric evaporation and condensation processes become isothermal processes. It is assumed that at the end of the condensation, state point 3 is saturated liquid, and after evaporation, state point 1 is at saturated

Table A3.7 Effect of internal irreversibility parameter on working fluid temperatures, power input, and performance of heat pump system in alternatively and continuously connected cases

$R_{\Delta s}$	Alternatively connected				Continuously connected			
–	$(T_e)_o$ K	$(T_c)_o$ K	COP_o	P_{in} kW	$(T_e)_o$ K	$(T_c)_o$ K	COP_o	P_{in} kW
0.50	278.75	316.83	1.79	1.68	281.21	314.00	1.81	1.66
0.55	278.46	316.97	1.93	1.55	281.03	314.00	1.97	1.52
0.60	278.18	317.10	2.11	1.42	280.85	314.00	2.16	1.39
0.65	277.90	317.22	2.32	1.29	280.68	314.00	2.39	1.26
0.70	277.62	317.35	2.58	1.16	280.50	314.00	2.67	1.12
0.75	277.35	317.46	2.90	1.03	280.32	314.00	3.03	0.99
0.80	277.09	317.58	3.31	0.91	280.15	314.00	3.49	0.86
0.85	276.82	317.69	3.86	0.78	279.97	314.00	4.13	0.73
0.90	276.56	317.79	4.61	0.65	279.79	314.00	5.05	0.59
0.95	276.31	317.90	5.74	0.52	279.62	314.00	6.49	0.46
1.00	276.06	318.00	7.58	0.40	279.44	314.00	9.09	0.33

Table A3.8 Effect of heat sink inlet fluid temperature on working fluid temperatures, power input, and performance of refrigeration/airconditioning system in alternatively and continuously connected cases

T_{C1}	Alternatively connected				Continuously connected			
K	$(T_c)_o$ K	$(T_e)_o$ K	COP_o	P_{in} kW	$(T_e)_o$ K	$(T_c)_o$ K	COP_o	P_{in} kW
310	319.58	267.00	5.08	0.59	271.00	314.64	6.21	0.48
312	321.64	267.00	4.89	0.61	271.00	316.67	5.93	0.51
314	323.70	267.00	4.71	0.64	271.00	318.70	5.68	0.53
316	325.76	267.00	4.54	0.66	271.00	320.73	5.45	0.55
318	327.82	267.00	4.39	0.68	271.00	322.76	5.24	0.57
318	327.82	267.00	4.39	0.68	271.00	322.76	5.24	0.57
320	329.88	267.00	4.25	0.71	271.00	324.79	5.04	0.60
322	331.95	267.00	4.11	0.73	271.00	326.82	4.85	0.62
324	334.01	267.00	3.98	0.75	271.00	328.85	4.68	0.64
326	336.07	267.00	3.87	0.78	271.00	330.88	4.53	0.66
328	338.13	267.00	3.75	0.80	271.00	332.91	4.38	0.69

vapour. It can be seen from Fig. A4.1 that during condensation, temperature of state point 2′ is higher than condensation temperature, so first it will desuperheat to state point 2, and then condensation will take place at constant temperature. For simplicity of theoretical analysis and to modify an ideal vapour compression cycle to an equivalent Carnot cycle to achieve the theoretical formula/analysis, we have used the concept of entropic average temperature with little loss of accuracy. Since the area under isobaric process 2′–2–3 in the *T*–*s* diagram of Fig. A4.1 represents the amount of heat rejection by VC cycle, we can make this area equal to the area under an isothermal process (2′–3) of Fig. A4.2 (a horizontal line), with an entropic

Table A3.9 Effect of heat source inlet fluid temperature on working fluid temperatures, power input, and performance of refrigeration/airconditioning system in alternatively and continuously connected cases

T_{E1}	Alternatively connected				Continuously connected			
K	$(T_c)_o$ K	$(T_e)_o$ K	COP_o	P_{in} kW	$(T_e)_o$ K	$(T_c)_o$ K	COP_o	P_{in} kW
265	328.22	257.00	3.61	0.83	261.00	322.95	4.21	0.71
267	328.14	259.00	3.75	0.80	263.00	322.91	4.39	0.68
269	328.06	261.00	3.89	0.77	265.00	322.87	4.58	0.66
271	327.98	263.00	4.05	0.74	267.00	322.84	4.78	0.63
273	327.90	265.00	4.21	0.71	269.00	322.80	5.00	0.60
275	327.82	267.00	4.39	0.68	271.00	322.76	5.24	0.57
277	327.75	269.00	4.58	0.66	273.00	322.73	5.49	0.55
279	327.67	271.00	4.78	0.63	275.00	322.69	5.77	0.52
281	327.60	273.00	5.00	0.60	277.00	322.66	6.07	0.49
283	327.53	275.00	5.24	0.57	279.00	322.63	6.40	0.47

Table A3.10 Effect of heat capacitance of heat sink fluid on working fluid temperatures, power input, and performance of refrigeration/airconditioning system in alternatively and continuously connected cases

C_C	Alternatively connected				Continuously connected			
kW/K	$(T_c)_o$ K	$(T_e)_o$ K	COP_o	P_{in} kW	$(T_e)_o$ K	$(T_c)_o$ K	COP_o	P_{in} kW
0.60	332.81	265.84	3.97	0.76	271.00	326.02	4.93	0.61
0.70	331.05	266.22	4.11	0.73	271.00	324.85	5.03	0.60
0.80	329.72	266.53	4.22	0.71	271.00	323.98	5.12	0.59
0.90	328.67	266.78	4.31	0.70	271.00	323.30	5.18	0.58
1.00	327.82	267.00	4.39	0.68	271.00	322.76	5.24	0.57
1.10	327.12	267.19	4.46	0.67	271.00	322.33	5.28	0.57
1.20	326.53	267.35	4.52	0.66	271.00	321.96	5.32	0.56
1.30	326.03	267.49	4.57	0.66	271.00	321.65	5.35	0.56
1.40	325.59	267.62	4.62	0.65	271.00	321.39	5.38	0.56
1.50	325.21	267.73	4.66	0.64	271.00	321.16	5.40	0.56

Table A3.11 Effect of heat capacitance of heat source fluid on working fluid temperatures, power input, and performance of refrigeration/airconditioning system in alternatively and continuously connected cases

C_E	Alternatively connected				Continuously connected			
kW/K	$(T_c)_o$ K	$(T_e)_o$ K	COP_o	P_{in} kW	$(T_e)_o$ K	$(T_c)_o$ K	COP_o	P_{in} kW
0.60	329.47	263.17	3.97	0.76	268.33	322.81	4.93	0.61
0.70	328.92	264.50	4.11	0.73	269.29	322.79	5.03	0.60
0.80	328.48	265.53	4.22	0.71	270.00	322.78	5.12	0.59
0.90	328.12	266.34	4.31	0.70	270.56	322.77	5.18	0.58
1.00	327.82	267.00	4.39	0.68	271.00	322.76	5.24	0.57
1.10	327.57	267.55	4.46	0.67	271.36	322.76	5.28	0.57
1.20	327.35	268.02	4.52	0.66	271.67	322.75	5.32	0.56
1.30	327.15	268.41	4.57	0.66	271.92	322.75	5.35	0.56
1.40	326.98	268.76	4.62	0.65	272.14	322.74	5.38	0.56
1.50	326.83	269.07	4.66	0.64	272.33	322.74	5.40	0.56

Table A3.12 Effect of effectiveness of sink side heat exchanger on working fluid temperatures, power input, and performance of refrigeration/airconditioning system in alternatively and continuously connected cases

ε_C	Alternatively connected				Continuously connected			
–	$(T_c)_o$ K	$(T_e)_o$ K	COP_o	P_{in} kW	$(T_e)_o$ K	$(T_c)_o$ K	COP_o	P_{in} kW
0.20	348.07	263.25	3.10	0.97	271.00	336.63	4.13	0.73
0.30	338.90	264.68	3.57	0.84	271.00	330.18	4.58	0.66
0.40	334.34	265.52	3.86	0.78	271.00	327.05	4.83	0.62
0.50	331.58	266.10	4.06	0.74	271.00	325.20	5.00	0.60
0.60	329.72	266.53	4.22	0.71	271.00	323.98	5.12	0.59
0.70	328.37	266.86	4.34	0.69	271.00	323.11	5.20	0.58
0.75	327.82	267.00	4.39	0.68	271.00	322.76	5.24	0.57
0.80	327.34	267.13	4.44	0.68	271.00	322.46	5.27	0.57
0.90	326.53	267.35	4.52	0.66	271.00	321.96	5.32	0.56
1.00	325.87	267.54	4.59	0.65	271.00	321.56	5.36	0.56

Table A3.13 Effect of effectiveness of source side heat exchanger on working fluid temperatures, power input, and performance of refrigeration/airconditioning system in alternatively and continuously connected cases

ε_E	Alternatively connected				Continuously connected			
–	$(T_c)_o$ K	$(T_e)_o$ K	COP_o	P_{in} kW	$(T_e)_o$ K	$(T_c)_o$ K	COP_o	P_{in} kW
0.20	333.53	252.25	3.10	0.97	260.00	322.97	4.13	0.73
0.30	331.22	258.68	3.57	0.84	265.00	322.87	4.58	0.66
0.40	329.93	262.02	3.86	0.78	267.50	322.83	4.83	0.62
0.50	329.09	264.10	4.06	0.74	269.00	322.80	5.00	0.60
0.60	328.48	265.53	4.22	0.71	270.00	322.78	5.12	0.59
0.70	328.02	266.57	4.34	0.69	270.71	322.77	5.20	0.58
0.75	327.82	267.00	4.39	0.68	271.00	322.76	5.24	0.57
0.80	327.65	267.38	4.44	0.68	271.25	322.76	5.27	0.57
0.90	327.35	268.02	4.52	0.66	271.67	322.75	5.32	0.56
1.00	327.09	268.54	4.59	0.65	272.00	322.75	5.36	0.56

Table A3.14 Effect of internal irreversibility parameter on working fluid temperatures, power input, and performance of refrigeration/airconditioning system in alternatively and continuously connected cases

$R_{\Delta s}$	Alternatively connected				Continuously connected			
–	$(T_c)_o$ K	$(T_e)_o$ K	COP_o	P_{in} kW	$(T_e)_o$ K	$(T_c)_o$ K	COP_o	P_{in} kW
0.50	335.26	265.34	0.65	4.58	271.00	327.67	0.71	4.25
0.55	333.92	265.61	0.78	3.86	271.00	326.77	0.84	3.58
0.60	332.81	265.84	0.92	3.26	271.00	326.02	0.99	3.02
0.65	331.87	266.04	1.09	2.76	271.00	325.39	1.18	2.54
0.70	331.05	266.22	1.29	2.33	271.00	324.85	1.40	2.14
0.75	330.34	266.38	1.53	1.96	271.00	324.38	1.68	1.79
0.80	329.72	266.53	1.83	1.64	271.00	323.98	2.02	1.48
0.85	329.16	266.66	2.21	1.36	271.00	323.62	2.47	1.21
0.90	328.67	266.78	2.71	1.11	271.00	323.30	3.07	0.98
0.95	328.22	266.90	3.40	0.88	271.00	323.02	3.93	0.76
1.00	327.82	267.00	4.39	0.68	271.00	322.76	5.24	0.57

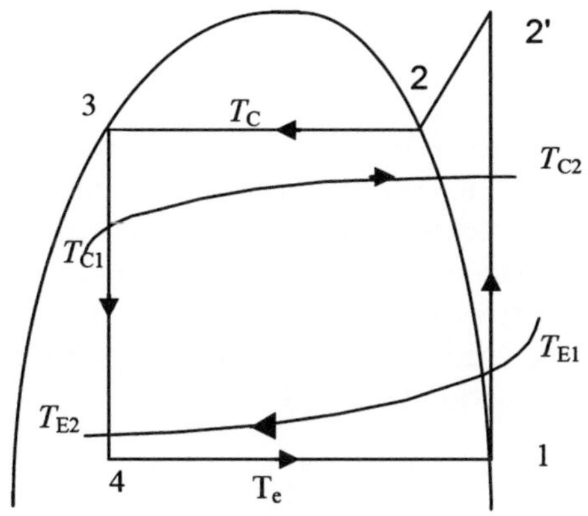

Fig. A4.1 *T–s* diagram of ideal vapour compression refrigeration/heat pump cycle

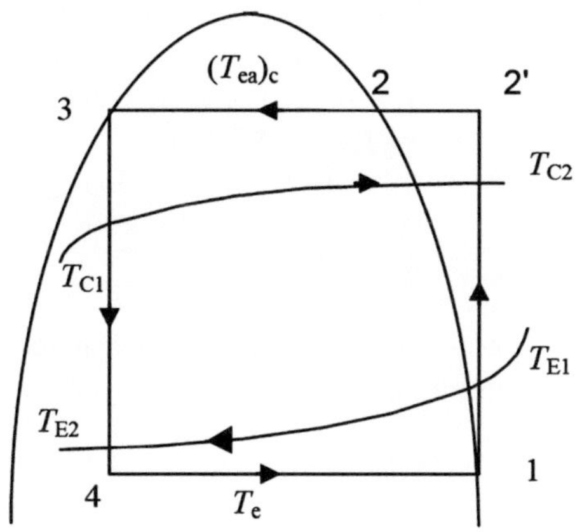

Fig. A4.2 *T–s* diagram of equivalent Carnot refrigeration/heat pump cycle

average temperature of heat rejection. The entropic average temperature T_{ea} can be defined as follows:

$$T_{ea} = \frac{\delta Q}{\Delta s}$$

where δQ and Δs are, respectively, the heat transfer and change in entropy from state point $2'$ to 3.

Entropic Average Temperature (Considering Isentropic Compression)

From the above equation and considering refrigeration/heat pump cycle 1–2′–2–3–4 (with isentropic compression) in Fig. A4.1, the entropic average temperature of heat rejection becomes:

$$(T_{ea})_c = \frac{h'_2 - h_3}{s'_2 - s_3}$$

Thus, for refrigerants (viz. R-12 and R-134a) after calculating entropy and enthalpy values at various state points, entropic average temperatures are calculated and tabulated in Tables A4.1 and A4.2. A sample calculation of entropic average temperature for R-12 is given below:

Let us consider, $T_c = 40\ °C = 313$ K, $T_e = 10\ °C = 283$ K.

Using computerized thermodynamic refrigerant tables (Sonntag et al. 1998) and considering Fig. A4.1:

$$\begin{aligned}
&P_c = P(T_c) = 0.9607 \text{ MPa}, \qquad P_e = P(T_e) = 0.4233 \text{ MPa} \\
&h_3 = h_f(T_c) = 74.59 \text{ kJ/kg} \\
&h_2 = h_g(T_c) = 203.2 \text{ kJ/kg} \\
&s_3 = s_f(T_c) = 0.2718 \text{ kJ/kg-K} \\
&s_2 = s_g(T_c) = 0.6825 \text{ kJ/kg-K} \\
&s_1 = s_g(T_e) = s'_2 = 0.6921 \text{ kJ/kg-K} \\
&h_f(T_e) = 45.37 \text{ kJ/kg} \\
&h_1 = h_g(T_e) = 191.7 \text{ kJ/kg} \\
&s_f(T_e) = 0.1752 \text{ kJ/kg-K} \\
&h'_2 = 206.2 \text{ kJ/kg}
\end{aligned}$$

Entropic average temperature is given by $(T_{ea})_c = (h'_2 - h_3)/(s'_2 - s_3)$; now, on substituting the values of enthalpy and entropy, we have:

$$(T_{ea})_c = 40.13^{\circ}\text{C} = 313.13\text{K}, \quad (T_{ea})_e = T_e = 10.0^{\circ}\text{C} = 283\text{K}$$

Thus it is found that entropic average temperature [$(T_{ea})_c$ = 313.13 K] is very close to the condensation temperature [T_c = 313.0 K], whereas entropic average

Table A4.1 Effect of condensation temperature on entropic average temperature with isentropic compression for R-12

T_c (K)	T_e (K)	$(T_{ea})_c$ (K)
313	283	313.1
315	283	315.2
317	283	317.3
319	283	319.2
321	283	321.4

Table A4.2 Effect of condensation temperature on entropic average temperature with isentropic compression for R-134a

T_c (K)	T_e (K)	$(T_{ea})_c$ (K)
313	283	313.0
315	283	315.3
317	283	317.2
319	283	319.3
321	283	321.3

temperature on evaporation side is same as evaporation temperature as no superheating has been assumed in the evaporator. Similarly, entropic average temperature for R-12 and R-134a with varying condensation temperature is calculated and tabulated below.

It can be seen from Tables A4.1 and A4.2 that entropic average temperatures are quite close to the condensation temperature, and it is only about 0.1% higher than the corresponding condensation temperature. So, the assumption of entropic average temperature as equivalent to condensation temperature simplifies the analysis considerably with little loss of accuracy.

Entropic Average Temperature (Considering Non-isentropic Compression)

If internal irreversibility due to friction, non-isentropic expansion/compression is accounted for the refrigeration/heat pump system, then the two isentropic processes become adiabatic processes with entropy generations which are shown by dotted lines (1–2″ and 3–4′) in Fig. A4.3.

On *T–s* diagram, the four processes of such a refrigeration/heat pump cycle constitute the cycle 1–2″–2–3–4′ as shown in Fig. A4.3. Now, entropic average temperature becomes:

$$(T_{\mathrm{ea}})_{\mathrm{c}} = \frac{h_2'' - h_3}{s_2'' - s_3}$$

With the help of this entropic average temperature, an isobaric process 2″–2′–2–3 of irreversible cycle (Fig. A4.3) can be replaced by an isothermal process 2″–2′–2–3 as shown in equivalent cycle (Fig. A4.4). A sample calculation of the entropic

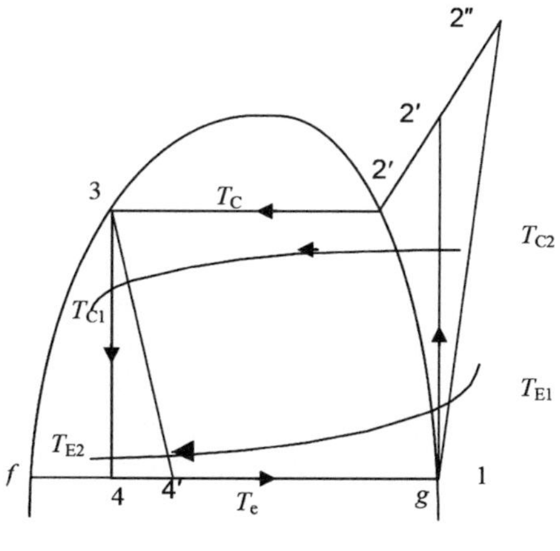

Fig. A4.3 *T–s* diagram of vapour compression refrigeration/heat pump cycle with non-isentropic compression and expansion

Fig. A4.4 *T–s* diagram of equivalent refrigeration/heat pump cycle with non-isentropic compression and expansion

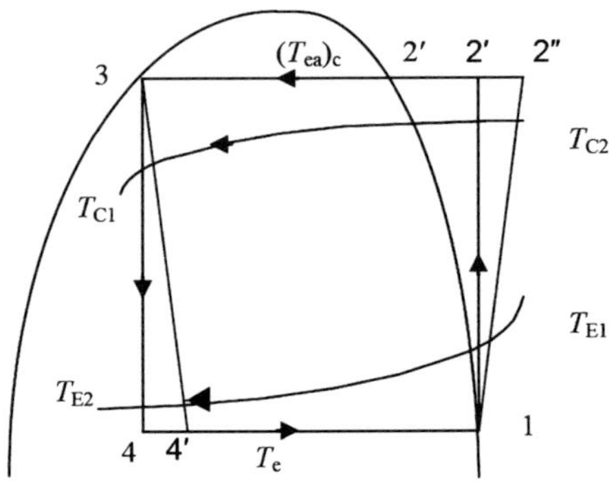

average temperature of heat rejection with non-isentropic compression/expansion for R-12 is given below:

Let us consider, $T_c = 40\ °C = 313$ K, $T_e = 10\ °C = 283$ K.

Following earlier Sect. 4.1 and considering Fig. A4.3:

$$\begin{aligned} h_1 &= h_g(T_e) = 191.7\ \text{kJ/kg} \\ &= 206.2\ \text{kJ/kg} \end{aligned}$$

Isentropic compression efficiency may be given as:

$$\eta_c = \frac{h_2' - h_1}{h_2'' - h_1}$$

Let, $\eta_c = 0.7$:

$$\begin{aligned} h_2'' &\quad 212.41\ \ \text{kJ/kg} \\ s_2'' &= 0.7113\ \ \text{kJ/kg} \end{aligned}$$

Entropic average temperature in non-isentropic case is given by $(T_{ea})_c = \dfrac{h_2'' - h_3}{s_2'' - s_3}$.

On substituting the values of enthalpy and entropy, we have:

$$(T_{ea})_c = 40.59°\text{C} = 313.6\ \text{K}, \quad (T_{ea})_e = T_e = 10.0°\text{C} = 283\ \text{K}$$

Entropic average temperature with varying condensation temperature and compression efficiency for R-12 and R-134a are calculated and tabulated below.

Tables A4.3, A4.4, A4.5, and A4.6 show that entropic average temperature in case of non-isentropic compression is found to be slightly higher than the case of isentropic process, but it is still only 0.2% higher than the corresponding condensation temperature. So, entropic average temperature can be considered as equivalent to condensation temperature, for the sake of simplicity of theoretical analysis with a little loss of accuracy. Further, the condensation pressure has also been taken as discharge pressure with some loss of accuracy.

Table A4.3 Effect of condensation temperature on entropic average temperature with non-isentropic compression for R-12

T_c (K)	T_e (K)	h_c	$(T_{ea})_c$ (K)
313	283	0.90	313.2
315	283	0.90	315.3
317	283	0.90	317.3
319	283	0.90	319.3
321	283	0.90	321.4

Table A4.4 Effect of compression efficiency on entropic average temperature with non-isentropic compression for R-12

T_c (K)	T_e (K)	h_c	$(T_{ea})_c$ (K)
313	283	0.70	313.6
313	283	0.75	313.4
313	283	0.80	313.4
313	283	0.85	313.3
313	283	0.90	313.2

Table A4.5 Effect of condensation temperature on entropic average temperature with non-isentropic compression for R-134a

T_c (K)	T_e (K)	h_c	$(T_{ea})_c$ (K)
313	283	0.90	312.7
315	283	0.90	313.5
317	283	0.90	317.3
319	283	0.90	319.7
321	283	0.90	321.2

Table A4.6 Effect of compression efficiency on entropic average temperature with non-isentropic compression for R-134a

T_c (K)	T_e (K)	h_c	$(T_{ea})_c$ (K)
313	283	0.70	313.2
313	283	0.75	313.4
313	283	0.80	312.9
313	283	0.85	313.3
313	283	0.90	312.7

Internal Irreversibility Parameter

Considering Fig. A4.4, heat absorbed by reversible cycle (process 4–1), $Q_E = T_e(S_1 - S_4)$ and heat absorbed by irreversible cycle (process 4′–1), $Q'_E = T_e(S_1 - S'_4)$. Since $S_4 < S'_4$, so $Q'_E < Q_E$ (primes are added to quantities associated with the irreversible cycle). One can define an irreversible heat absorption parameter C_1, such that:

$$Q'_E = C_1 Q_E \quad \text{with} \quad C_1 = \frac{(s_1 - s'_4)}{(s_1 - s_4)} < 1 \tag{A4.1}$$

Similarly, an irreversible heat rejection parameter C_2 may be defined as:

$$Q_c' = C_2 Q_c \quad \text{with} \quad C_2 = \frac{\left(s_2'' - s_3\right)}{\left(s_2' - s_3\right)} > 1 \tag{A4.2}$$

Assuming that the irreversible cycle operates between the same source and sink temperatures and that the effective thermal conductance remains the same, the second law requires: $\int \frac{\mathrm{d}Q}{T} = \frac{Q_E'}{T_e} - \frac{Q_c'}{T_c} < 0.$

We can write the above inequality as:

$$\frac{Q_c'}{T_e} = R_{\Delta s} \frac{Q_c'}{T_c} \tag{A4.3}$$

where

$$R_{\Delta s} = \frac{C_1}{C_2} = \frac{s_1 - s_4'}{s_2'' - s_3} \tag{A4.4}$$

Thus, it is seen that internal irreversibility parameter $R_{\Delta S}$ represents the ratio of entropy differences and it can be calculated for any refrigerant, by calculating these entropy values at various state points of the cycle. A set of sample calculation of internal irreversibility for R-12 is given below.

Let us consider, $T_c = 40\ °\text{C} = 313\ \text{K} \quad T_e = 10\ °\text{C} = 283\ \text{K}$

Internal irreversibility parameter is calculated, using computerized thermodynamic refrigerant tables (Sonntag et al. 1998) and considering Figs. A4.3 and A4.4:

$$\begin{aligned}
h_3 &= h_f(T_c) = 74.59\,\text{kJ/kg} \\
s_3 &= s_f(T_c) = 0.2718\,\text{kJ/kg-K} \\
s_1 &= s_g(T_e) = 0.6921\,\text{kJ/kg-K} \\
h_f(T_e) &= 45.37\,\text{kJ/kg} \\
h_1 &= h_g(T_e) = 191.7\text{kJ/kg} \\
s_f(T_e) &= 0.1752\,\text{kJ/kg-K}
\end{aligned}$$

Since expansion process is isenthalpic process, therefore:

$$\begin{aligned}
h_4' &= h_3 = 74.59\,\text{kJ/kg} \\
h_4' &= h_f(T_e) + x_4' h_{fg}(T_e) \\
x_4' &= 0.1997 \\
s_4' &= s_f(T_e) + x_4' S_{fg}(T_e) \\
s_4' &= 0.2784\,\text{kJ/kg-K}
\end{aligned}$$

In Sect. 4.2, it is seen with $h_c = 0.7$, $s_2'' = 0.7113\,\text{kJ/kg}$

$$R_{\Delta s} = \frac{s_1 - s_4'}{s_2'' - s_3}$$

On substituting the entropy values, we have:

$$R_{\Delta s} = 0.9413$$

Similar calculations have been carried out with varying compression efficiency and condensation temperature for R-12 and R-134a, and corresponding internal irreversibility parameters are calculated and tabulated below.

It can be seen clearly from Tables A4.7, A4.8, A4.9, and A4.10 that for both refrigerants (viz. R-12 and R-134a), internal irreversibility parameter increases and

Table A4.7 Effect of compression efficiency on internal irreversibility parameter for R-12

T_c (K)	T_e (K)	h_C	$R_{\Delta S}$
313	283	0.70	0.9413
313	283	0.75	0.9503
313	283	0.80	0.9587
313	283	0.85	0.9661
313	283	0.90	0.9727

Table A4.8 Effect of condensation temperature on internal irreversibility parameter for R-12

T_c (K)	T_e (K)	h_C	$R_{\Delta S}$
313	283	0.90	0.9727
315	283	0.90	0.9696
317	283	0.90	0.9660
319	283	0.90	0.9630
321	283	0.90	0.9590

Table A4.9 Effect of compression efficiency on internal irreversibility parameter for R-134a

T_c (K)	T_e (K)	h_C	$R_{\Delta S}$
313	283	0.70	0.9385
313	283	0.75	0.9487
313	283	0.80	0.9556
313	283	0.85	0.9644
313	283	0.90	0.9698

Table A4.10 Effect of condensation temperature on internal irreversibility parameter for R-134a

T_c (K)	T_e (K)	h_C	$R_{\Delta S}$
313	283	0.90	0.9698
315	283	0.90	0.9626
317	283	0.90	0.9645
319	283	0.90	0.9619
321	283	0.90	0.9561

approaches unity (endoreversible) with increasing compression efficiency/condensation temperature. For most practical cases, it will be between 0.94 and 0.97.

Appendix 5: Entropic Average Temperature and Internal Irreversibility Parameter for Modified VCR Cycle

Entropic Average Temperature for Irreversible VC R/AC and HP System with Liquid–Vapour Heat Exchanger

A sample calculation of entropic average temperature of heat rejection/evaporation for R-12 is given below.

Let us consider, $T_c = 40\ °C = 313\ K$, $T_e = 10\ °C = 283\ K$.

Considering Fig. 7.5b and using computerized thermodynamic refrigerant tables (Sanntag et al. 1998):

$$\begin{aligned}
P_c &= P(T_c) = 0.9607\ \text{MPa}\\
P_e &= P(T_e) = 0.4233\ \text{MPa}\\
h_3 &= h_f(T_c) = 74.59\ \text{kJ/kg}\\
s_3 &= s_f(T_c) = 0.2718\ \text{kJ/kg-K}\\
s_1 &= s_g(T_e) = 0.6921\ \text{kJ/kg-K}\\
h_f(T_e) &= 45.37\ \text{kJ/kg}\\
h_1 &= h_g(T_e) = 191.7\ \text{kJ/kg}\\
s_f(T_e) &= 0.1752\ \text{kJ/kg-K}
\end{aligned}$$

Specific heat of saturated vapour of R-12 refrigerant (Prasad 1989) is given by:

$$C_{pg}(T) = 0.624 + 0.00319T + 3.353T^2 10^{-5}$$

Effectiveness of liquid–vapour heat exchanger may be defined as:

$$\varepsilon = \frac{\text{actual heat transfer}}{\text{maximum possible heat transfer}}\varepsilon = \frac{h_3 - h_3'}{C_{\min}(T_3 - T_1)}C_{\min} = C_{pg}(T = 10°\text{C})$$
$$= 0.659253\,\text{kJ/kg-K}$$

Let us take $\varepsilon = 0.9$:

$$\begin{aligned}
h_3' &= 56.79\,\text{kJ/kg}\\
s_3' &= 0.2142\,\text{kJ/kg-K}
\end{aligned}$$

Effectiveness of L–V heat exchanger may also be written as:

$$\begin{aligned}
\varepsilon &= \frac{h_1' - h_1}{C_{\min}(T_3 - T_1)} \\
h_1' &= 209.5\,\text{kJ/kg} \\
s_1' &= 0.752\,\text{kJ/kg-K} \\
s_2' &= s_1' = 0.752\,\text{kJ/kg-K} \\
h_2' &= 226.0\,\text{kJ/kg} \\
h_4' &= h_3' = 56.79\,\text{kJ/kg} \\
h_4' &= h_\text{f}(T_\text{e}) + x_4' h\text{fg}(T_\text{e}) \\
x_4' &= 0.0780427 \\
s_4' &= s_\text{f}(T_\text{e}) + x_4'\,\text{sfg}(T_\text{e}) \\
s_4' &= 0.2155\,\text{kJ/kg-K}
\end{aligned}$$

Entropic average temperature on rejection side considering isentropic compression is given as:

$(T_\text{ea})_\text{c} = \dfrac{h_2' - h_3'}{s_2' - s_3'}$. On substituting the values of enthalpy and entropy, we have:

$$(T_\text{ea})_\text{c} = 41.63\,^\circ\text{C} = 314.63\ \text{K}$$

And entropic average temperature on evaporation side is given as:

$$\begin{aligned}
(T_\text{ea})_\text{e} &= \frac{h_1' - h_4'}{s_1' - s_4'} \\
(T_\text{ea})_\text{e} &= 11.66\,^\circ\text{C} \quad = \quad 284.66\ \text{K}
\end{aligned}$$

If non-isentropic compression is considered, entropic average temperature of evaporation will remain unchanged, whereas entropic average temperature of heat rejection becomes

$$(T_\text{ea})_\text{c} = \frac{h_2'' - h_3'}{s_2'' - s_3'}$$

Isentropic compression efficiency may be given as $\eta_c = \dfrac{h_2' - h_1'}{h_2'' - h_1'}$.

Let $\eta_\text{c} = 0.9$:

$$\begin{aligned}
h_2'' &= 227.83\,\text{kJ/kg} \\
\text{s}_2'' &= 0.7573\,\text{kJ/kg-K} \\
(\text{T}_\text{ea})_\text{c} &= 41.94\,^\circ\text{C} = 314.94\ \text{K}
\end{aligned}$$

Similarly, entropic average temperature of condensation and evaporation for R-12 and R-134a is calculated and tabulated in Tables A5.1, A5.2, and A5.3.

Table A5.1 Entropic average temperature and internal irreversibility parameter for R-12 and R-134a with varying condensation temperature

T_c	T_e			*R-12*			*R-134a*		
				$(T_{ea})_e$	$(T_{ea})_c$		$(T_{ea})_e$	$(T_{ea})_c$	
K	K	Effectiveness	h_C	K	K	$R_{\Delta S}$	K	K	$R_{\Delta S}$
313	283	0.90	0.90	284.66	314.94	0.9878	283.21	313.05	0.9751
315	283	0.90	0.90	284.81	317.07	0.9870	282.90	315.48	0.9741
317	283	0.90	0.90	285.07	319.29	0.9862	283.09	317.58	0.9714
319	283	0.90	0.90	285.27	321.49	0.9850	283.28	319.56	0.9680
321	283	0.90	0.90	285.48	323.76	0.9845	282.95	321.60	0.9653

Table A5.2 Entropic average temperature and internal irreversibility parameter for R-12 and R-134a with varying compression efficiency

T_c	T_e			*R-12*			*R-134a*		
				$(T_{ea})_e$	$(T_{ea})_c$		$(T_{ea})_e$	$(T_{ea})_c$	
K	K	Effectiveness	h_C	K	K	$R_{\Delta S}$	K	K	$R_{\Delta S}$
313	283	0.90	0.70	284.66	315.86	0.9612	283.21	313.82	0.9456
313	283	0.90	0.75	284.66	315.53	0.9689	283.21	313.38	0.9536
313	283	0.90	0.80	284.66	315.27	0.9757	283.21	313.32	0.9617
313	283	0.90	0.85	284.66	315.07	0.9820	283.21	313.05	0.9684
313	283	0.90	0.90	284.66	314.94	0.9878	283.21	313.05	0.9751

Table A5.3 Entropic average temperature and internal irreversibility parameter for R-12 and R-134a with varying effectiveness of L–V heat exchanger

T_c	T_e			R-12			R-134a		
				$(T_{ea})_e$	$(T_{ea})_c$		$(T_{ea})e$	$(T_{ea})_c$	
K	K	Effectiveness	h_C	K	K	$R_{\Delta S}$	K	K	$R_{\Delta S}$
313	283	0.70	0.90	284.15	314.52	0.9855	282.93	312.99	0.9740
313	283	0.75	0.90	284.24	314.69	0.9864	283.00	312.96	0.9743
313	283	0.80	0.90	284.39	314.73	0.9867	283.07	312.92	0.9745
313	283	0.85	0.90	284.53	314.84	0.9873	283.14	313.44	0.9765
313	283	0.90	0.90	284.66	314.94	0.9878	283.21	313.05	0.9751

Internal Irreversibility Parameter for Irreversible VC R/AC and HP System with Liquid–Vapour Heat Exchanger

Similar to Appendix 4, internal irreversibility parameter can also be calculated for the present case with liquid–vapour heat exchanger. A sample calculation for R-12 is given below.

Internal irreversibility parameter for a system with L–V heat exchanger can be defined (following Appendix 4) as:

$$\frac{S'_1\ S'_4}{S''_2\ S'_3}$$

On substituting the entropy values, we have:

$$\mathbf{R_{\Delta S}} = 0.9878$$

With varying compression efficiency, effectiveness, and condensation temperature, similar calculations have been carried out for R-12 and R-134a, and corresponding internal irreversibility parameters are found out and tabulated below.

It can be seen from Tables A5.1, A5.2, and A5.3 that even with liquid–vapour heat exchanger within a VC R/AC system, entropic average temperature of heat rejection and of heat extraction is, respectively, quite close to the condensation and evaporation temperatures. It is found that in this case, it is slightly higher than the case without liquid–vapour heat exchanger, but still it is only about 1.0% higher than the corresponding condensation/evaporation temperature. So, with a little loss of accuracy, it can be considered as equal to condensation and evaporation temperature, for the sake of simplicity in theoretical analysis.

It is further seen from the results that with increasing compression efficiency and effectiveness of liquid–vapour heat exchanger, the internal irreversibility parameter for both cases (viz. R-12, and R-134a) increases and reaches towards endoreversible case (i.e. $R_{\Delta s} = 1$). But it is found to be decreasing with increasing condensation temperature for both R-12 and R-134a.

Appendix 6: Comparison of Predicted and Reported Experimental Performance

Vapour Compression Systems

In order to have a comparison of theoretical results with the available experimental performance data, a simple comparative study for same set of operating conditions is given below.

Internal irreversibility parameter for a vapour compression system can be given as (considering Fig. A4.3 of Appendix 4):

$$R_{\Delta s} = \frac{s_1 - s'_4}{s''_2 - s_3} \tag{A6.1}$$

A set of sample calculation of actual internal irreversibility parameter for commercial R-22 chiller (Chua et al. 1996) is given below:

Inlet condenser temperature $T_{C1} = 297.09$ K
Inlet evaporator temperature $T_{E1} = 281.02$ K

Thermal conductance of heat exchanger on evaporator side $EE = 0.608$ kW/K
Thermal conductance of heat exchanger on condenser side $CE = 0.840$ kW/K
Cooling load $P_L = 9.8$ kW
Heat rejection rate on condenser side $P_H = 13.8$ kW

Heat transfer rates on evaporator and condenser sides are given by:

$$P_L = EE(T_{E1} - T_e) \tag{A6.2}$$

$$P_H = CE(T_c - T_{C1}) \tag{A6.3}$$

Using Eqs. (A6.2) and (A6.3), we have evaporation and condensation temperature as given by:

$$T_e = T_{E1} - \frac{P_L}{EE} \tag{A6.4}$$

$$T_c = T_{C1} + \frac{P_H}{CE} \tag{A6.5}$$

Now, substituting input parameters in Eqs. (A6.4) and (A6.5), we have:

$$T_e = -8.10^{\circ}\mathrm{C} \quad T_c = 40.52^{\circ}\mathrm{C}$$

Heat extraction and rejection rates may also be written as:

$$P_L = \dot{m}_r(h_1 - h_4') \tag{A6.6}$$

$$P_H = \dot{m}_r(h_2'' - h_3) \tag{A6.7}$$

where $\dot{m}_r$ is mass flow rate of refrigerant. With an assumption that state point 1 (saturated vapour) and state point 3 (saturated liquid) are saturated states (see Fig. A4.3 of Appendix 4), enthalpy and entropy values at various points (following computerized refrigerant tables, Sanntag et al. 1998) are given by:

$$\begin{aligned}
h_1 &= h_g = 246.9 \text{ kJ/kg} \\
h_f &= 35.19 \text{ kJ/kg} \\
h_3 &= h_4' = 94.96 \text{ kJ/kg} \\
s_3 &= 0.3438 \text{ kJ/kg-K} \\
s_1 &= 0.9392 \text{ kJ/kg-K} \\
s_f &= 0.1405 \text{ kJ/kg-K}
\end{aligned}$$

Using Eqs. (A6.6) and (A6.7), we have:

$$
\begin{aligned}
h_2'' &= 308.92\,\mathrm{kJ/kg} \\
s_2'' &= 1.014\,\mathrm{kJ/kg\text{-}K} \\
h_3 &= h_4' = 94.96\,\mathrm{kJ/kg} \quad \text{gives} \\
x_4' &= 0.2823 \\
s_4' &= 0.3660\,\mathrm{kJ/kg\text{-}K} \\
R_{\Delta s} &= \frac{s_1 - s_4'}{s_2'' - s_3} = 0.8552
\end{aligned}
$$

The coefficient of performance of a VCR system continuously connected to the heat source/sink thermal reservoirs (see Chap. 7) is given by:

$$
\mathrm{COP} = \left(\frac{\frac{T_{\mathrm{C1}}}{R_{\Delta S}}}{T_{\mathrm{E1}} - \frac{P_{\mathrm{L}}}{K}} - 1 \right)^{-1} \tag{A6.8}
$$

where

$$
K = \frac{CE\; EE\; R_{\Delta S}}{R_{\Delta S} CE + EE}
$$

Using Eq. (A6.8) performance is calculated which is found to be as COP = 2.62, (1/COP = 0.382). However, for the same input parameters, experimental value (Chua et al. 1996) is COP = 2.68, (1/COP = 0.373).

Our predicted performance is coming only about 2.4% lower than the experimental value which may be because of our calculated value of internal irreversibility parameter which may be lower than the actual value because of the simplifying assumption of taking saturated states at points 1 and 3. For different set of input parameters, internal irreversibility parameters are calculated, and corresponding predicted irreversible performance and experimental performance (Chua et al. 1996) from a commercial R-22 chiller are tabulated in Table A6.1.

It can be seen from Table A6.1 that for the same input parameters (viz. inlet external fluid temperatures of evaporator and condenser, cooling load, and thermal conductance on evaporator and condenser side), predicted performance in endoreversible case ($R_{\Delta S} = 1.0$) is quite higher than experimental value, as expected, and in actual irreversible case ($R_{\Delta S}$ has been calculated for actual given input data), performance predicted by finite time thermodynamic model (when system is continuously connected to thermal reservoirs) is found to be very close (maximum difference is about 5%) to the experimental performance as reported in Chua et al. (1996). Optimal input power and heat rejection rates are also found to be comparable with the experimental values. If an alternatively connected model is considered, the predicted performance is found to be much lower than the experimental performance.

Table A6.1 A comparison between measured performance of a commercial reciprocating R-22 chillers as reported in Chua et al. (1996) and the performance predicted by endoreversible ($R_{\Delta s} = 1.0$) and irreversible (actual calculated $R_{\Delta s}$) model

T_{C1}	T_{E1}	P_L	EE	CE	Predicted				Predicted				Experimental (data from Chua et al. 1996)		
K	K	kW	kW/K	kW/K	Endoreversible model				Irreversible model						
						P_H	P_o			P_H	P_o		P_H	P_o	
					$R_{\Delta s}$	kW	kW	$1/COP_o$	$R_{\Delta s}$	kW	kW	$1/COP_o$	kW	kW	$1/COP_o$
297.09	281.02	9.8	0.608	0.840	1.0	11.5	1.70	0.173	0.8553	13.5	3.75	0.382	13.8	3.66	0.373
296.89	283.00	10.5	0.622	0.884	1.0	12.3	1.76	0.168	0.8648	14.3	3.78	0.360	14.5	3.71	0.353
296.81	285.37	11.4	0.643	0.918	1.0	13.3	1.86	0.163	0.8834	15.1	3.70	0.325	15.3	3.81	0.336
296.94	286.96	11.7	0.639	0.905	1.0	13.6	1.89	0.161	0.8872	15.4	3.71	0.317	15.6	3.84	0.329
296.90	288.99	12.4	0.646	0.907	1.0	14.4	1.97	0.159	0.8981	16.1	3.70	0.299	16.3	3.90	0.315
296.88	291.00	12.9	0.649	0.917	1.0	14.9	2.00	0.155	0.9008	16.6	3.74	0.290	16.8	3.92	0.302
299.75	281.00	9.8	0.618	0.869	1.0	11.6	1.77	0.181	0.8634	13.5	3.70	0.377	13.7	3.74	0.380
299.75	283.00	10.5	0.640	0.900	1.0	12.3	1.85	0.176	0.8754	14.2	3.70	0.352	14.4	3.81	0.364
299.75	285.41	11.3	0.672	0.939	1.0	13.2	1.90	0.168	0.8798	15.1	3.80	0.336	15.3	3.94	0.348
299.67	287.00	11.9	0.689	0.953	1.0	13.9	1.96	0.165	0.8877	15.7	3.81	0.320	15.9	3.98	0.335
299.74	289.01	12.7	0.716	0.992	1.0	14.7	2.03	0.160	0.8915	16.6	3.92	0.309	16.8	4.10	0.322
299.67	290.99	13.4	0.734	1.007	1.0	15.5	2.08	0.155	0.8939	17.4	4.02	0.300	17.6	4.14	0.308

References

Andresen, B., Salamon, P. and Berry, R.S. (1984). Thermodynamics in finite time. *Physics Today*, **37**, 62–70.

Badescu, V. (1992). Optimum operation of solar convertor in combustion with a Stirling or Ericsson heat engine. *Energy*, **17**, 601–607.

Band, Y.B., Kafri, O. and Salamon, P. (1982). Finite time thermodynamics optimal expansion of a heated working fluid. *J. Appl. Phys*, **53(1)**, 8–28.

Bejan, A. (1982). Entropy generation through heat and fluid flow. John Wiley, New York, 181.

Bejan, A. (1993). Power and refrigeration plants for minimum heat exchanger inventory. *ASME, I. Energy Res. Tech*, **115**, 148–150.

Bejan, A. (1995). Theory of heat transfer-irreversible power plants-II. The optimal allocation of heat exchange equipment. *Int. J. Heat Mass Transfer*, **38(3)**, 433–444.

Bejan, A. (1996a). Models of power plants that generate minimum entropy while operating at maximum power. *American Journal of Physics*, **64(8)**, 1054–1059.

Bejan, A. (1996b). Entropy generation minimization: The new thermodynamics of finite-size devices and finite time processes. *J. Appl. Phys*, **79(3)** 1191–1215.

Berry, R.S., Kazakov, V., Sieniutycz, S., Szwast, Z. and Tsirlin, A.M. (1999). Thermodynamic optimization of finite-time processes. John Wiley & Sons, Ltd. ISBN 0 471 967521.

Bhardwaj, P.K., Kaushik, S.C. and Jain, S. (2001). Finite Time heat transfer and thermodynamic optimization of an irreversible Rankine cycle airconditioning system. Published in International Conference on "Emerging Technologies in Airconditioning and Refrigeration" organized by ISHRAE and ASHRAE in New Delhi, ACRECONF-2001 September 26–28, New Delhi, India, 398–408.

Bhardwaj, P.K. Kaushik, S.C. and Jain, S. (2003a). Finite Time Thermodynamic Optimization of an Irreversible Rankine Cycle Heat Pump System. *Indian Journal of Pure and Applied Physics*, **41**, 515–521.

Bhardwaj, P.K., Kaushik, S.C. and Jain, S. (2003b). Finite Time Optimization of an Endoreversible and Irreversible Vapour-Absorption Refrigeration System. *Energy Conversion and Management*, **44**, 1131–1144.

Bhardwaj, P.K., Kaushik, S.C. and Jain, S. (2003c). Finite Time Optimization of an Irreversible Vapour-Absorption Heat Pump System. *International Journal of Ambient Energy*, **24(4)**, 207–219.

Bhardwaj, P.K., Kaushik, S.C. and Jain, S. (2005). General Performance Characteristics of an Irreversible Vapour-Absorption Refrigeration System using Finite Time Thermodynamic Approach. *International Journal of Thermal Science*, **44(2)**, 189–196.

S.C. Kaushik et al., *Finite Time Thermodynamics of Power and Refrigeration Cycles*,
https://doi.org/10.1007/978-3-319-62812-7

Bhargava, R. and Peretto, A. (2002). A unique approach for thermo-economic optimization of an intercooled, reheat, and recuperated gas turbine for cogeneration applications. *J. Eng. for Gas Turbine & Power*, **124**, 881–891.

Blanchard, C.H. (1980). Coefficient of performance for finite speed heat pump. *J. Appl. Phys*, **51 (5)**, 2471–2472.

Blank, D.A., Davis, G.W. and Wu, C. (1994). Power optimization of an endoreversible Stirling cycle with regeneration. *Energy*, **19**, 125–133.

Blank, D.A. and Wu, C. (1995). Power optimization of an extra-terrestrial solar radiant Stirling heat engine, *Energy*, **20(6)**, 523–530.

Blank, D.A. and Wu, C. (1996a). Power limit of an endoreversible Ericsson cycle with regeneration. *Energy Convs. Mgmt*, **37**, 59–66.

Blank, D.A. and Wu, C. (1996b). Finite time power limit for solar-radiant Ericsson engines for space applications. *Applied Thermal Engineering*, **18**, 1347–1357.

Carnot, S. (1824). Reflections on the Motive Power of Fire. Bachelier, Paris.

Cengel, Y.A. and Boles, M.A. (2006). Thermodynamics. An Engineering approach. 5th edition. McGraw Hill.

Chambadal, P. (1957). Les centrales nuclearies. Paris: Armand Colin, 41–58.

Chen, J. (1994). The maximum power and maximum efficiency of an irreversible Carnot heat engine. *J. Phys. D: Appl. Phys*, **27**, 1144–1149.

Chen, J. (1995). The equivalent cycle system of an endoreversible absorption refrigerator and its general performance characteristics. *Energy*, **20(10)**, 995–1003.

Chen, J. and Andresen, B. (1995). Optimal analysis of primary performance parameters for an endoreversible absorption heat pump. *Heat Recovery System & CHP*, **15(8)**, 723–731.

Chen, J. (1997a). Optimal performance analysis of irreversible cycles used as heat pumps and refrigerators. *J. Phys. D: Appl. Phys*, **30**, 582–587.

Chen, J. (1997b). Effect of regenerative losses on the efficiency of Stirling heat engine at maximum power output. *Int. Journal of Ambient Energy*, **18**, 107–112.

Chen, J. (1998). Minimum power input of irreversible Stirling refrigerator for given cooling rate. *Energy Conv. & Mgmt*, **39(12)**, 1255–1263.

Chen, J. and Schouten, J.A. (1998). Optimum performance characteristic of an irreversible absorption refrigeration system. *Energy Conv. & Mgmt*, **39(10)**, 999–1007.

Chen, J. and Schouten, J.A. (1999). The comprehensive influence of several major irreversibilities on the performance of an Ericsson heat engine. *Applied Thermal Engineering*, **19**, 555.

Chen, J., Tyagi, S.K. and Wu, C. (2003). Optimal design on the performance parameters of an irreversible Carnot heat engine based on the thermoeconomic approach. *Int. Journal of Ambient Energy*, **24**, 201–206.

Chen, J. and Yan, Z. (1988). Optimal performance of an endoreversible-combined refrigeration cycle. *J. Appl. Phys*, **63(10)**, 4795–4798.

Chen, J. and Yan, Z. (1989). Unified description of endoreversible cycles. *Physical Review A*, **39 (8)**, 4140–4146.

Chen, J., Yan, Z., Chen, L. and Andresen, B. (1998). Efficiency bound of solar-driven Stirling heat engine system. *Int. Journal Energy Res*, **22**, 805–812.

Chen, L., Wu, C. and Sun, F. (1997). Optimal coefficient of performance and heating load relationship of three-heat-reservoir endoreversible heat pump. *Energy Conv. & Mgmt*. **38(8)**, 727–733.

Chen, L., Wu, C. and Sun, F. (1999a). Finite time thermodynamic optimization or entropy generation minimization of energy systems. *J. Non-Equilibrium Thermodynamics*, **24**, 327–359.

Chen, L., Wu, C., Sun, F. and Cao, S. (1999b). Maximum profit performance of a three-heat-reservoir heat pump. *Int. J. Energy Research*, **23**, 773–777.

Chen, L., Zheng, J., Sun, F. and Wu, C. (2001). Optimum distribution of heat exchanger inventory for power density optimization of an endoreversible closed Brayton cycle. *J. Phys. D: Appl. Phys*, **34**, 422–427.

Chen, W.Z., Sun, F.R., Cheng, S.M. and Chen, L.G. (1995). Study on optimal performance and working temperatures of endoreversible forward and reverse Carnot cycles. *Int. J. Energy Research*, **19**, 751–759.

Chua, H.T., Ng, K.C. and Gordon, J.M. (1996). Experimental study of the fundamental properties of reciprocating chillers and their relation to thermodynamic modeling and chiller design. *Int. J. Heat Mass Transfer*, **39(11)**, 2195–2204.

Curzon, F.L. and Ahlborn, B. (1975). Efficiency of a Carnot engine at maximum power output. *American Journal of Physics*, **43**, 22–24.

Davis, G.W. and Wu, C. (1997). Finite time analysis of a geothermal heat engine driven airconditioning system. *Energy Conv. & Mgmt*. **38(3)**, 263–268.

Erbay, L.B., Göktun, S. and Yavuz, H. (2001). Optimal design of the regenerative gas turbine engine with isothermal heat addition. *Applied Energy*, **68**, 249–269.

Erbay, L.B. and Yavuz, H. (1997). Analysis of Stirling heat engine at maximum power conditions. *Energy*, **22**, 645–650.

Goktun, S. (1997). Optimal performance of an irreversible refrigerator with three heat source (IRWTHS). *Energy*, **22(1)**, 27–31.

Goktun, S. and Ozkaynak, S. (1997). Optimum performance of a corrugated collector driven irreversible Carnot heat engine and absorption refrigerator. *Energy*, **22(5)**, 481–485.

Göktun, S. and Yavuz, H. (1999). Thermal efficiency of a regenerative Brayton cycle with isothermal heat addition. *Energy Conv. Mangt*, **40**, 1259–1266.

Hoffmann, K.H., Burzler, J.M. and Schubert, S. (1997). Endoreversible thermodynamics. *J Non-Equilib. Thermodynamics*, **22**, 311–355.

Holman, J.P. (1992). Heat Transfer 1963 (7th ed. in SI units, 1992). McGraw-Hill, ISBN 0-07-112644-9, 539–584.

Ibrahim, O.M., Klein, S.A. and Mitchell, J.W. (1991). Optimum heat power cycles for specified boundary conditions. *J. Engg. Gas Turbines Power*, **113**, 514–521.

Kaushik, S.C. (1999). State-of-the-art on finite time thermodynamics. Internal Report CES, IIT Delhi, India.

Kaushik, S.C., Singh, N. and Kumar, S. (1999a). Thermodynamic evaluation of a modified steam regenerative Brayton heat engine for solar thermal power generation. *Solar Energy Society of India*, **9**, 63–75.

Kaushik, S.C., Kumar, P. and Khaliq, A. (1999b). Analysis of a endoreversible Rankine Cycle Cooling System. *Energy Opportunities*, **xiv(1&2)**, 20–29.

Kaushik, S.C. and Kumar, S. (2000a). Finite time thermodynamic analysis of an endoreversibble Stirling heat engine with regenerative losses. *Energy*, **25**, 989–1003.

Kaushik, S.C. and Kumar, S. (2000b). Finite time thermodynamic evaluation of irreversible Ericsson and Stirling heat pump cycles, Proceedings of 4th Minsk International Seminar on Heat Pipes, Heat Pumps & Refrigerators, Minsk, Belarus 2000b, 113–126.

Kaushik, S.C., Kumar, P. and Jain, S. (2000). A Parametric Study based on Finite Time Thermodynamic Analysis of a Solar-Engine driven Airconditioning System. *SESI Journal*, **10(2)**, 63–78.

Kaushik, S.C., Kumar, P. and Jain, S. (2001). Finite Time Thermodynamic Optimisation of an Irreversible Heat Pump System Using the Lagrangian Multiplier Method. *International Journal of Ambient Energy*, **22(2)**, 105–112.

Kaushik, S.C. and Kumar, S. (2001). Finite time thermodynamic evaluation of irreversible Ericsson and Stirling heat engines. *Energy Convers. Mgmt*, **42**, 295–312.

Kaushik, S.C. and Tyagi, S.K. (2002). Finite time thermodynamic analysis of a nonisentropic regenerative Brayton heat engine. *Int. J. Solar Energy*, **22**, 141–151.

Kaushik, S.C., Tyagi, S.K., Bose, S.K. and Singhal, M.K. (2002a). Performance evaluation of irrerversible Ericsson and Stirling heat pump cycles. *Int. J. Thermal Sciences*, **41**, 193–200.

Kaushik, S.C., Kumar, P. and Jain, S. (2002b). Finite Time Optimization of Irreversible Airconditioning System Using Method of Lagrangian Multiplier *Journal of Energy and Environment* **2**, 53–61.

Kaushik, S.C., Kumar, P. and Jain, S. (2002c). Performance Evaluation of Irreversible Cascaded Refrigeration and Heat Pump Cycles. *Energy Conversion and Management*, **43(17)**, 2405–2424.

Kaushik, S.C., Bhardwaj, P.K. and Jain, S. (2002d). Finite Time Thermodynamics in Energy Conversion Processes. Published in National Conference on "Advances in Contemporary Physics and Energy-2002", Feb. 8–9, IIT Delhi, New Delhi, India, 464–486.

Kaushik, S.C., Tyagi, S.K. and Singhal, M.K. (2003). Parametric study of an irreversible regenerative Brayton heat engine with isothermal heat addition. *Energy Convers Mgmt*, **44**, 2013–2025.

Kays, W.M. and London, A.L. (1964). Compact heat exchangers. Second edition, McGraw-Hill, New York.

Kodal, A., Sahin, B., Ekmekei, I. and Yilmaz, T. (2003). Thermoeconomic optimization for irreversible absorption refrigerators and heat pumps. *Energy Convers. Mgmt*, **44**, 109–123.

Kodal, A., Sahin, B. and Erdil, A. (2002). Performance analysis of two-stage irreversible heat pump under maximum heating load per unit total cost conditions. *Exergy—An Int. Journal*, **2**, 159–166.

Kodal, A., Sahin, B. and Yilmaz, T. (2000). Effect of internal irreversibility and heat leakage on the finite time thermoeconomic performance of refrigerators and heat pumps, *Energy Conv. & Mgmt*, **41**, 607–619.

Kumar, S. (2000). Finite time thermodynamic analysis and second law evaluation of thermal energy conversion systems. Ph.D. Thesis, C.C.S. University, Meerut India.

Kumar, P. (2002). Finite time thermodynamic analysis of refrigeration airconditioning and heat pump systems. Ph.D. Thesis, CES, IIT Delhi, India.

Ladas, H.G. and Ibrahim, O.M. (1994). Finite time view of Stirling heat engine. *Energy*, **19**, 837–843.

Lee, W.Y. and Kim, S.S. (1992). Finite time optimization of a Rankine heat engine. *Energy Conv. & Mgmt*, **33(1)**, 59–67.

Leff, H.S. (1987). Thermal efficiency at maximum work output: New results for old heat engines. *American Journal of Physics*, **55(7)**, 602–610.

Leff, H.S. and Teeters, W.D. (1978). EER, COP and second law efficiency for airconditioner. *Am. J. Phys*, **41(1)**, 19–22.

Medina, A., Rocco, J.M.M. and Hernandez, C. (1996). Regenerative gas turbine at maximum power density conditions. *J. Phys. D: Appl. Phys*, **29**, 2802–2805.

Moran, M.J. and Shapiro, H.N. (eds.) (2007). Fundamentals of Engineering Thermodynamics. Wiley, New York.

Negri-di, M.G., Gambini, M. and Peretto, A. (1995). Reheat and regenerative gas turbine for feed water repowering of steam power plant. ASME Turbo Expo. Houston, June 5–8, 1995.

Novikov, I.I. (1957). The efficiency of atomic power stations (A review). *Attommaya Energia*, **3**, 409.

Prasad, M. (1989). Refrigeration and airconditioning data book. Wiley Eastern Limited, ISBN 81-224-0104-x.

Redcenco, V., Vargas, J.V.C. and Bejan, A. (1998). Theoretical optimization of a gas turbine power plant with pressure drop irreversibilities. *J. Energy Res. Tech*, **129**, 233–240.

Sahin, B., Kodal, A., Yilmaz, T. and Yavuz, H. (1996). Maximum power density analysis of irreversible Joule-Brayton engine. *J. Phys. D: Appl. Phys*, **29**, 1162–1167.

Sahin, B., Kodal, A. and Kaya, S.S. (1998). A comparative performance analysis of irreversible regenerative reheating Joule-Brayton heat engine under maximum power density and maximum power condition. *J. Phys D: Appl. Phys*. **31**, 2125–2131.

Sahin, B. and Kodal, A. (1999). Finite time thermoeconomic optimization for endoreversible refrigerators and heat pumps. *Energy Conv. & Mgmt*, **40**, 951–960.

Salamon, P. and Nitzan, A. (1981). Finite time optimizations of a Newton's law Carnot cycle. *J. Chem Phys*, **74(6)**, 3546–3560.

Senft, J.R. (1998). Theoretical limits on the performance of Stirling engine. *Int Journal Energy Res*, **22**, 991–1000.

Sonntag, R.E., Borgnakke, C. and Van Wylen, G.J. (1998). Fundamentals of thermodynamics. Fifth Edition. John Wiley & Sons Inc.

Tolle, H. (1975). Optimization methods. Springer, Berlin.

Trukhow, V.S., Tursunbaev, I.A., Lezhebokov, I.A. and Kenzhaev, I.G. (1997). Energy balance of autonomous solar power plant, with the Stirling engine *Appl Solar Energy*, **33**, 17–23.

Tyagi, S.K., Kaushik, S.C. and Tyagi, B.K. (2000). Thermodynamic analysis of a regenerative Brayton cycle with isothermal heat addition, NREC-2000, 419–424, Nov. 30-Dec. 2, 2000, IIT Bombay, India.

Tyagi, S.K., Kaushik, S.C. and Salhotra, R. (2002). Ecological optimization and parametric study of irreversible Ericsson and Stirling heat engines. *Journal of Phys D: Appl. Phys*, **35**, 2668–2675.

Tyagi, S.K., Chen, J. and Hua, B. (2004a). Performance evaluation and parametric study of irreversible Carnot heat pump and refrigerator cycles. Proceedings of 3rd International Symposium on Heat Transfer Enhancement and Energy Conservation, Guangzhou, China.

Tyagi, S.K., Chen, J. and Kaushik, S.C. (2004b). Thermoeconomic optimization and parametric study of an irreversible Stirling heat pump cycle. *Int. J. of Thermal Sciences*, **43**, 105–112.

Tyagi, S.K., Zhou, Y. and Chen, J. (2004c). Optimum criteria on the performance of an irreversible Braysson heat engine based on the new thermoeconomic approach. *Entropy—An Int. Journal*, **6**, 244–256.

Tyagi, S.K. and Kaushik, S.C. (2005). Ecological optimization of an irreversible regenerative intercooled Brayton heat engine with direct heat loss. *Int. Journal of Ambient Energy*, **26**, 81–92.

Tyagi, S.K., Chen, G.M., Wang, Q. and Kaushik, S.C. (2006a). A new thermoeconomic approach and parametric study of irreversible regenerative Brayton refrigeration cycle. *Int. Journal of Refrigeration*, **29**, 1167–1174.

Tyagi, S.K., Chen, G.M., Wang, Q. and Kaushik, S.C. (2006b). Thermodynamic analysis and parametric study of an irreversible regenerative-intercooled-reheat Brayton cycle heat engine Int. Journal of Thermal Sciences, **45**, 829–840.

Tyagi, S.K., Chen, J. and Kaushik, S.C. (2007). Effects of the intercooling on the performance of an irreversible regenerative modified Brayton cycle. *Int. Journal of Power and Energy Systems*, **27**, 56–64.

Tyagi, S.K., Wang, S.W. and Park, S.R. (2008). Performance criteria on different pressure ratios of an irreversible modified complex Brayton cycle. *Indian Journal of Pure & Applied Physics*, **46**, 565–574.

Tyagi, S.K. (2009). Effects of intercooling on the performance of a realistic regenerative Brayton heat engine cycle. *Int. Journal of Sustainable Energy*, **28**, 231–245.

Tyagi, S.K., Wang, Q., Xia, P. and Chen, G.M. (2010). Optimization of an irreversible Carnot refrigerator working between two heat reservoirs. *Int. Journal of Exergy*, **7**, 76–88.

Vecchiarelli, J., Kawall, J.G. and Wallace, J.S. (1997). Analysis of a concept for increasing the efficiency of a Brayton cycle via isothermal heat addition. *Int. Journal of Energy Research*, **21**, 113–127.

Wang, W., Chen, L., Sun, F. and Wu, C. (2003). Performance analysis of and irreversible variable temperature heat reservoir closed intercooled regenerated Brayton cycle. *Energy Convers Mgmt*, **44**, 2713–2732.

Wu, C. (1988). Power optimization of a finite time Carnot heat engine. *Energy*, **13(9)**, 681–687.

Wu, C. (1993a). Cooling capacity optimization of a waste heat absorption refrigeration cycle. *Heat Recovery Systems & CHP*, **13(2)**, 161–166.

Wu, C. (1993b). Maximum cooling load of a heat-engine driven refrigerator. *Energy Conv. & Mgmt*, **34(8)**, 691–696.

Wu, C. (1993c). Performance of a solar-engine driven airconditioning system. *Int. J. Ambient Energy*. **14(2)**, 77–82.

Wu, C. (1993d). Specific heating load of an endoreversible Carnot heat pump. *Int. J. Ambient Energy*. **14(1)**, 25–28.

Wu, C. (1995). Maximum obtainable specific cooling load of a refrigerator. *Energy Conv. & Mgmt*, **36(1)**, 7–10.

Wu, C., Chen, L. and Chen, J. (eds.), (1999). Recent advances in finite time thermodynamics. New York: Nova Science Publishing, 560.

Wu, C., Chen, L. and Sun, F. (1996). Performance of regenerative Brayton heat engine. *Energy*, **21**, 71–76.

Wu, C., Chen, L. and Sun, F. (1998a). Optimization of steady flow heat pumps. *Energy Conv. & Mgmt*, **39(5,6)**, 445–453.

Wu, C., Chen, L. and Sun, F. (1998b). Effect of heat transfer law on finite time exergoeconomic performance of Carnot heat pump. *Energy Conv. & Mgmt*, **39(7)**, 579–588.

Wu, C., Chen, L., Sun, F. and Cao, S. (1998c). Optimal collector temperature for solar driven heat pumps. *Energy Conv. & Mgmt*, **39(1–2)**, 143–147.

Wu, C. and Kiang, R.L. (1990). Work and power optimization of a finite time Brayton cycle. *Int. J. Ambient Energy*, **11**, 129–136.

Wu, C. and Kiang, R.L. (1991). Power performance of a nonisentropic Brayton cycle. *J. Engg. Gas Turbines Power*, **113**, 501–504.

Wu, C. and Kiang, R.L. (1992). Finite time thermodynamic analysis of a Carnot engine with internal irreversibility. *Energy*, **17(12)**, 1173–1178.

Yan, Z. and Chen, J. (1989). An optimal endoreversible three-heat-source refrigerator. *J. Appl. Phys*, **65(1)**, 1–4.

Zhou Y., Tyagi, S.K. and Chen, J. (2004). The optimal performance of an irreversible Braysson heat engine cycle, *Int. J. Therm. Sci.*, **43**, 1101–1106.

Index

S.C. Kaushik et al., *Finite Time Thermodynamics of Power and Refrigeration Cycles*,
https://doi.org/10.1007/978-3-319-62812-7

MIX
Papier aus verantwortungsvollen Quellen
Paper from responsible sources
FSC® C105338
FSC
www.fsc.org

If you have any concerns about our products,
you can contact us on
ProductSafety@springernature.com

In case Publisher is established outside the EU,
the EU authorized representative is:
Springer Nature Customer Service Center GmbH
Europaplatz 3, 69115 Heidelberg, Germany

Printed by Libri Plureos GmbH
in Hamburg, Germany